Dynamic Programming and Optimal Control

Volume I

Dimitri P. Bertsekas

Massachusetts Institute of Technology

Athena Scientific, Belmont, Massachusetts

Athena Scientific
Post Office Box 391
Belmont, Mass. 02178-9998
U.S.A.

Email: athenasc@world.std.com

Cover Design: *Ann Gallager*

Portions of this volume are adapted and reprinted from the author's *Dynamic Programming: Deterministic and Stochastic Models*, Prentice-Hall, 1987, by permission of Prentice-Hall, Inc.

Publisher's Cataloging-in-Publication Data

Bertsekas, Dimitri P.
Dynamic Programming and Optimal Control
Includes Bibliography and Index
1. Mathematical Optimization. 2. Dynamic Programming. I. Title.
QA402.5 .B465 1995 519.703 95-075941

ISBN 1-886529-12-4 (Vol. I)
ISBN 1-886529-13-2 (Vol. II)
ISBN 1-886529-11-6 (Vol. I and II)

Contents

CONTENTS OF VOLUME II

ABOUT THE AUTHOR

Dimitri Bertsekas studied Mechanical and Electrical Engineering at the National Technical University of Athens, Greece, and obtained his Ph.D. in system science from the Massachusetts Institute of Technology. He has held faculty positions with the Engineering-Economic Systems Dept., Stanford University and the Electrical Engineering Dept. of the University of Illinois, Urbana. He is currently Professor of Electrical Engineering and Computer Science at the Massachusetts Institute of Technology. He consults regularly with private industry and has held editorial positions in several journals. He has been elected Fellow of the IEEE.

Professor Bertsekas has done research in a broad variety of subjects from control theory, optimization theory, parallel and distributed computation, data communication networks, and systems analysis. He has written numerous papers in each of these areas. This book is his fourth on dynamic programming and optimal control.

Other books by the author:

1) *Dynamic Programming and Stochastic Control*, Academic Press, 1976.
2) *Stochastic Optimal Control: The Discrete-Time Case*, Academic Press, 1978 (with S. E. Shreve; translated in Russian).
3) *Constrained Optimization and Lagrange Multiplier Methods*, Academic Press, 1982 (translated in Russian).
4) *Dynamic Programming: Deterministic and Stochastic Models*, Prentice-Hall, 1987.
5) *Data Networks*, Prentice-Hall, 1987 (with R. G. Gallager; translated in Russian and Japanese); 2nd Edition 1992.
6) *Parallel and Distributed Computation: Numerical Methods*, Prentice-Hall, 1989 (with J. N. Tsitsiklis).
7) *Linear Network Optimization: Algorithms and Codes*, M.I.T. Press 1991.

Preface

This two-volume book is based on a first-year graduate course on dynamic programming and optimal control that I have taught for over twenty years at Stanford University, the University of Illinois, and the Massachusetts Institute of Technology. The course has been typically attended by students from engineering, operations research, economics, and applied mathematics. Accordingly, a principal objective of the book has been to provide a unified treatment of the subject, suitable for a broad audience. In particular, problems with a continuous character, such as stochastic control problems, popular in modern control theory, are simultaneously treated with problems with a discrete character, such as Markovian decision problems, popular in operations research. Furthermore, many applications and examples, drawn from a broad variety of fields, are discussed.

The book may be viewed as a greatly expanded and pedagogically improved version of my 1987 book "Dynamic Programming: Deterministic and Stochastic Models," published by Prentice-Hall. I have included much new material on deterministic and stochastic shortest path problems, as well as a new chapter on continuous-time optimal control problems and the Pontryagin Maximum Principle, developed from a dynamic programming viewpoint. I have also added a fairly extensive exposition of simulation-based approximation techniques for dynamic programming. These techniques, which are often referred to as "neuro-dynamic programming" or "reinforcement learning," represent a breakthrough in the practical application of dynamic programming to complex problems that involve the dual curse of large dimension and lack of an accurate mathematical model. Other material was also augmented, substantially modified, and updated.

With the new material, however, the book grew so much in size that it became necessary to divide it into two volumes: one on finite horizon, and the other on infinite horizon problems. This division was not only natural in terms of size, but also in terms of style and orientation. The first volume is more oriented towards modeling, and the second is more oriented towards mathematical analysis and computation. To make the first volume self-contained for instructors who wish to cover a modest amount of infinite horizon material in a course that is primarily oriented towards modeling, conceptualization, and finite horizon problems, I have added a final chapter that provides an introductory treatment of infinite horizon problems.

Many topics in the book are relatively independent of the others. For example Chapter 2 of Vol. I on shortest path problems can be skipped without loss of continuity, and the same is true for Chapter 3 of Vol. I, which deals with continuous-time optimal control. As a result, the book can be used to teach several different types of courses.

(a) A two-semester course that covers both volumes.

(b) A one-semester course primarily focused on finite horizon problems that covers most of the first volume.

(c) A one-semester course focused on stochastic optimal control that covers Chapters 1, 4, 5, and 6 of Vol. I, and Chapters 1, 2, and 4 of Vol. II.

(c) A one-semester course that covers Chapter 1, about 50% of Chapters 2 through 6 of Vol. I, and about 70% of Chapters 1, 2, and 4 of Vol. II. This is the course I usually teach at MIT.

(d) A one-quarter engineering course that covers the first three chapters and parts of Chapters 4 through 6 of Vol. I.

(e) A one-quarter mathematically oriented course focused on infinite horizon problems that covers Vol. II.

The mathematical prerequisite for the text is knowledge of advanced calculus, introductory probability theory, and matrix-vector algebra. A summary of this material is provided in the appendixes. Naturally, prior exposure to dynamic system theory, control, optimization, or operations research will be helpful to the reader, but based on my experience, the material given here is reasonably self-contained.

The book contains a large number of exercises, and the serious reader will benefit greatly by going through them. Solutions to all exercises are compiled in a manual that is available to instructors from Athena Scientific or from the author. Many thanks are due to the several people who spent long hours contributing to this manual, particularly Steven Shreve, Eric Loiederman, Lakis Polymenakos, and Cynara Wu.

Dynamic programming is a conceptually simple technique that can be adequately explained using elementary analysis. Yet a mathematically rigorous treatment of general dynamic programming requires the complicated machinery of measure-theoretic probability. My choice has been to bypass the complicated mathematics by developing the subject in generality, while claiming rigor only when the underlying probability spaces are countable. A mathematically rigorous treatment of the subject is carried out in my monograph "Stochastic Optimal Control: The Discrete Time Case," Academic Press, 1978, coauthored by Steven Shreve. This monograph complements the present text and provides a solid foundation for the subjects developed somewhat informally here.

Finally, I am thankful to a number of individuals and institutions for their contributions to the book. My understanding of the subject was sharpened while I worked with Steven Shreve on our 1978 monograph. My interaction and collaboration with John Tsitsiklis on stochastic shortest paths and approximate dynamic programming have been most valuable. Michael Caramanis, Emmanuel Fernandez-Gaucherand, Pierre Humblet, Lennart Ljung, and John Tsitsiklis taught from versions of the book, and contributed several substantive comments and homework problems. A number of colleagues offered valuable insights and information, particularly David Castanon, Eugene Feinberg, and Krishna Pattipati. NSF provided research support. Prentice-Hall graciously allowed the use of material from my 1987 book. Teaching and interacting with the students at MIT have kept up my interest and excitement for the subject.

Dimitri P. Bertsekas

bertsekas@lids.mit.edu

1

The Dynamic Programming Algorithm

Contents

Life can only be understood going backwards,
but it must be lived going forwards.
Kierkegaard

1.1 INTRODUCTION

This book deals with situations where decisions are made in stages. The outcome of each decision is not fully predictable but can be anticipated to some extent before the next decision is made. The objective is to minimize a certain cost – a mathematical expression of what is considered an undesirable outcome.

A key aspect of such situations is that decisions cannot be viewed in isolation since one must balance the desire for low present cost with the undesirability of high future costs. The dynamic programming technique captures this tradeoff. At each stage, decisions are ranked based on the sum of the present cost and the expected future cost, assuming optimal decision making for subsequent stages.

There is a very broad variety of practical problems that can be treated by dynamic programming. In this book, we try to keep the main ideas uncluttered by irrelevant assumptions on problem structure. To this end, we formulate in this section a broadly applicable model of optimal control of a dynamic system over a finite number of stages (a finite horizon). This model will occupy us for the first six chapters; its infinite horizon version will be the subject of the last chapter and Vol. II.

Our basic model has two principal features: (1) an underlying *discrete-time dynamic system*, and (2) a *cost function that is additive over time*. The dynamic system is of the form

$$x_{k+1} = f_k(x_k, u_k, w_k), \qquad k = 0, 1, \ldots, N-1,$$

where

k indexes discrete time,

x_k is the state of the system and summarizes past information that is relevant for future optimization,

u_k is the control or decision variable to be selected at time k,

w_k is a random parameter (also called disturbance or noise depending on the context),

N is the horizon or number of times control is applied.

The cost function is additive in the sense that the cost incurred at time k, denoted by $g_k(x_k, u_k, w_k)$, accumulates over time. The total cost is

$$g_N(x_N) + \sum_{k=0}^{N-1} g_k(x_k, u_k, w_k),$$

where $g_N(x_N)$ is a terminal cost incurred at the end of the process. However, because of the presence of w_k, cost is generally a random variable and cannot be meaningfully optimized. We therefore formulate the problem as an optimization of the *expected cost*

$$E\left\{ g_N(x_N) + \sum_{k=0}^{N-1} g_k(x_k, u_k, w_k) \right\},$$

where the expectation is with respect to the joint distribution of the random variables involved. The optimization is over the controls $u_0, u_1, \ldots, u_{N-1}$, but some qualification is needed here; each control u_k is selected with some knowledge of the current state x_k (either its exact value or some other information relating to it).

A more precise definition of the terminology just used will be given shortly. We first provide some orientation by means of examples.

Example 1.1 (Inventory Control)

Consider a problem of ordering a quantity of a certain item at each of N periods so as to meet a stochastic demand. Let us denote

x_k stock available at the beginning of the kth period,

u_k stock ordered (and immediately delivered) at the beginning of the kth period,

w_k demand during the kth period with given probability distribution.

We assume that $w_0, w_1, \ldots, w_{N-1}$ are independent random variables, and that excess demand is backlogged and filled as soon as additional inventory becomes available. Thus, stock evolves according to the discrete-time equation

$$x_{k+1} = x_k + u_k - w_k,$$

where negative stock corresponds to backlogged demand (see Fig. 1.1.1).

The cost incurred in period k consists of two components:

(a) A cost $r(x_k)$ representing a penalty for either positive stock x_k (holding cost for excess inventory) or negative stock x_k (shortage cost for unfilled demand).

(b) The purchasing cost cu_k, where c is cost per unit ordered.

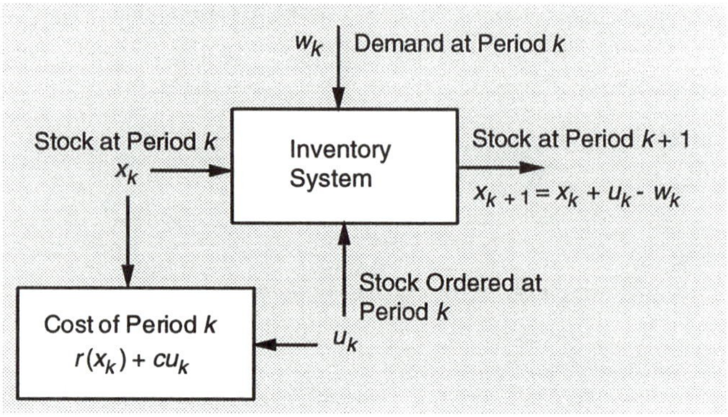

Figure 1.1.1 Inventory control example. At period k, the current stock (state) x_k, the stock ordered (control) u_k, and the demand (random disturbance) w_k determine the cost $r(x_k) + cu_k$ and the stock $x_{k+1} = x_k + u_k - w_k$ at the next period.

There is also a terminal cost $R(x_N)$ for being left with inventory x_N at the end of N periods. Thus, the total cost over N periods is

$$E\left\{ R(x_N) + \sum_{k=0}^{N-1} \big(r(x_k) + cu_k\big) \right\}.$$

We want to minimize this cost by proper choice of the orders $u_0, \ldots, u_{N-1}$, subject to the natural constraint $u_k \geq 0$ for all k.

At this point we need to distinguish between *closed-loop* and *open-loop* minimization of the cost. In open-loop minimization we select all orders $u_0, \ldots, u_{N-1}$ at once at time 0, without waiting to see the subsequent demand levels. In closed-loop minimization we postpone placing the order u_k until the last possible moment (time k) when the current stock x_k will be known. The idea is that since there is no penalty for delaying the order u_k up to time k, we can take advantage of information that becomes available between times 0 and k (the demand and stock level in past periods).

Closed-loop optimization is of central importance in dynamic programming and is the type of optimization that we will consider almost exclusively in this book. Thus, in our basic formulation, decisions are made in stages while gathering information between stages that will be used to enhance the quality of the decisions. The effect of this on the structure of the resulting optimization problem is quite profound. In particular, in closed-loop inventory optimization we are not interested in finding optimal numerical values of the orders but rather we want to find an *optimal rule for selecting at each period k an order u_k for each possible value of stock x_k that can occur*. This is an "action versus strategy" distinction.

Mathematically, in closed-loop inventory optimization, we want to find a sequence of functions μ_k, $k = 0, \ldots, N-1$, mapping stock x_k into order u_k so as to minimize the expected cost. The meaning of μ_k is that, for each k

and each possible value of x_k,

$\mu_k(x_k) =$ amount that should be ordered at time k if the stock is x_k.

The sequence $\pi = \{\mu_0, \ldots, \mu_{N-1}\}$ will be referred to as a *policy* or *control law*. For each π, the corresponding cost for a fixed initial stock x_0 is

$$J_\pi(x_0) = E\left\{ R(x_N) + \sum_{k=0}^{N-1} \big(r(x_k) + c\mu_k(x_k) \big) \right\},$$

and we want to minimize $J_\pi(x_0)$ for a given x_0 over all π that satisfy the constraints of the problem. This is a typical dynamic programming problem. We will analyze this problem in various forms in subsequent sections. For example, we will show in Section 4.2 that for a reasonable choice of the cost function, the optimal ordering policy is of the form

$$\mu_k(x_k) = \begin{cases} S_k - x_k & \text{if } x_k < S_k, \\ 0 & \text{otherwise,} \end{cases}$$

where S_k is a suitable threshold level determined by the data of the problem. In other words, when stock falls below the threshold S_k, order just enough to bring stock up to S_k.

The preceding example illustrates the main ingredients of the basic problem formulation:

(a) A *discrete-time system* of the form

$$x_{k+1} = f_k(x_k, u_k, w_k),$$

where f_k is some function; in the inventory example $f_k(x_k, u_k, w_k) = x_k + u_k - w_k$.

(b) *Independent random parameters* w_k. This will be generalized by allowing the probability distribution of w_k to depend on x_k and u_k; in the context of the inventory example, we can think of a situation where the level of demand w_k is influenced by the current stock level x_k.

(c) A *control constraint*; in the example, we have $u_k \geq 0$. In general, the constraint set will depend on x_k and the time index k, that is, $u_k \in U_k(x_k)$. To see how constraints dependent on x_k can arise in the inventory context, think of a situation where there is an upper bound B on the level of stock that can be accommodated, so $u_k \leq B - x_k$.

(d) An *additive cost* of the form

$$E\left\{ g_N(x_N) + \sum_{k=0}^{N-1} g_k(x_k, u_k, w_k) \right\},$$

where g_k are some functions; in the example, we have $g_N(x_N) = R(x_N)$, and $g_k(x_k, u_k, w_k) = r(x_k) + cu_k$.

(e) *Optimization over (closed-loop) policies*, that is, rules for choosing u_k for each k and possible value of x_k.

Discrete-State and Finite-State Systems

In the preceding example, the state x_k was a continuous real variable, and it is easy to think of multidimensional generalizations where the state is an n-dimensional vector of real variables. It is also possible, however, that the state takes values from a discrete set, such as the integers.

A version of the inventory problem where a discrete viewpoint is more natural arises when stock is measured in whole units (such as cars), each of which is a significant fraction of x_k, u_k, or w_k. It is more appropriate then to take as state space the set of all integers rather than the set of real numbers. The form of the system equation and the cost per period will, of course, stay the same.

In other systems the state is naturally discrete and there is no continuous counterpart of the problem. Such systems are often conveniently specified in terms of the probabilities of transition between the states. What we need to know is $p_{ij}(u,k)$, which is the probability at time k that the next state will be j, given that the current state is i, and the control selected is u, i.e.,

$$p_{ij}(u,k) = P\{x_{k+1} = j \mid x_k = i, u_k = u\}.$$

Such a system can be described alternatively in terms of the discrete-time system equation

$$x_{k+1} = w_k,$$

where the probability distribution of the random parameter w_k is

$$P\{w_k = j \mid x_k = i, u_k = u\} = p_{ij}(u,k).$$

Conversely, given a discrete-state system in the form

$$x_{k+1} = f_k(x_k, u_k, w_k),$$

together with the probability distribution $P_k(w_k \mid x_k, u_k)$ of w_k, we can provide an equivalent transition probability description. The corresponding transition probabilities are given by

$$p_{ij}(u,k) = P_k\{W_k(i,u,j) \mid x_k = i, u_k = u\},$$

where $W(i,u,j)$ is the set

$$W_k(i,u,j) = \{w \mid j = f_k(i,u,w)\}.$$

Thus a discrete-state system can equivalently be described in terms of a difference equation and in terms of transition probabilities. Depending on the given problem, it may be notationally more convenient to use one description over the other.

The following three examples illustrate discrete-state systems.

Example 1.2 (Machine Replacement)

Consider a problem of operating efficiently over N time periods a machine that can be in any one of n states, denoted $1, 2, \ldots, n$. The implication here is that state i is better than state $i+1$, and state 1 corresponds to a machine in perfect condition. In particular, we denote by $g(i)$ the operating cost per period when the machine is in state i, and we assume that

$$g(1) \leq g(2) \leq \cdots \leq g(n).$$

During a period of operation, the state of the machine can become worse or it may stay unchanged. We thus assume that the transition probabilities

$$p_{ij} = P\{\text{next state will be } j \mid \text{current state is } i\}$$

satisfy

$$p_{ij} = 0 \qquad \text{if } j < i.$$

We assume that at the start of each period we know the state of the machine and we must choose one of the following two options:

(a) Let the machine operate one more period in the state it currently is.

(b) Repair the machine and bring it to the perfect state 1 at a cost R.

We assume that the machine, once repaired, is guaranteed to stay in state 1 for one period. In subsequent periods, it may deteriorate to states $j > 1$ according to the transition probabilities p_{1j}.

Thus the objective here is to decide on the level of deterioration (state) at which it is worth paying the cost of machine repair, thereby obtaining the benefit of smaller future operating costs. Note that the decision should also be affected by the period we are in. For example, we would be less inclined to repair the machine when there are few periods left.

The system equation for this problem is represented by the graphs of Fig. 1.1.2. These graphs depict the transition probabilities between various pairs of states for each value of the control and are known as *transition probability graphs* or simply *transition graphs*. Note that there is a different graph for each control; in the present case there are two controls (repair or not repair).

Example 1.3 (Control of a Queue)

Consider a queueing system with room for n customers operating over N time periods. We assume that service of a customer can start (end) only at the beginning (end) of the period and that the system can serve only one customer at a time. The probability p_m of m customer arrivals during a period is given, and the numbers of arrivals in two different periods are independent. Customers finding the system full depart without attempting to enter later. The system offers two kinds of service, *fast* and *slow*, with cost per period c_f and c_s, respectively. Service can be switched between fast and

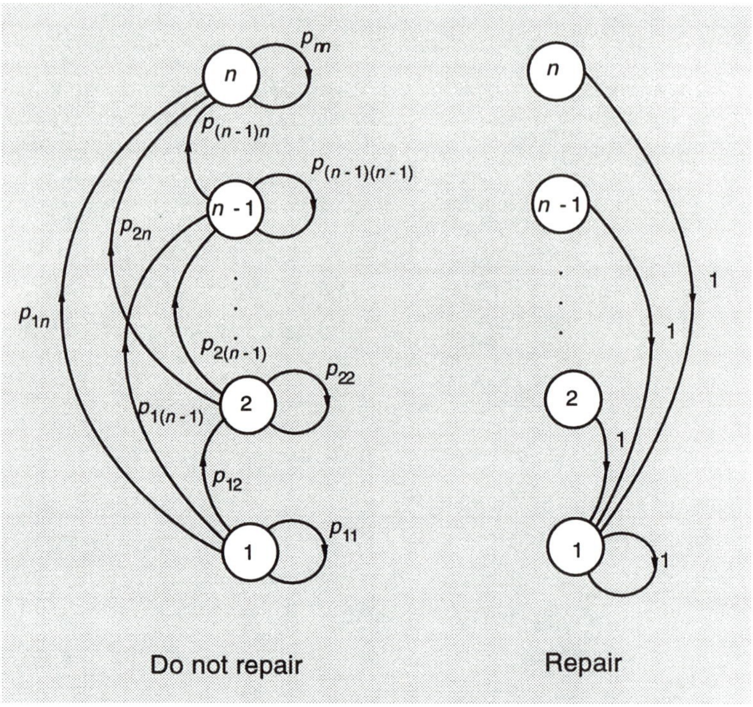

Figure 1.1.2 Machine replacement example. Transition probability graphs for each of the two possible controls (repair or not repair). At each stage and state i, the cost of repairing is $R+g(1)$, and the cost of not repairing is $g(i)$. The terminal cost is 0.

slow at the beginning of each period. With fast (slow) service, a customer in service at the beginning of a period will terminate service at the end of the period with probability q_f (respectively, q_s) independently of the number of periods the customer has been in service and the number of customers in the system ($q_f > q_s$). There is a cost $r(i)$ for each period for which there are i customers in the system. There is also a terminal cost $R(i)$ for i customers left in the system at the end of the last period. The problem is to choose, at each period, the type of service as a function of the number of customers in the system so as to minimize the expected total cost over N periods.

It is appropriate to take as state here the number i of customers in the system at the start of a period and as control the type of service provided. The cost per period then is $r(i)$ plus c_f or c_s depending on whether fast or slow service is provided. We derive the transition probabilities of the system.

When the system is empty at the start of the period, the probability that the next state is j is independent of the type of service provided. It equals the given probability of j customer arrivals when $j < n$,

$$p_{0j}(u_f) = p_{0j}(u_s) = p_j, \qquad j = 0, 1, \ldots, n-1,$$

and it equals the probability of n or more customer arrivals when $j = n$,

$$p_{0n}(u_f) = p_{0n}(u_s) = \sum_{m=n}^{\infty} p_m.$$

When there is at least one customer in the system ($i > 0$), we have

$$p_{ij}(u_f) = 0, \qquad \text{if } j < i - 1,$$

$$p_{ij}(u_f) = q_f p_0, \qquad \text{if } j = i - 1,$$

$$\begin{aligned} p_{ij}(u_f) &= P\{j - i + 1 \text{ arrivals, service completed}\} \\ &\quad + P\{j - i \text{ arrivals, service not completed}\} \\ &= q_f p_{j-i+1} + (1 - q_f) p_{j-i}, \qquad \text{if } i - 1 < j < n - 1, \end{aligned}$$

$$p_{i(n-1)}(u_f) = q_f \sum_{m=n-i}^{\infty} p_m + (1 - q_f) p_{n-1-i},$$

$$p_{in}(u_f) = (1 - q_f) \sum_{m=n-i}^{\infty} p_m.$$

The transition probabilities when slow service is provided are also given by these formulas with u_f and q_f replaced by u_s and q_s, respectively.

Example 1.4 (Optimizing a Chess Match Strategy)

A player is about to play a two-game chess match with an opponent, and wants to maximize his winning chances. Each game can have one of two outcomes:

(a) A win by one of the players (1 point for the winner and 0 for the loser).

(b) A draw (1/2 point for each of the two players).

If the score is tied at 1-1 at the end of the two games, the match goes into sudden-death mode, whereby the players continue to play until the first time one of them wins a game (and the match). The player has two playing styles and he can choose one of the two at will in each game, independently of the style he chose in previous games.

(1) *Timid play* with which he draws with probability $p_d > 0$, and he loses with probability $(1 - p_d)$.

(2) *Bold play* with which he wins with probability p_w, and he loses with probability $(1 - p_w)$.

Thus, in a given game, timid play never wins, while bold play never draws. The player wants to find a style selection strategy that maximizes his probability of winning the match. Note that once the match gets into sudden death, the player should play bold, since with timid play he can at best prolong the sudden death play, while running the risk of losing. Therefore, there are only two decisions for the player to make, the selection of the playing strategy in the first two games. Thus, we can model the problem as one with two stages, and with states the possible scores at the start of each of the first two stages (games), as shown in Fig. 1.1.3. The initial state is the initial score 0-0. The transition probabilities for each of the two different controls (playing styles) are also shown in Fig. 1.1.3. There is a cost at the terminal states: a cost of -1 at the winning scores 2-0 and 1.5-0.5, a cost of 0 at the losing scores 0-2 and 0.5-1.5, and a cost of $-p_w$ at the tied score 1-1 (since the probability of winning in sudden death is p_w). Note that to maximize the probability P of winning the match, we must minimize $-P$.

This problem has an interesting feature. One would think that if $p_w < 1/2$, the player would have a less than 50-50 chance of winning the match, even with optimal play, since his probability of losing is greater than his probability of winning any one game, regardless of his playing style. This is not so, however, because the player can adapt his playing style to the current score, but his opponent does not have that option. In other words, the player can use a closed-loop strategy, and it will be seen later that with optimal play, as determined by the dynamic programming algorithm, he has a better than 50-50 chance of winning the match provided p_w is higher than a threshold value $\overline{p}$, which, depending on the value of p_d, may satisfy $\overline{p} < 1/2$.

1.2 THE BASIC PROBLEM

We now formulate a general problem of decision under stochastic uncertainty over a finite number of stages. This problem, which we call *basic*, is the central problem in this book. We will discuss solution methods for this problem based on dynamic programming in the first six chapters, and we will extend our analysis to versions of this problem involving an infinite number of stages in the last chapter and in Vol. II of this work.

The basic problem is very general. In particular, we will not require that the state, control, or random parameter take a finite number of values or belong to a space of n-dimensional vectors. A surprising aspect of dynamic programming is that its applicability depends very little on the nature of the state, control, and random parameter spaces. For this reason it is convenient to proceed without any assumptions on the structure of these spaces; indeed such assumptions would become a serious impediment later.

Basic Problem

We are given a discrete-time dynamic system

$$x_{k+1} = f_k(x_k, u_k, w_k), \qquad k = 0, 1, \ldots, N-1, \tag{2.1}$$

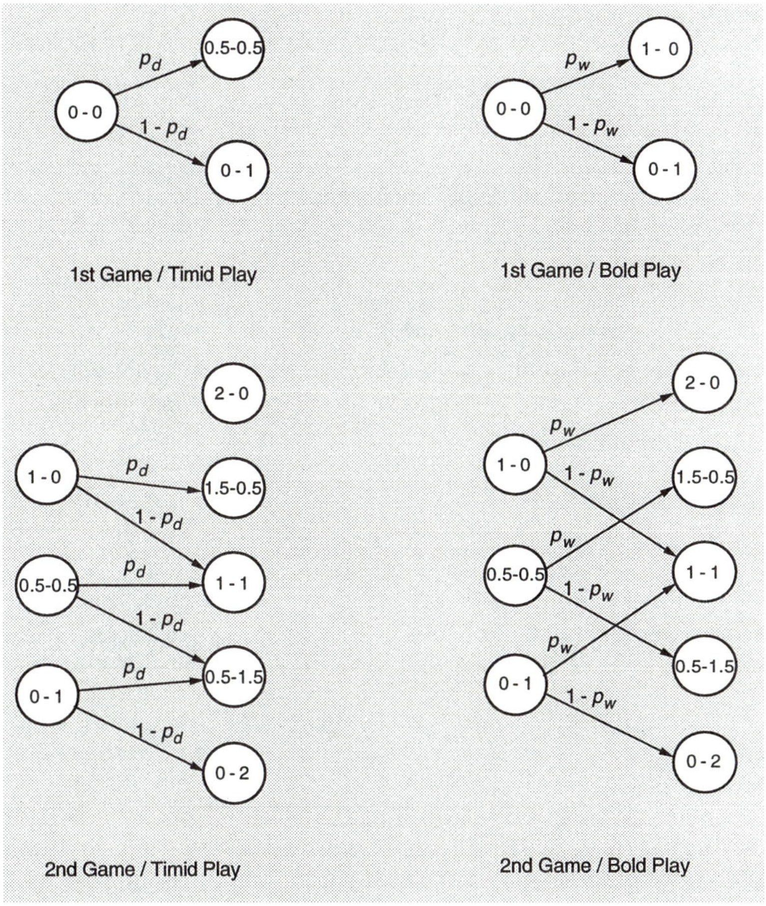

Figure 1.1.3 Chess match example. Transition probability graphs for each of the two possible controls (timid or bold play). Note here that the state space is not the same at each stage. The terminal cost is -1 at the winning final scores 2-0 and 1.5-0.5, 0 at the losing final scores 0-2 and 0.5-1.5, and $-p_w$ at the tied score 1-1.

where the state x_k is an element of a space S_k, the control u_k is an element of a space C_k, and the random "disturbance" w_k is an element of a space D_k.

The control u_k is constrained to take values in a given nonempty subset $U(x_k) \subset C_k$, which depends on the current state x_k; that is, $u_k \in U_k(x_k)$ for all $x_k \in S_k$ and k.

The random disturbance w_k is characterized by a probability distribution $P_k(\cdot \mid x_k, u_k)$ that may depend explicitly on x_k and u_k but not on values of prior disturbances $w_{k-1}, \ldots, w_0$.

We consider the class of policies (also called control laws) that consist of a sequence of functions

$$\pi = \{\mu_0, \ldots, \mu_{N-1}\},$$

where μ_k maps states x_k into controls $u_k = \mu_k(x_k)$ and is such that $\mu_k(x_k) \in U_k(x_k)$ for all $x_k \in S_k$. Such policies will be called *admissible*.

Given an initial state x_0 and an admissible policy $\pi = \{\mu_0, \ldots, \mu_{N-1}\}$, the system equation

$$x_{k+1} = f_k\big(x_k, \mu_k(x_k), w_k\big), \qquad k = 0, 1, \ldots, N-1, \tag{2.2}$$

makes x_k and w_k random variables with well-defined distributions. Thus, for given functions g_k, $k = 0, 1, \ldots, N$, the expected cost

$$J_\pi(x_0) = \mathop{E}_{\substack{w_k \\ k=0,1,\ldots,N-1}} \left\{ g_N(x_N) + \sum_{k=0}^{N-1} g_k\big(x_k, \mu_k(x_k), w_k\big) \right\} \tag{2.3}$$

is a well-defined quantity. For a given initial state x_0, an optimal policy π^* is one that minimizes this cost; that is,

$$J_{\pi^*}(x_0) = \min_{\pi \in \Pi} J_\pi(x_0),$$

where Π is the set of all admissible policies.

Note that the optimal policy π^* is associated with a fixed initial state x_0. However, an interesting aspect of the basic problem and of dynamic programming is that it is typically possible to find a policy π^* that is simultaneously optimal for all initial states.

The optimal cost depends on x_0 and is denoted by $J^*(x_0)$; that is,

$$J^*(x_0) = \min_{\pi \in \Pi} J_\pi(x_0).$$

It is useful to view J^* as a function that assigns to each initial state x_0 the optimal cost $J^*(x_0)$ and call it the *optimal cost function* or *optimal value function*.

For the benefit of the mathematically oriented reader we note that in the preceding equation, "min" denotes the greatest lower bound (or infimum) of the set of numbers $\{J_\pi(x_0) \mid \pi \in \Pi\}$. A notation more in line with normal mathematical usage would be to write $J^*(x_0) = \inf_{\pi \in \Pi} J_\pi(x_0)$. However (as discussed in Appendix B), we find it convenient to use "min" in place of "inf" even when the infimum is not attained. It is less distracting, and it will not lead to any confusion.

The Role and Value of Information

We mentioned earlier the distinction between open-loop minimization, where we select all controls $u_0, \ldots, u_{N-1}$ at once at time 0, and closed-loop minimization, where we select a policy $\{\mu_0, \ldots, \mu_{N-1}\}$ that applies the control $\mu_k(x_k)$ at time k with knowledge of the current state x_k (see Fig. 1.2.1). With closed-loop policies, it is possible to achieve lower cost, essentially by taking advantage of the extra information (the value of the current state). The reduction in cost may be called the *value of the information* and can be significant indeed. If the information is not available, the controller cannot adapt appropriately to unexpected values of the state, and as a result the cost can be adversely affected. For example, in the inventory control example of the preceding section, the information that becomes available at the beginning of each period k is the inventory stock x_k. Clearly, this information is very important to the inventory manager, who will want to adjust the amount u_k to be purchased depending on whether the current stock x_k is running high or low.

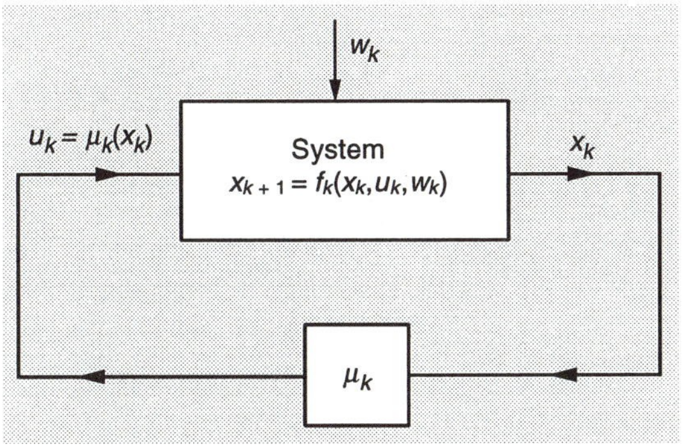

Figure 1.2.1 Information gathering in the basic problem. At each time k the controller observes the current state x_k and applies a control $u_k = \mu_k(x_k)$ that depends on that state.

Example 2.1

To illustrate the benefits of the proper use of information, let us consider the chess match example of the preceding section. There, a player can select timid play (probabilities p_d and $1 - p_d$ for a draw and a loss, respectively) or bold play (probabilities p_w and $1 - p_w$ for a win and a loss, respectively) in each of the two games of the match. Suppose the player chooses a policy of playing timid if and only if he is ahead in the score, as illustrated in Fig. 1.2.2; we will see in the next section that this policy is optimal, assuming $p_d > p_w$. Then after the first game (in which he plays bold), the score is 1-0 with probability p_w and 0-1 with probability $1 - p_w$. In the second game, he plays timid in the former case and bold in the latter case. Thus after two games, the probability of a match win is $p_w p_d$, the probability of a match loss is $(1-p_w)^2$, and the probability of a tied score is $p_w(1-p_d)+(1-p_w)p_w$, in which case he has a probability p_w of winning the subsequent sudden-death game. Thus the probability of winning the match with the given strategy is

$$p_w p_d + p_w\big(p_w(1-p_d)+(1-p_w)p_w\big),$$

which, with some rearrangement, gives

$$\text{Probability of a match win } = p_w^2(2-p_w)+p_w(1-p_w)p_d. \tag{2.4}$$

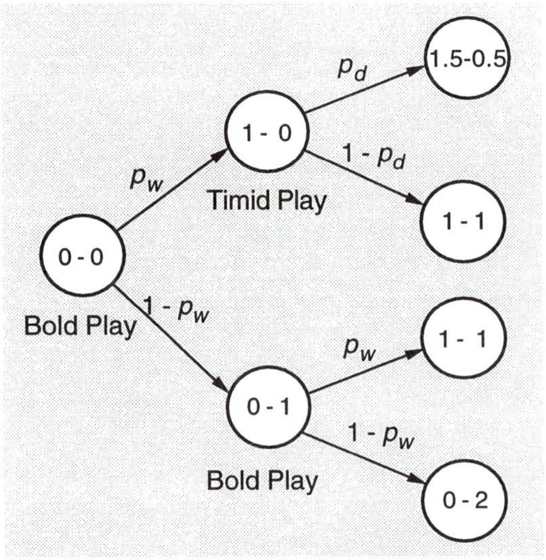

Figure 1.2.2 Illustration of the policy used in Example 2.1 to obtain a greater than 50-50 chance of winning the chess match and associated transition probabilities. The player chooses a policy of playing timid if and only if he is ahead in the score.

Suppose now that $p_w < 1/2$. Then the player has a greater probability of losing than winning any one game, regardless of the type of play he uses. From this we can infer that no open-loop strategy can give the player a greater than 50-50 chance of winning the match. Yet from Eq. (2.4) it can be seen that with the closed-loop strategy of playing timid if and only if the player is ahead in the score, the chance of a match win can be greater than 50-50, provided that p_w is close enough to $1/2$ and p_d is close enough to 1. As an example, for $p_w = 0.45$ and $p_d = 0.9$, Eq. (2.4) gives a match win probability of roughly 0.53.

To calculate the value of information, let us consider the four open-loop policies, whereby we decide on the type of play to be used without waiting to see the result of the first game. These are:

(1) Play timid in both games; this has a probability $p_d^2 p_w$ of winning the match.

(2) Play bold in both games; this has a probability $p_w^2 + p_w^2(1-p_w) = p_w^2(2-p_w)$ of winning the match.

(3) Play bold in the first game and timid in the second game; this has a probability $p_w p_d + p_w^2(1-p_d)$ of winning the match.

(4) Play timid in the first game and bold in the second game; this also has a probability $p_w p_d + p_w^2(1-p_d)$ of winning the match.

The first policy is always dominated by the others, and the optimal open-loop probability of winning the match is

$$\begin{aligned}\text{Open-loop probability of win } &= \max\big(p_w^2(2-p_w),\, p_w p_d + p_w^2(1-p_d)\big)\\ &= p_w^2 + p_w(1-p_w)\max(p_w, p_d).\end{aligned}$$

By making the reasonable assumption $p_d > p_w$, we see that the optimal open-loop policy is to play timid in one of the two games and play bold in the other, and that

$$\text{Open-loop probability of a match win } = p_w p_d + p_w^2(1-p_d). \tag{2.5}$$

For $p_w = 0.45$ and $p_d = 0.9$, Eq. (2.5) gives an optimal open-loop match win probability of roughly 0.425. Thus, the value of the information (the outcome of the first game) is the difference of the optimal closed-loop and open-loop values, which is approximately $0.53 - 0.425 = 0.105$. More generally, by subtracting Eqs. (2.4) and (2.5), we see that

$$\begin{aligned}\text{Value of information } &= p_w^2(2-p_w) + p_w(1-p_w)p_d - \big(p_w p_d + p_w^2(1-p_d)\big)\\ &= p_w^2(1-p_w).\end{aligned}$$

It should be noted, however, that whereas availability of the state information cannot hurt, it may not result in an advantage either. For instance, in deterministic problems, where no random disturbances are present, one can predict the future states given the initial state and the sequence of controls. Thus, optimization over all sequences $\{u_0, u_1, \ldots, u_{N-1}\}$ of controls leads to the same optimal cost as optimization over all admissible policies. The same can be true even in some stochastic problems (see for example Exercise 1.13). This brings up a related issue. Assuming no information is forgotten, the controller actually knows the prior states and controls $x_0, u_0, \ldots, x_{k-1}, u_{k-1}$ as well as the current state x_k. Therefore, the question arises whether policies that use the entire system history can be superior to policies that use just the current state. The answer turns out to be negative although the proof is technically complicated (see [BeS78]). The intuitive reason is that, for a given time k and state x_k, all future expected costs depend explicitly just on x_k and not on prior history.

1.3 THE DYNAMIC PROGRAMMING ALGORITHM

The dynamic programming (DP) technique rests on a very simple idea, the *principle of optimality*. The name is due to Bellman, who contributed a great deal to the popularization of DP and to its transformation into a systematic tool. Roughly, the principle of optimality states the following rather obvious fact.

Principle of Optimality

Let $\pi^* = \{\mu_0^*, \mu_1^*, \ldots, \mu_{N-1}^*\}$ be an optimal policy for the basic problem, and assume that when using π^*, a given state x_i occurs at time i with positive probability. Consider the subproblem whereby we are at x_i at time i and wish to minimize the "cost-to-go" from time i to time N

$$E\left\{ g_N(x_N) + \sum_{k=i}^{N-1} g_k\big(x_k, \mu_k(x_k), w_k\big) \right\}.$$

Then the truncated policy $\{\mu_i^*, \mu_{i+1}^*, \ldots, \mu_{N-1}^*\}$ is optimal for this subproblem.

The intuitive justification of the principle of optimality is very simple. If the truncated policy $\{\mu_i^*, \mu_{i+1}^*, \ldots, \mu_{N-1}^*\}$ were not optimal as stated, we would be able to reduce the cost further by switching to an optimal policy for the subproblem once we reach x_i. For an auto travel analogy, suppose that the fastest route from Los Angeles to Boston passes through Chicago. The principle of optimality translates to the obvious fact that the Chicago to Boston portion of the route is also the fastest route for a trip that starts from Chicago and ends in Boston.

The principle of optimality suggests that an optimal policy can be constructed in piecemeal fashion, first constructing an optimal policy for the "tail subproblem" involving the last stage, then extending the optimal policy to the "tail subproblem" involving the last two stages, and continuing in this manner until an optimal policy for the entire problem is constructed. The DP algorithm is based on this idea. We introduce the algorithm by means of an example.

The DP Algorithm for the Inventory Control Example

Consider the inventory control example of the previous section and the following procedure for determining the optimal ordering policy starting with the last period and proceeding backward in time.

Period $N-1$: Assume that at the beginning of period $N-1$ the stock is x_{N-1}. Clearly, no matter what happened in the past, the inventory manager should order the amount of inventory that minimizes over $u_{N-1} \geq 0$ the sum of the ordering cost and the expected terminal holding/shortage cost $cu_{N-1} + E\{R(x_N)\}$, which can be written as

$$cu_{N-1} + \underset{w_{N-1}}{E}\big\{R(x_{N-1} + u_{N-1} - w_{N-1})\big\}.$$

Adding the holding/shortage cost of period $N-1$, the optimal cost for the last period (plus the terminal cost) is

$$J_{N-1}(x_{N-1}) = r(x_{N-1}) + \min_{u_{N-1}\geq 0}\left[cu_{N-1} + \underset{w_{N-1}}{E}\big\{R(x_{N-1} + u_{N-1} - w_{N-1})\big\}\right].$$

Naturally, J_{N-1} is a function of the stock x_{N-1}. It is calculated either analytically or numerically (in which case a table is used for computer storage of the function J_{N-1}). In the process of calculating J_{N-1}, we obtain the optimal inventory policy $\mu^*_{N-1}(x_{N-1})$ for the last period; $\mu^*_{N-1}(x_{N-1})$ is the value of u_{N-1} that minimizes the right-hand side of the preceding equation for a given value of x_{N-1}.

Period $N-2$: Assume that at the beginning of period $N-2$ the stock is x_{N-2}. It is clear that the inventory manager should order the amount of inventory that minimizes not just the expected cost of period $N-2$ but rather the

(expected cost of period $N-2$) + (expected cost of period $N-1$,
given that an optimal policy will be used at period $N-1$),

which is equal to

$$r(x_{N-2}) + cu_{N-2} + E\big\{J_{N-1}(x_{N-1})\big\}.$$

Using the system equation $x_{N-1} = x_{N-2} + u_{N-2} - w_{N-2}$, the last term is also written as $J_{N-1}(x_{N-2} + u_{N-2} - w_{N-2})$.

Thus the optimal cost for the last two periods given that we are at state x_{N-2}, denoted $J_{N-2}(x_{N-2})$, is given by

$$J_{N-2}(x_{N-2}) = r(x_{N-2}) + \min_{u_{N-2}\geq 0}\left[cu_{N-2} + \underset{w_{N-2}}{E}\big\{J_{N-1}(x_{N-2} + u_{N-2} - w_{N-2})\big\}\right]$$

Again $J_{N-2}(x_{N-2})$ is calculated for every x_{N-2}. At the same time, the optimal policy $\mu^*_{N-2}(x_{N-2})$ is also computed.

Period k: Similarly, we have that at period k, when the stock is x_k, the inventory manager should order u_k to minimize

(expected cost of period k) + (expected cost of periods $k+1, \ldots, N-1$, given that an optimal policy will be used for these periods).

By denoting by $J_k(x_k)$ the optimal cost, we have

$$J_k(x_k) = r(x_k) + \min_{u_k \geq 0} \left[cu_k + \mathop{E}_{w_k} \big\{ J_{k+1}(x_k + u_k - w_k) \big\} \right], \tag{3.1}$$

which is actually the dynamic programming equation for this problem.

The functions $J_k(x_k)$ denote the optimal expected cost for the remaining periods when starting at period k and with initial inventory x_k. These functions are computed recursively backward in time, starting at period $N-1$ and ending at period 0. The value $J_0(x_0)$ is the optimal expected cost for the process when the initial stock at time 0 is x_0. During the calculations, the optimal policy is simultaneously computed from the minimization in the right-hand side of Eq. (3.1).

The example illustrates the main advantage offered by DP. While the original inventory problem requires an optimization over the set of policies, the DP algorithm of Eq. (3.1) decomposes this problem into a sequence of minimizations carried out over the set of controls. Each of these minimizations is much simpler than the original problem.

The DP Algorithm

We now state the DP algorithm for the basic problem and show its optimality by translating in mathematical terms the heuristic argument given above for the inventory example.

Proposition 3.1: For every initial state x_0, the optimal cost $J^*(x_0)$ of the basic problem is equal to $J_0(x_0)$, where the function J_0 is given by the last step of the following algorithm, which proceeds backward in time from period $N-1$ to period 0:

$$J_N(x_N) = g_N(x_N), \tag{3.2}$$

$$J_k(x_k) = \min_{u_k \in U_k(x_k)} \mathop{E}_{w_k} \Big\{ g_k(x_k, u_k, w_k) + J_{k+1}\big(f_k(x_k, u_k, w_k)\big) \Big\}, \quad k = 0, 1, \ldots, N-1, \tag{3.3}$$

where the expectation is taken with respect to the probability distribution of w_k, which depends on x_k and u_k. Furthermore, if $u_k^* = \mu_k^*(x_k)$ minimizes the right side of Eq. (3.3) for each x_k and k, the policy $\pi^* = \{\mu_0^*, \ldots, \mu_{N-1}^*\}$ is optimal.

Proof: † For any admissible policy $\pi = \{\mu_0, \mu_1, \ldots, \mu_{N-1}\}$ and each $k = 0, 1, \ldots, N-1$, denote $\pi^k = \{\mu_k, \mu_{k+1}, \ldots, \mu_{N-1}\}$. For $k = 0, 1, \ldots, N-1$, let $J_k^*(x_k)$ be the optimal cost for the $(N-k)$-stage problem that starts at state x_k and time k, and ends at time N; that is,

$$J_k^*(x_k) = \min_{\pi^k} \mathop{E}_{w_k, \ldots, w_{N-1}} \left\{ g_N(x_N) + \sum_{i=k}^{N-1} g_i\big(x_i, \mu_i(x_i), w_i\big) \right\}.$$

For $k = N$, we define $J_N^*(x_N) = g_N(x_N)$. We will show by induction that the functions J_k^* are equal to the functions J_k generated by the DP algorithm, so that for $k = 0$, we will obtain the desired result.

Indeed, we have by definition $J_N^* = J_N = g_N$. Assume that for some k and all x_{k+1}, we have $J_{k+1}^*(x_{k+1}) = J_{k+1}(x_{k+1})$. Then, since $\pi^k = (\mu_k, \pi^{k+1})$, we have for all x_k

$$\begin{aligned}
J_k^*(x_k) &= \min_{(\mu_k, \pi^{k+1})} \mathop{E}_{w_k, \ldots, w_{N-1}} \Bigg\{ g_k\big(x_k, \mu_k(x_k), w_k\big) \\
&\qquad\qquad + g_N(x_N) + \sum_{i=k+1}^{N-1} g_i\big(x_i, \mu_i(x_i), w_i\big) \Bigg\} \\
&= \min_{\mu_k} \mathop{E}_{w_k} \Bigg\{ g_k\big(x_k, \mu_k(x_k), w_k\big) \\
&\qquad + \min_{\pi^k} \Bigg[\mathop{E}_{w_{k+1}, \ldots, w_{N-1}} \Bigg\{ g_N(x_N) + \sum_{i=k+1}^{N-1} g_i\big(x_i, \mu_i(x_i), w_i\big) \Bigg\} \Bigg] \Bigg\} \\
&= \min_{\mu_k} \mathop{E}_{w_k} \Big\{ g_k\big(x_k, \mu_k(x_k), w_k\big) + J_{k+1}^*\Big(f_k\big(x_k, \mu_k(x_k), w_k\big)\Big) \Big\} \\
&= \min_{\mu_k} \mathop{E}_{w_k} \Big\{ g_k\big(x_k, \mu_k(x_k), w_k\big) + J_{k+1}\Big(f_k\big(x_k, \mu_k(x_k), w_k\big)\Big) \Big\} \\
&= \min_{u_k \in U_k(x_k)} \mathop{E}_{w_k} \Big\{ g_k(x_k, u_k, w_k) + J_{k+1}\big(f_k(x_k, u_k, w_k)\big) \Big\} \\
&= J_k(x_k),
\end{aligned}$$

---

† For a strictly rigorous proof, some technical mathematical issues must be addressed; see Section 1.5. These issues do not arise if the disturbance w_k takes a finite or countable number of values and the expected values of all terms in the expression of the cost function (2.2) are well defined and finite for every admissible policy π.

completing the induction. In the second equation above, we moved the minimum over π^{k+1} inside the braced expression, using the assumption that the probability distribution of w_i, $i = k+1, \ldots, N-1$, depends only on x_i and u_i. In the third equation, we used the definition of J^*_{k+1}, and in the fourth equation we used the induction hypothesis. In the fifth equation, we converted the minimization over μ_k to a minimization over u_k, using the fact that for any function F of x and u, we have

$$\min_{\mu \in M} F\big(x, \mu(x)\big) = \min_{u \in U(x)} F(x, u),$$

where M is the set of all functions $\mu(x)$ such that $\mu(x) \in U(x)$ for all x. **Q.E.D.**

The argument of the preceding proof provides an interpretation of $J_k(x_k)$ as the optimal cost for an $(N-k)$-stage problem starting at state x_k and time k, and ending at time N. We consequently call $J_k(x_k)$ the *cost-to-go* at state x_k and time k, and refer to J_k as the *cost-to-go function* at time k.

Ideally, we would like to use the DP algorithm to obtain closed-form expressions for J_k or an optimal policy. In this book, we will consider a large number of models that admit analytical solution by DP. Even if such models encompass oversimplified assumptions, they are often very useful. They may provide valuable insights about the structure of the optimal solution of more complex models, and they may form the basis for suboptimal control schemes. Furthermore, the broad collection of analytically solvable models provides helpful guidelines for modeling; when faced with a new problem it is worth trying to pattern its model after one of the principal analytically tractable models.

Unfortunately, in many practical cases an analytical solution is not possible, and one has to resort to numerical execution of the DP algorithm. This may be quite time-consuming since the minimization in the DP Eq. (3.3) must be carried out for each value of x_k. This means that the state space must be discretized in some way (if it is not already a finite set). The computational requirements are proportional to the number of discretization points, so for complex problems the computational burden may be excessive. Nonetheless, DP is the only general approach for sequential optimization under uncertainty, and even when it is computationally prohibitive, it can serve as the basis for more practical suboptimal approaches, which will be discussed in Chapter 6.

The following examples illustrate some of the analytical and computational aspects of DP.

Example 3.1

A certain material is passed through a sequence of two ovens (see Fig. 1.3.1). Denote

x_0: initial temperature of the material,

x_k, $k = 1, 2$: temperature of the material at the exit of oven k,

u_{k-1}, $k = 1, 2$: prevailing temperature in oven k.

We assume a model of the form

$$x_{k+1} = (1 - a)x_k + au_k, \qquad k = 0, 1,$$

where a is a known scalar from the interval $(0, 1)$. The objective is to get the final temperature x_2 close to a given target T, while expending relatively little energy. This is expressed by a cost function of the form

$$r(x_2 - T)^2 + u_0^2 + u_1^2,$$

where $r > 0$ is a given scalar. We assume no constraints on u_k. (In reality, there are constraints, but if we can solve the unconstrained problem and verify that the solution satisfies the constraints, everything will be fine.) The problem is deterministic; that is, there is no stochastic uncertainty. However, such problems can be placed within the basic framework by introducing a fictitious disturbance taking a unique value with probability one.

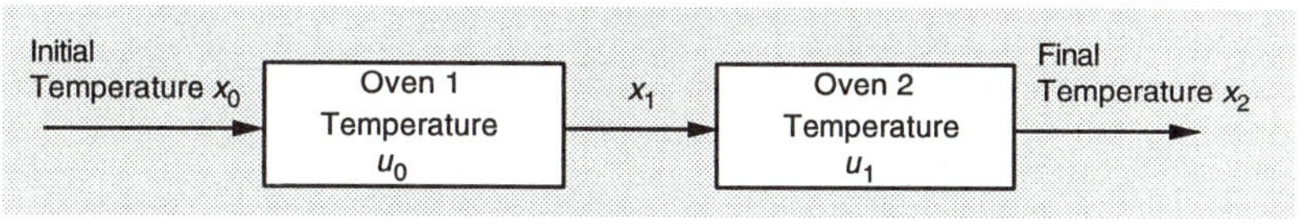

Figure 1.3.1 Problem of Example 3.1. The temperature of the material evolves according to $x_{k+1} = (1 - a)x_k + au_k$, where a is some scalar with $0 < a < 1$.

We have $N = 2$ and a terminal cost $g_2(x_2) = r(x_2 - T)^2$, so the initial condition for the DP algorithm is [cf. Eq. (3.2)]

$$J_2(x_2) = r(x_2 - T)^2.$$

For the next-to-last stage, we have [cf. Eq. (3.3)]

$$\begin{aligned} J_1(x_1) &= \min_{u_1} \big[u_1^2 + J_2(x_2)\big] \\ &= \min_{u_1} \Big[u_1^2 + J_2\big((1 - a)x_1 + au_1\big)\Big]. \end{aligned}$$

Substituting the previous form of J_2, we obtain

$$J_1(x_1) = \min_{u_1} \Big[u_1^2 + r\big((1 - a)x_1 + au_1 - T\big)^2\Big]. \tag{3.4}$$

This minimization will be done by setting to zero the derivative with respect to u_1. This yields

$$0 = 2u_1 + 2ra\big((1 - a)x_1 + au_1 - T\big),$$

and by collecting terms and solving for u_1, we obtain the optimal temperature for the last oven:

$$\mu_1^*(x_1) = \frac{ra\big(T - (1-a)x_1\big)}{1+ra^2}.$$

Note that this is not a single control but rather a control function, a rule that tells us the optimal oven temperature $u_1 = \mu_1^*(x_1)$ for each possible state x_1.

By substituting the optimal u_1 in the expression (3.4) for J_1, we obtain

$$\begin{aligned} J_1(x_1) &= \frac{r^2a^2\big((1-a)x_1 - T\big)^2}{(1+ra^2)^2} + r\left((1-a)x_1 + \frac{ra^2\big(T-(1-a)x_1\big)}{1+ra^2} - T\right)^2 \\ &= \frac{r^2a^2\big((1-a)x_1 - T\big)^2}{(1+ra^2)^2} + r\left(\frac{ra^2}{1+ra^2} - 1\right)^2 \big((1-a)x_1 - T\big)^2 \\ &= \frac{r\big((1-a)x_1 - T\big)^2}{1+ra^2}. \end{aligned}$$

We now go back one stage. We have [cf. Eq. (3.3)]

$$J_0(x_0) = \min_{u_0}\big[u_0^2 + J_1(x_1)\big] = \min_{u_0}\big[u_0^2 + J_1\big((1-a)x_0 + au_0\big)\big],$$

and by substituting the expression already obtained for J_1, we have

$$J_0(x_0) = \min_{u_0}\left[u_0^2 + \frac{r\big((1-a)^2x_0 + (1-a)au_0 - T\big)^2}{1+ra^2}\right].$$

We minimize with respect to u_0 by setting the corresponding derivative to zero. We obtain

$$0 = 2u_0 + \frac{2r(1-a)a\big((1-a)^2x_0 + (1-a)au_0 - T\big)}{1+ra^2}.$$

This yields, after some calculation, the optimal temperature of the first oven:

$$\mu_0^*(x_0) = \frac{r(1-a)a\big(T - (1-a)^2x_0\big)}{1+ra^2\big(1+(1-a)^2\big)}.$$

The optimal cost is obtained by substituting this expression in the formula for J_0. This leads to a straightforward but lengthy calculation, which in the end yields the rather simple formula

$$J_0(x_0) = \frac{r\big((1-a)^2x_0 - T\big)^2}{1+ra^2\big(1+(1-a)^2\big)}.$$

This completes the solution of the problem.

One noteworthy feature in the preceding example is the facility with which we obtained an analytical solution. A little thought while tracing the steps of the algorithm will convince the reader that what simplifies the solution is the quadratic nature of the cost and the linearity of the system equation. In Section 4.1 we will see that, generally, when the system is linear and the cost is quadratic then, regardless of the number of stages N, the optimal policy is given by a closed-form expression.

Another noteworthy feature of the example is that the optimal policy remains unaffected when a zero-mean stochastic disturbance is added in the system equation. To see this, assume that the material' s temperature evolves according to

$$x_{k+1} = (1-a)x_k + au_k + w_k, \qquad k = 0, 1,$$

where w_0, w_1 are independent random variables with given distribution, zero mean

$$E\{w_0\} = E\{w_1\} = 0,$$

and finite variance. Then the equation for J_1 [cf. Eq. (3.3)] becomes

$$\begin{aligned} J_1(x_1) &= \min_{u_1} \mathop{E}_{w_1} \left\{ u_1^2 + r\big((1-a)x_1 + au_1 + w_1 - T\big)^2 \right\} \\ &= \min_{u_1} \big[u_1^2 + r\big((1-a)x_1 + au_1 - T\big)^2 \\ &\qquad + 2rE\{w_1\}\big((1-a)x_1 + au_1 - T\big) + rE\{w_1^2\}\big]. \end{aligned}$$

Since $E\{w_1\} = 0$, we obtain

$$J_1(x_1) = \min_{u_1} \left[u_1^2 + r\big((1-a)x_1 + au_1 - T\big)^2 \right] + rE\{w_1^2\}.$$

Comparing this equation with Eq. (3.4), we see that the presence of w_1 has resulted in an additional inconsequential term, $rE\{w_1^2\}$. Therefore, the optimal policy for the last stage remains unaffected by the presence of w_1, while $J_1(x_1)$ is increased by the constant term $rE\{w_1^2\}$. It can be seen that a similar situation also holds for the first stage. In particular, the optimal cost is given by the same expression as before except for an additive constant that depends on $E\{w_0^2\}$ and $E\{w_1^2\}$.

If the optimal policy is unaffected when the disturbances are replaced by their means, we say that *certainty equivalence* holds. We will derive certainty equivalence results for several types of problems involving a linear system and a quadratic cost (see Sections 4.1, 5.2, and 5.3).

Example 3.2

To illustrate the computational aspects of DP, consider an inventory control problem that is slightly different from the one of Sections 1.1 and 1.2. In particular, we assume that inventory u_k and the demand w_k are nonnegative integers, and that the excess demand $(w_k - x_k - u_k)$ is lost. As a result, the stock equation takes the form

$$x_{k+1} = \max(0, x_k + u_k - w_k).$$

We also assume that there is an upper bound of 2 units on the stock that can be stored, i.e. there is a constraint $x_k + u_k \leq 2$. The holding/storage cost for the kth period is given by

$$(x_k + u_k - w_k)^2,$$

implying a penalty both for excess inventory and for unmet demand at the end of the kth period. The ordering cost is 1 per unit stock ordered. Thus the cost per period is

$$g_k(x_k, u_k, w_k) = u_k + (x_k + u_k - w_k)^2.$$

The terminal cost is assumed to be 0,

$$g_N(x_N) = 0.$$

The planning horizon N is 3 periods, and the initial stock x_0 is 0. The demand w_k has the same probability distribution for all periods, given by

$$p(w_k = 0) = 0.1, \qquad p(w_k = 1) = 0.7, \qquad p(w_k = 2) = 0.2.$$

The system can also be represented in terms of the transition probabilities $p_{ij}(u)$ between the three possible states, for the different values of the control (see Fig. 1.3.2).

The starting equation for the DP algorithm is

$$J_3(x_3) = 0,$$

since the terminal state cost is 0 [cf. Eq. (3.2)]. The algorithm takes the form [cf. Eq. (3.3)]

$$J_k(x_k) = \min_{\substack{0 \leq u_k \leq 2 - x_k \\ u_k = 0,1,2}} \mathop{E}_{w_k} \Big\{ u_k + (x_k + u_k - w_k)^2 + J_{k+1}\big(\max(0, x_k + u_k - w_k)\big) \Big\},$$

where $k = 0, 1, 2$, and x_k, u_k, w_k can take the values 0, 1, and 2.

Period 2: We compute $J_2(x_2)$ for each of the three possible states. We have

$$\begin{aligned} J_2(0) &= \min_{u_2 = 0,1,2} \mathop{E}_{w_2} \big\{ u_2 + (u_2 - w_2)^2 \big\} \\ &= \min_{u_2 = 0,1,2} \big[u_2 + 0.1(u_2)^2 + 0.7(u_2 - 1)^2 + 0.2(u_2 - 2)^2 \big]. \end{aligned}$$

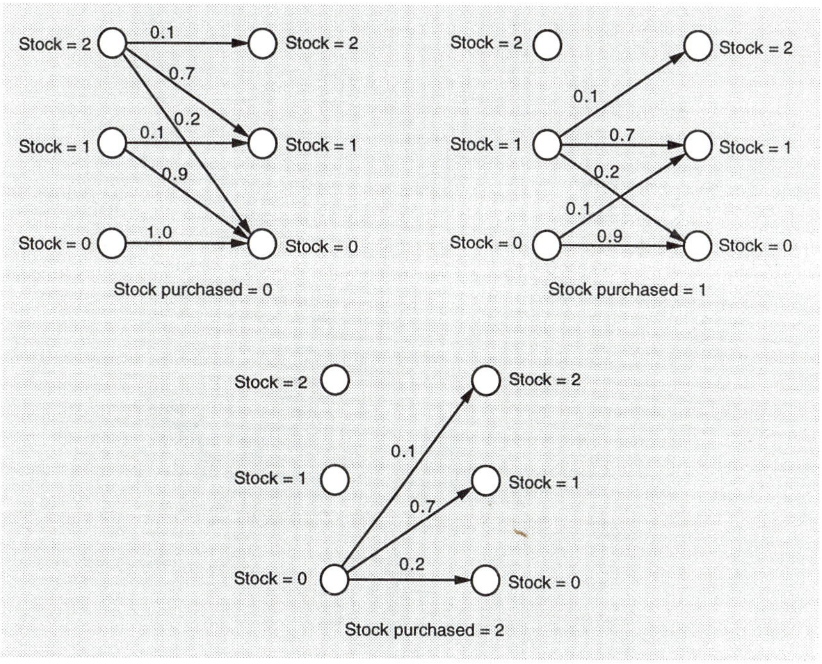

Stock	Stage 0 Cost-to-go	Stage 0 Optimal stock to purchase	Stage 1 Cost-to-go	Stage 1 Optimal stock to purchase	Stage 2 Cost-to-go	Stage 2 Optimal stock to purchase
0	3.67	1	2.5	1	1.3	1
1	2.67	0	1.2	0	0.3	0
2	2.608	0	1.68	0	1.1	0

Figure 1.3.2 System and DP results for Example 3.2. The transition probability diagrams for the different values of stock purchased (control) are shown. The numbers next to the arcs are the transition probabilities. The control $u = 1$ is not available at state 2 because of the limitation $x_k + u_k \leq 2$. Similarly, the control $u = 2$ is not available at states 1 and 2. The results of the DP algorithm are given in the table.

We calculate the expectation of the right side for each of the three possible values of u_2:

$$u_2 = 0 : E\{\cdot\} = 0.7 \cdot 1 + 0.2 \cdot 4 = 1.5,$$
$$u_2 = 1 : E\{\cdot\} = 1 + 0.1 \cdot 1 + 0.2 \cdot 1 = 1.3,$$
$$u_2 = 2 : E\{\cdot\} = 2 + 0.1 \cdot 4 + 0.7 \cdot 1 = 3.1.$$

Hence we have, by selecting the minimizing u_2,

$$J_2(0) = 1.3, \qquad \mu_2^*(0) = 1.$$

For $x_2 = 1$, we have

$$\begin{aligned} J_2(1) &= \min_{u_2=0,1} \mathop{E}_{w_2}\left\{u_2 + (1 + u_2 - w_2)^2\right\} \\ &= \min_{u_2=0,1} \left[u_2 + 0.1(1 + u_2)^2 + 0.7(u_2)^2 + 0.2(u_2 - 1)^2\right]. \end{aligned}$$

$$\begin{aligned} u_2 = 0 &: E\{\cdot\} = 0.1 \cdot 1 + 0.2 \cdot 1 = 0.3, \\ u_2 = 1 &: E\{\cdot\} = 1 + 0.1 \cdot 4 + 0.7 \cdot 1 = 2.1. \end{aligned}$$

Hence

$$J_2(1) = 0.3, \qquad \mu_2^*(1) = 0.$$

For $x_2 = 2$, the only admissible control is $u_2 = 0$, so we have

$$J_2(2) = \mathop{E}_{w_2}\left\{(2 - w_2)^2\right\} = 0.1 \cdot 4 + 0.7 \cdot 1 = 1.1,$$

$$J_2(2) = 1.1, \qquad \mu_2^*(2) = 0.$$

Period 1: Again we compute $J_1(x_1)$ for each of the three possible states $x_2 = 0, 1, 2$, using the values $J_2(0)$, $J_2(1)$, $J_2(2)$ obtained in the previous period. For $x_1 = 0$, we have

$$J_1(0) = \min_{u_1=0,1,2} \mathop{E}_{w_1}\left\{u_1 + (u_1 - w_1)^2 + J_2\big(\max(0, u_1 - w_1)\big)\right\},$$

$$\begin{aligned} u_1 = 0 &: E\{\cdot\} = 0.1 \cdot J_2(0) + 0.7\big(1 + J_2(0)\big) + 0.2\big(4 + J_2(0)\big) = 2.8, \\ u_1 = 1 &: E\{\cdot\} = 1 + 0.1\big(1 + J_2(1)\big) + 0.7 \cdot J_2(0) + 0.2\big(1 + J_2(0)\big) = 2.5, \\ u_1 = 2 &: E\{\cdot\} = 2 + 0.1\big(4 + J_2(2)\big) + 0.7\big(1 + J_2(1)\big) + 0.2 \cdot J_2(0) = 3.68, \end{aligned}$$

$$J_1(0) = 2.5, \qquad \mu_1^*(0) = 1.$$

For $x_1 = 1$, we have

$$J_1(1) = \min_{u_1=0,1} \mathop{E}_{w_1}\left\{u_1 + (1 + u_1 - w_1)^2 + J_2\big(\max(0, 1 + u_1 - w_1)\big)\right\},$$

$$\begin{aligned} u_1 = 0 &: E\{\cdot\} = 0.1\big(1 + J_2(1)\big) + 0.7 \cdot J_2(0) + 0.2\big(1 + J_2(0)\big) = 1.2, \\ u_1 = 1 &: E\{\cdot\} = 1 + 0.1\big(4 + J_2(2)\big) + 0.7\big(1 + J_2(1)\big) + 0.2 \cdot J_2(0) = 2.68, \end{aligned}$$

$$J_1(1) = 1.2, \qquad \mu_1^*(1) = 0.$$

For $x_1 = 2$, the only admissible control is $u_1 = 0$, so we have

$$\begin{aligned} J_1(2) &= \mathop{E}_{w_1}\left\{(2 - w_1)^2 + J_2\big(\max(0, 2 - w_1)\big)\right\} \\ &= 0.1\big(4 + J_2(2)\big) + 0.7\big(1 + J_2(1)\big) + 0.2 \cdot J_2(0) \\ &= 1.68, \end{aligned}$$

$$J_1(2) = 1.68, \qquad \mu_1^*(2) = 0.$$

Period 0: Here we need to compute only $J_0(0)$ since the initial state is known to be 0. We have

$$J_0(0) = \min_{u_0=0,1,2} \underset{w_0}{E}\Big\{u_0 + (u_0 - w_0)^2 + J_1\big(\max(0, u_0 - w_0)\big)\Big\},$$

$$u_0 = 0 : E\{\cdot\} = 0.1 \cdot J_1(0) + 0.7\big(1 + J_1(0)\big) + 0.2\big(4 + J_1(0)\big) = 4.0,$$
$$u_0 = 1 : E\{\cdot\} = 1 + 0.1\big(1 + J_1(1)\big) + 0.7 \cdot J_1(0) + 0.2\big(1 + J_1(0)\big) = 3.67,$$
$$u_0 = 2 : E\{\cdot\} = 2 + 0.1\big(4 + J_1(2)\big) + 0.7\big(1 + J_1(1)\big) + 0.2 \cdot J_1(0) = 5.108,$$

$$J_0(0) = 3.67, \qquad \mu_0^*(0) = 1.$$

If the initial state were not known a priori, we would have to compute in a similar manner $J_0(1)$ and $J_0(2)$, as well as the minimizing u_0. The reader may verify (Exercise 1.2) that these calculations yield

$$J_0(1) = 2.67, \qquad \mu_0^*(1) = 0,$$

$$J_0(2) = 2.608, \qquad \mu_0^*(2) = 0.$$

Thus the optimal ordering policy for each period is to order one unit if the current stock is zero and order nothing otherwise. The results of the DP algorithm are given in tabular form in Fig. 1.3.1.

Example 3.3 (Optimizing a Chess Match Strategy)

Consider the chess match example considered in Section 1.1. There, a player can select timid play (probabilities p_d and $1 - p_d$ for a draw or loss, respectively) or bold play (probabilities p_w and $1-p_w$ for a win or loss, respectively) in each game of the match. We want to formulate a DP algorithm for finding the policy that maximizes the player's probability of winning the match. Note that here we are dealing with a maximization problem. We can convert the problem to a minimization problem by changing the sign of the cost function, but a simpler alternative, which we will generally adopt, is to replace the minimization in the DP algorithm with maximization.

Let us consider the general case of an N-game match, and let the state be the *net score*, that is, the difference between the points of the player minus the points of the opponent (so a state of 0 corresponds to an even score). The optimal cost-to-go function at the start of the kth game is given by the dynamic programming recursion

$$\begin{aligned} J_k(x_k) = \max\big[&p_d J_{k+1}(x_k) + (1 - p_d) J_{k+1}(x_k - 1), \\ &p_w J_{k+1}(x_k + 1) + (1 - p_w) J_{k+1}(x_k - 1)\big]. \end{aligned} \tag{3.5}$$

The maximum above is taken over the two possible decisions:

(a) Timid play, which keeps the score at x_k with probability p_d, and changes x_k to $x_k - 1$ with probability $1 - p_d$.

(b) Bold play, which changes x_k to $x_k + 1$ or to $x_k - 1$ with probabilities p_w or $(1 - p_w)$, respectively.

It is optimal to play bold when

$$p_w J_{k+1}(x_k + 1) + (1 - p_w) J_{k+1}(x_k - 1) \geq p_d J_{k+1}(x_k) + (1 - p_d) J_{k+1}(x_k - 1)$$

or equivalently, if

$$\frac{p_w}{p_d} \geq \frac{J_{k+1}(x_k) - J_{k+1}(x_k - 1)}{J_{k+1}(x_k + 1) - J_{k+1}(x_k - 1)}. \tag{3.6}$$

The dynamic programming recursion is started with

$$J_N(x_N) = \begin{cases} 1 & \text{if } x_N > 0, \\ p_w & \text{if } x_N = 0, \\ 0 & \text{if } x_N < 0. \end{cases} \tag{3.7}$$

We have $J_N(0) = p_w$ because when the score is even after N games ($x_N = 0$), it is optimal to play bold in the first game of sudden death.

By executing the DP algorithm (3.5) starting with the terminal condition (3.7), and using the criterion (3.6) for optimality of bold play, we find the following, assuming that $p_d > p_w$:

$$\begin{aligned}
J_{N-1}(x_{N-1}) &= 1 \text{ for } x_{N-1} > 1; \quad \text{optimal play: either} \\
J_{N-1}(1) &= \max[p_d + (1 - p_d)p_w,\ p_w + (1 - p_w)p_w] \\
&= p_d + (1 - p_d)p_w; \quad \text{optimal play: timid} \\
J_{N-1}(0) &= p_w; \quad \text{optimal play: bold} \\
J_{N-1}(-1) &= p_w^2; \quad \text{optimal play: bold} \\
J_{N-1}(x_{N-1}) &= 0 \text{ for } x_{N-1} < -1; \quad \text{optimal play: either.}
\end{aligned}$$

Also, given $J_{N-1}(x_{N-1})$, and Eqs. (3.5) and (3.6) we obtain

$$\begin{aligned}
J_{N-2}(0) &= \max\big[p_d p_w + (1 - p_d)p_w^2,\ p_w\big(p_d + (1 - p_d)p_w\big) + (1 - p_w)p_w^2\big] \\
&= p_w\big(p_w + (p_w + p_d)(1 - p_w)\big)
\end{aligned}$$

and that if the score is even with 2 games remaining, it is optimal to play bold. Thus for a 2-game match, the optimal policy for both periods is to play timid if and only if the player is ahead in the score. The region of pairs (p_w, p_d) for which the player has a better than 50-50 chance to win a 2-game match is

$$R_2 = \big\{(p_w, p_d) \mid J_0(0) = p_w\big(p_w + (p_w + p_d)(1 - p_w)\big) > 1/2\big\},$$

and, as noted in the preceding section, it includes points where $p_w < 1/2$.

Example 3.4 (Finite-State Systems)

We mentioned earlier (cf. the examples in Section 1.1) that systems with a finite number of states can be represented either in terms of a discrete-time system equation or in terms of the probabilities of transition between the states. Let us work out the DP algorithm corresponding to the latter case. We will assume for the sake of the following discussion that the problem is stationary (i.e., the transition probabilities, the cost per stage, and the control constraint sets do not change from one stage to the next). Then, if

$$p_{ij}(u) = P\{x_{k+1} = j \mid x_k = i, u_k = u\}$$

are the transition probabilities, we can alternatively represent the system by the system equation (cf. the discussion of the previous section)

$$x_{k+1} = w_k,$$

where the probability distribution of the disturbance w_k is

$$P\{w_k = j \mid x_k = i, u_k = u\} = p_{ij}(u).$$

Using this system equation and denoting by $g(i,u)$ the expected cost per stage at state i when control u is applied, the DP algorithm can be rewritten as

$$J_k(i) = \min_{u \in U(i)} \left[g(i,u) + E\big\{J_{k+1}(w_k)\big\}\right]$$

or equivalently (in view of the distribution of w_k given previously)

$$J_k(i) = \min_{u \in U(i)} \left[g(i,u) + \sum_j p_{ij}(u) J_{k+1}(j)\right].$$

As an illustration, in the machine replacement example of Section 1.1, this algorithm takes the form

$$J_N(i) = 0, \qquad i = 1, \ldots, n,$$

$$J_k(i) = \min \left[R + g(1) + J_{k+1}(1),\ g(i) + \sum_{j=i}^{n} p_{ij} J_{k+1}(j)\right].$$

The two expressions in the above minimization correspond to the two available decisions (replace or not replace the machine).

In the queueing example of Section 1.1, the DP algorithm takes the form

$$J_N(i) = R(i), \qquad i = 0, 1, \ldots, n,$$

$$J_k(i) = \min \left[r(i) + c_f + \sum_{j=0}^{n} p_{ij}(u_f) J_{k+1}(j),\ r(i) + c_s + \sum_{j=0}^{n} p_{ij}(u_s) J_{k+1}(j)\right].$$

The two expressions in the above minimization correspond to the two possible decisions (fast and slow service).

1.4 STATE AUGMENTATION

We now discuss how to deal with situations where some of the assumptions of the basic problem are violated. Generally, in such cases the problem can be reformulated into the basic problem format. This process is called *state augmentation* because it typically involves the enlargement of the state space. The general guideline in state augmentation is to *include in the enlarged state at time k all the information that is known to the controller at time k and can be used with advantage in selecting u_k*. Unfortunately, state augmentation often comes at a price: the reformulated problem may have very complex state and/or control spaces. We provide some examples.

Time Lags

In many applications the system state x_{k+1} depends not only on the preceding state x_k and control u_k but also on earlier states and controls. In other words, states and controls influence future states with some time lag. Such situations can be handled by state augmentation; the state is expanded to include an appropriate number of earlier states and controls.

For simplicity, assume that there is at most a single period time lag in the state and control; that is, the system equation has the form

$$x_{k+1} = f_k(x_k, x_{k-1}, u_k, u_{k-1}, w_k), \qquad k = 1, 2, \ldots, N-1, \tag{4.1}$$

$$x_1 = f_0(x_0, u_0, w_0).$$

Time lags of more than one period can be handled similarly.

If we introduce additional state variables y_k and s_k, and we make the identifications $y_k = x_{k-1}$, $s_k = u_{k-1}$, the system equation (4.1) yields

$$\begin{pmatrix} x_{k+1} \\ y_{k+1} \\ s_{k+1} \end{pmatrix} = \begin{pmatrix} f_k(x_k, y_k, u_k, s_k, w_k) \\ x_k \\ u_k \end{pmatrix}. \tag{4.2}$$

By defining $\tilde{x}_k = (x_k, y_k, s_k)$ as the new state, we have

$$\tilde{x}_{k+1} = \tilde{f}_k(\tilde{x}_k, u_k, w_k),$$

where the system function $\tilde{f}_k$ is defined from Eq. (4.2). By using the preceding equation as the system equation and by expressing the cost function in terms of the new state, the problem is reduced to the basic problem without time lags. Naturally, the control u_k should now depend on the new state $\tilde{x}_k$, or equivalently a policy should consist of functions μ_k of the current state x_k, as well as the preceding state x_{k-1} and the preceding control u_{k-1}.

When the DP algorithm for the reformulated problem is translated in terms of the variables of the original problem, it takes the form

$$J_N(x_N) = g_N(x_N),$$

$$\begin{aligned} &J_{N-1}(x_{N-1}, x_{N-2}, u_{N-2}) \\ &\quad = \min_{u_{N-1}\in U_{N-1}(x_{N-1})} \mathop{E}_{w_{N-1}} \Big\{ g_{N-1}(x_{N-1}, u_{N-1}, w_{N-1}) \\ &\qquad\qquad + J_N\big(f_{N-1}(x_{N-1}, x_{N-2}, u_{N-1}, u_{N-2}, w_{N-1})\big)\Big\}, \end{aligned}$$

$$\begin{aligned} J_k(x_k, x_{k-1}, u_{k-1}) &= \min_{u_k\in U_k(x_k)} \mathop{E}_{w_k} \Big\{ g_k(x_k, u_k, w_k) \\ &+ J_{k+1}\big(f_k(x_k, x_{k-1}, u_k, u_{k-1}, w_k), x_k, u_k\big)\Big\}, \quad k = 1, \ldots, N-2, \end{aligned}$$

$$J_0(x_0) = \min_{u_0\in U_0(x_0)} \mathop{E}_{w_0} \big\{ g_0(x_0, u_0, w_0) + J_1\big(f_0(x_0, u_0, w_0), x_0, u_0\big)\big\}.$$

Similar reformulations are possible when time lags appear in the cost; for example, in the case where the cost is of the form

$$E\left\{ g_N(x_N, x_{N-1}) + g_0(x_0, u_0, w_0) + \sum_{k=1}^{N-1} g_k(x_k, x_{k-1}, u_k, w_k) \right\}.$$

The extreme case of time lags in the cost arises in the nonadditive form

$$E\big\{g_N(x_N, x_{N-1}, \ldots, x_0, u_{N-1}, \ldots, u_0, w_{N-1}, \ldots, w_0)\big\}.$$

Then, the problem can be reduced to the basic problem format, by taking as augmented state

$$\tilde{x}_k = \big(x_k, x_{k-1}, \ldots, x_0, u_{k-1}, \ldots, u_0, w_{k-1}, \ldots, w_0\big)$$

and $E\{g_N(\tilde{x}_N)\}$ as reformulated cost. Policies consist of functions μ_k of the present and past states $x_k, \ldots, x_0$, the past controls $u_{k-1}, \ldots, u_0$, and the past disturbances $w_{k-1}, \ldots, w_0$. Naturally, we must assume that the past disturbances are known to the controller. Otherwise, we are faced with a problem where the state is imprecisely known to the controller. Such problems are known as problems with imperfect state information and will be discussed in Chapter 5.

Correlated Disturbances

We turn now to the case where the disturbances w_k are correlated over time. A common situation that can be handled efficiently by state augmentation arises when the process $w_0, \ldots, w_{N-1}$ can be represented as the output of a linear system driven by independent random variables. As an example, suppose that by using statistical methods, we determine that the evolution of w_k can be modeled by an equation of the form

$$w_k = \lambda w_{k-1} + \xi_k,$$

where λ is a given scalar and $\{\xi_k\}$ is a sequence of independent random vectors with given distribution. Then we can introduce an additional state variable

$$y_k = w_{k-1}$$

and obtain a new system equation

$$\begin{pmatrix} x_{k+1} \\ y_{k+1} \end{pmatrix} = \begin{pmatrix} f_k(x_k, u_k, \lambda y_k + \xi_k) \\ \lambda y_k + \xi_k \end{pmatrix},$$

where the new state is the pair $\tilde{x}_k = (x_k, y_k)$ and the new disturbance is the vector ξ_k.

More generally, suppose that w_k can be modeled by

$$w_k = C_k y_{k+1},$$

where

$$y_{k+1} = A_k y_k + \xi_k, \qquad k = 0, \ldots, N-1,$$

A_k, C_k are known matrices of appropriate dimension, and ξ_k are independent random vectors with given distribution (see Fig. 1.4.1). By viewing y_k as an additional state variable, we obtain the new system equation

$$\begin{pmatrix} x_{k+1} \\ y_{k+1} \end{pmatrix} = \begin{pmatrix} f_k\big(x_k, u_k, C_k(A_k y_k + \xi_k)\big) \\ A_k y_k + \xi_k \end{pmatrix}.$$

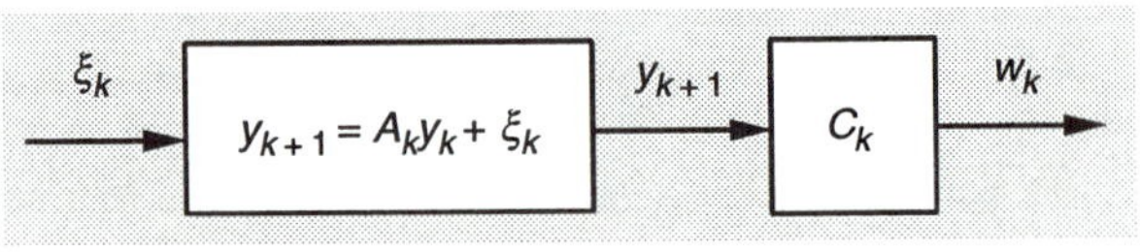

Figure 1.4.1 Representing correlated disturbances as the output of a linear system driven by independent random vectors.

Note that in order to have perfect state information, the controller must be able to observe y_k. Unfortunately, this is true only in the minority of practical cases; for example when C_k is the identity matrix and w_{k-1} is observed before u_k is applied. In the case of perfect state information, the DP algorithm takes the form

$$J_N(x_N, y_N) = g_N(x_N),$$

$$J_k(x_k, y_k) = \min_{u_k \in U_k(x_k)} \mathop{E}_{\xi_k}\big\{g_k\big(x_k, u_k, C_k(A_k y_k + \xi_k)\big) \\ + J_{k+1}\big(f_k\big(x_k, u_k, C_k(A_k y_k + \xi_k)\big), A_k y_k + \xi_k\big)\big\}.$$

Forecasts

Finally, consider the case where at time k the controller has access to a forecast y_k that results in a reassessment of the probability distribution of w_k and possibly of future disturbances. For example, y_k may be an exact prediction of w_k or an exact prediction that the probability distribution of w_k is a specific one out of a finite collection of distributions. Forecasts of interest in practice are, for example, probabilistic predictions on the state of the weather, the interest rate for money, and the demand for inventory.

Generally, forecasts can be handled by state augmentation although the reformulation into the basic problem format may be quite complex. We will treat here only a simple special case.

Assume that at the beginning of each period k, the controller receives an accurate prediction that the next disturbance w_k will be selected according to a particular probability distribution out of a given collection of distributions $\{Q_1, \ldots, Q_m\}$; that is, if the forecast is i, then w_k is selected according to Q_i. The a priori probability that the forecast will be i is denoted by p_i and is given.

As an example, suppose that in our earlier inventory example the demand w_k is determined according to one of three distributions Q_1, Q_2, and Q_3, corresponding to "small," "medium," and "large" demand. Each of the three types of demand occurs with a given probability at each time period, independently of the values of demand at previous time periods. However, the inventory manager, prior to ordering u_k, gets to know through a forecast the type of demand that will occur. (Note that it is the probability distribution of demand that becomes known through the forecast, not the demand itself.)

The forecasting process can be represented by means of the equation

$$y_{k+1} = \xi_k,$$

where y_{k+1} can take the values $1, \ldots, m$, corresponding to the m possible forecasts, and ξ_k is a random variable taking the value i with probability

p_i. The interpretation here is that when ξ_k takes the value i, then w_{k+1} will occur according to the distribution Q_i.

By combining the system equation with the forecast equation $y_{k+1} = \xi_k$, we obtain an augmented system given by

$$\begin{pmatrix} x_{k+1} \\ y_{k+1} \end{pmatrix} = \begin{pmatrix} f_k(x_k, u_k, w_k) \\ \xi_k \end{pmatrix}.$$

The new state is $\tilde{x}_k = (x_k, y_k)$, and because the forecast y_k is known at time k, perfect state information prevails. The new disturbance is $\tilde{w}_k = (w_k, \xi_k)$, and its probability distribution is determined by the distributions Q_i and the probabilities p_i, and depends explicitly on $\tilde{x}_k$ (via y_k) but not on the prior disturbances. Thus, by suitable reformulation of the cost, the problem can be cast into the basic problem format. Note that the control applied depends on both the current state and the current forecast.

The DP algorithm takes the form

$$J_N(x_N, y_N) = g_N(x_N), \tag{4.3}$$

$$\begin{aligned} J_k(x_k, y_k) = \min_{u_k \in U_k(x_k)} \underset{w_k}{E} \Big\{ & g_k(x_k, u_k, w_k) \\ & + \sum_{i=1}^{m} p_i J_{k+1}\big(f_k(x_k, u_k, w_k), i\big) \,\Big|\, y_k \Big\}, \end{aligned} \tag{4.4}$$

where y_k may take the values $1, \ldots, m$, and the expectation over w_k is taken with respect to the distribution Q_{y_k}.

There is a nice simplification of the above algorithm that allows DP to be executed over a smaller space. In particular, define

$$\hat{J}_k(x_k) = \sum_{i=1}^{m} p_i J_k(x_k, i), \qquad k = 0, 1, \ldots, N-1,$$

and

$$\hat{J}_N(x_N) = g_N(x_N).$$

Then from Eq. (4.4), we obtain the algorithm

$$\begin{aligned} \hat{J}_k(x_k) = \sum_{i=1}^{m} p_i \min_{u_k \in U_k(x_k)} \underset{w_k}{E} \Big\{ & g_k(x_k, u_k, w_k) \\ & + \hat{J}_{k+1}\big(f_k(x_k, u_k, w_k)\big) \,\Big|\, y_k = i \Big\}, \end{aligned}$$

which is executed over the space of x_k rather than x_k *and* y_k. This simplification arises also in other contexts where the state has a component that cannot be affected by the choice of control (see Exercise 1.22).

It should be clear that the preceding formulation admits several extensions; one example is the case where forecasts can be influenced by the control action and involve several future disturbances. However, the price for these extensions is increased complexity of the corresponding DP algorithm.

1.5 SOME MATHEMATICAL ISSUES

Let us now discuss some technical issues relating to the basic problem formulation. The reader who is not mathematically inclined need not be concerned about these issues and can skip this section without loss of continuity.

Once an admissible policy $\{\mu_0, \ldots, \mu_{N-1}\}$ is adopted, the following sequence of events is envisioned at the typical stage k:

1. The controller observes x_k and applies $u_k = \mu_k(x_k)$.
2. The disturbance w_k is generated according to the given distribution $P_k\big(\cdot \mid x_k, \mu_k(x_k)\big)$.
3. The cost $g_k\big(x_k, \mu_k(x_k), w_k\big)$ is incurred and added to previous costs.
4. The next state x_{k+1} is generated according to the system equation

$$x_{k+1} = f_k\big(x_k, \mu_k(x_k), w_k\big).$$

If this is the last stage ($k = N - 1$), the terminal cost $g_N(x_N)$ is added to previous costs and the process terminates. Otherwise, k is incremented, and the same sequence of events is repeated for the next stage.

For each stage, the above process is well-defined and is couched in precise probabilistic terms. Matters are, however, complicated by the need to view the cost as a well-defined random variable with well-defined expected value. The framework of probability theory requires that for each policy we define an underlying probability space, that is, a set Ω, a collection of events in Ω, and a probability measure on these events. In addition, the cost must be a well-defined random variable on this space in the sense of Appendix C (a measurable function from the probability space into the real line in the terminology of measure-theoretic probability theory). For this to be true, additional (measurability) assumptions on the functions f_k, g_k, and μ_k may be required, and it may be necessary to introduce additional structure on the spaces S_k, C_k, and D_k. Furthermore, these assumptions may restrict the class of admissible policies, since the functions μ_k may be constrained to satisfy additional (measurability) requirements.

Thus, unless these additional assumptions and structure are specified, the basic problem is formulated inadequately. Unfortunately, a rigorous formulation for general state, control, and disturbance spaces is well beyond the mathematical framework of this introductory book and will not be undertaken here. Nonetheless, it turns out that these difficulties are mainly technical and do not substantially affect the basic results to be obtained. For this reason, we find it convenient to proceed with informal derivations and arguments; this is consistent with most of the literature on the subject.

We would like to stress, however, that under at least one frequently satisfied assumption, the mathematical difficulties mentioned above disappear. In particular, let us assume that the disturbance spaces D_k are all countable and the expected values of all terms in the cost are finite for every admissible policy (this is true in particular if the spaces D_k are finite sets). Then, for every admissible policy, the expected values of all the cost terms can be written as (possibly infinite) sums involving the probabilities of the elements of the spaces D_k.

Alternatively, one may write the cost as

$$J_\pi(x_0) = \mathop{E}_{x_1,\ldots,x_N}\left\{ g_N(x_N) + \sum_{k=0}^{N-1} \tilde{g}_k\big(x_k, \mu_k(x_k)\big) \right\}, \tag{5.1}$$

where

$$\tilde{g}_k\big(x_k, \mu_k(x_k)\big) = \mathop{E}_{w_k}\big\{ g_k\big(x_k, \mu_k(x_k), w_k\big) \mid x_k, \mu_k(x_k)\big\},$$

with the preceding expectation taken with respect to the distribution $P_k\big(\cdot \mid x_k, \mu_k(x_k)\big)$ defined on the countable set D_k. Then one may take as the basic probability space the Cartesian product of the spaces $\tilde{S}_k$, $k = 1, \ldots, N$, given for all k by

$$\tilde{S}_{k+1} = \big\{ x_{k+1} \in S_{k+1} \mid x_{k+1} = f_k\big(x_k, \mu_k(x_k), w_k\big),\ x_k \in \tilde{S}_k,\ w_k \in D_k \big\},$$

where $\tilde{S}_0 = \{x_0\}$. The set $\tilde{S}_k$ is the subset of all states that can be reached at time k when the policy $\{\mu_0, \ldots, \mu_{N-1}\}$ is used. Because the disturbance spaces D_k are countable, the sets $\tilde{S}_k$ are also countable (this is true since the union of any countable collection of countable sets is a countable set). The system equation $x_{k+1} = f_k\big(x_k, \mu_k(x_k), w_k\big)$, the probability distributions $P_k\big(\cdot \mid x_k, \mu_k(x_k)\big)$, the initial state x_0, and the policy $\{\mu_0, \ldots, \mu_{N-1}\}$ define a probability distribution on the countable set $\tilde{S}_1 \times \cdots \times \tilde{S}_N$, and the expected value in the cost expression (5.1) is defined with respect to this latter distribution.

In conclusion, the basic problem has been formulated rigorously only when the disturbance spaces $D_0, \ldots, D_{N-1}$ are countable sets. In the absence of countability of these spaces, the reader should interpret subsequent results and conclusions as essentially correct but mathematically imprecise statements. In fact, when discussing infinite horizon problems (where the need for precision is greater), we will make the countability assumption explicit.

We note, however, that the advanced reader will have little difficulty in establishing rigorously most of our subsequent results concerning specific finite horizon applications, even if the countability assumption is not satisfied. This can be done by using the DP algorithm as a verification theorem. In particular, if one can find within a subset of policies $\tilde{\Pi}$ (such as those satisfying certain measurability restrictions) a policy that attains the minimum in the DP algorithm, then this policy can be readily shown to be optimal within $\tilde{\Pi}$. This result is developed in Exercise 1.12, and can be used by the mathematically oriented reader to establish rigorously many of our subsequent results concerning specific applications. For example, in linear-quadratic problems (Section 4.1) one determines from the DP algorithm a policy in closed form, which is linear in the current state. When w_k can take uncountably many values, it is necessary that admissible policies consist of Borel measurable functions μ_k. Since the linear policy obtained from the DP algorithm belongs to this class, the result of Exercise 1.12 guarantees that this policy is optimal. For a rigorous mathematical treatment of DP that resolves the associated measurability issues and supplements the present text, see [BeS78].

1.6 NOTES, SOURCES, AND EXERCISES

Dynamic programming is a simple mathematical technique that has been used for many years by engineers, mathematicians, and social scientists in a variety of contexts. It was Bellman, however, who realized in the early fifties that DP could be developed (in conjunction with the then appearing digital computer) into a systematic tool for optimization. In his influential books [Bel57], [BeD62], Bellman demonstrated the broad scope of DP and helped streamline its theory.

Following Bellman's works, there was much research on DP. In particular, the mathematical and algorithmic issues associated with infinite horizon problems were extensively investigated, extensions to continuous-time problems were formulated and analyzed, and the mathematical issues discussed in the preceding section relating to the formulation of DP were addressed. In addition, DP was used in a very broad variety of applications, ranging from many branches of engineering to statistics, economics, finance, and some of the social sciences. Samples of these applications will be given in subsequent chapters.

EXERCISES

1.1

Consider the system

$$x_{k+1} = x_k + u_k + w_k, \qquad k = 0, 1, 2, 3,$$

with initial state $x_0 = 5$, and the cost function

$$\sum_{k=0}^{3}(x_k^2 + u_k^2).$$

Apply the DP algorithm for the following three cases:

(a) The control constraint set $U_k(x_k)$ is $\{u \mid 0 \leq x_k + u \leq 5,\ u : \text{integer}\}$ for all x_k and k, and the disturbance w_k is equal to zero for all k.

(b) The control constraint and the disturbance w_k are as in part (a), but there is in addition a constraint $x_4 = 5$ on the final state. *Hint*: For this problem you need to define a state space for x_4 that consists of just the value $x_4 = 5$, and also to redefine $U_3(x_3)$. Alternatively, you may use a terminal cost $g_4(x_4)$ equal to a very large number for $x_4 \neq 5$.

(c) The control constraint is as in part (a) and the disturbance w_k takes the values -1 and 1 with equal probability $1/2$ for all x_k and u_k, except if $x_k + u_k$ is equal to 0 or 5, in which case $w_k = 0$ with probability 1.

1.2

Carry out the calculations needed to verify that $J_0(1) = 2.67$ and $J_0(2) = 2.608$ in Example 3.2.

1.3

Suppose we have a machine that is either running or is broken down. If it runs throughout one week, it makes a gross profit of \$100. If it fails during the week, gross profit is zero. If it is running at the start of the week and we perform preventive maintenance, the probability that it will fail during the week is 0.4. If we do not perform such maintenance, the probability of failure is 0.7. However, maintenance will cost \$20. When the machine is broken down at the start of the week, it may either be repaired at a cost of \$40, in which case it will fail during the week with a probability of 0.4, or it may be replaced at a cost of \$150 by a new machine that is guaranteed to run through its first week of operation. Find the optimal repair, replacement, and maintenance policy that maximizes total profit over four weeks, assuming a new machine at the start of the first week.

1.4

A game of the blackjack variety is played by two players as follows: Both players throw a die. The first player, knowing his opponent's result, may stop or may throw the die again and add the result to the result of his previous throw. He then may stop or throw again and add the result of the new throw to the sum of his previous throws. He may repeat this process as many times as he wishes. If his sum exceeds seven (i.e., he busts), he loses the game. If he stops before exceeding seven, the second player takes over and throws the die successively until the sum of his throws is four or higher. If the sum of the second player is over seven, he loses the game. Otherwise the player with the larger sum wins, and in case of a tie the second player wins. The problem is to determine a stopping strategy for the first player that maximizes his probability of winning for each possible initial throw of the second player. Formulate the problem in terms of DP and find an optimal stopping strategy for the case where the second player's initial throw is three. *Hint*: Take $N = 6$ and a state space consisting of the following 15 states:

$$x^1 : \text{busted}$$

$$x^{1+i} : \text{already stopped at sum } i \ (1 \leq i \leq 7),$$

$$x^{8+i} : \text{current sum is } i \text{ but the player has not yet stopped } (1 \leq i \leq 7).$$

The optimal strategy is to throw until the sum is four or higher.

1.5 (Min-Max Problems) Solution on www

In the framework of the basic problem, consider the case where the disturbances $w_0, w_1, \ldots, w_{N-1}$ do not have a probabilistic description but rather are known to belong to corresponding given sets $W_k(x_k, u_k) \subset D_k$, $k = 0, 1, \ldots, N-1$, which may depend on the current state x_k and control u_k. Consider the problem of finding a policy $\pi = \{\mu_0, \ldots, \mu_{N-1}\}$ with $\mu_k(x_k) \in U_k(x_k)$ for all x_k, k, which minimizes the cost function

$$J_\pi(x_0) = \max_{\substack{w_k \in W_k(x_k, \mu_k(x_k)) \\ k=0,1,\ldots,N-1}} \left[g_N(x_N) + \sum_{k=0}^{N-1} g_k\big(x_k, \mu_k(x_k), w_k\big) \right].$$

The DP algorithm for this problem takes the form

$$J_N(x_N) = g_N(x_N),$$

$$J_k(x_k) = \min_{u_k \in U(x_k)} \max_{w_k \in W_k(x_k, u_k)} \big[g_k(x_k, u_k, w_k) + J_{k+1}\big(f_k(x_k, u_k, w_k)\big) \big].$$

Assuming that $J_k(x_k) > -\infty$ for all x_k and k, show that the optimal cost equals $J_0(x_0)$. *Hint*: Imitate the proof for the stochastic case. Prove and use the following fact: If U, W, X are three sets, $f : W \to X$ is a function, and M is the set of all functions $\mu : X \to U$, then for any functions $G_0 : W \to (-\infty, \infty]$, $G_1 : X \times U \to (-\infty, \infty]$ such that

$$\min_{u \in U} G_1\big(f(w), u\big) > -\infty, \qquad \text{for all } w \in W$$

we have

$$\min_{\mu \in M} \max_{w \in W} \big[G_0(w) + G_1\big(f(w), \mu(f(w))\big) \big] = \max_{w \in W} \big[G_0(w) + \min_{u \in U} G_1\big(f(w), u\big) \big].$$

1.6 (Discounted Cost per Stage)

In the framework of the basic problem, consider the case where the cost is of the form

$$\underset{\substack{w_k \\ k=0,1,\ldots,N-1}}{E} \left\{ \alpha^N g_N(x_N) + \sum_{k=0}^{N-1} \alpha^k g_k(x_k, u_k, w_k) \right\},$$

where α is a discount factor with $0 < \alpha < 1$. Show that an alternate form of the DP algorithm is given by

$$V_N(x_N) = g_N(x_N),$$

$$V_k(x_k) = \min_{u_k \in U_k(x_k)} \underset{w_k}{E} \left\{ g_k(x_k, u_k, w_k) + \alpha V_{k+1}\big(f_k(x_k, u_k, w_k)\big) \right\}.$$

1.7 (Exponential Cost Function)

In the framework of the basic problem, consider the case where the cost is of the form

$$\underset{\substack{w_k \\ k=0,1,\ldots,N-1}}{E} \left\{ \exp\left(g_N(x_N) + \sum_{k=0}^{N-1} g_k(x_k, u_k, w_k) \right) \right\}.$$

(a) Show that the optimal cost and an optimal policy can be obtained from the DP-like algorithm

$$J_N(x_N) = \exp\big(g_N(x_N)\big),$$

$$J_k(x_k) = \min_{u_k \in U_k(x_k)} \underset{w_k}{E} \left\{ J_{k+1}\big(f_k(x_k, u_k, w_k)\big) \exp\big(g_k(x_k, u_k, w_k)\big) \right\}.$$

(b) Define the functions $V_k(x_k) = \ln J_k(x_k)$. Assume also that g_k is a function of x_k and u_k only (and not of w_k). Show that the above algorithm can be rewritten as

$$V_N(x_N) = g_N(x_N),$$

$$V_k(x_k) = \min_{u_k \in U_k(x_k)} \left\{ g_k(x_k, u_k) + \ln \underset{w_k}{E} \left\{ \exp\Big(V_{k+1}\big(f_k(x_k, u_k, w_k)\big)\Big) \right\} \right\}.$$

1.8 (Terminating Process)

In the framework of the basic problem, consider the case where the system evolution terminates at time i when a given value $\overline{w}_i$ of the disturbance at time i occurs, or when a termination decision u_i is made by the controller. If termination occurs at time i, the resulting cost is

$$T + \sum_{k=0}^{i} g_k(x_k, u_k, w_k),$$

where T is a termination cost. If the process has not terminated up to the final time N, the resulting cost is $g_N(x_N) + \sum_{k=0}^{N-1} g_k(x_k, u_k, w_k)$. Reformulate the problem into the framework of the basic problem. *Hint*: Augment the state space with a special termination state.

1.9 (Multiplicative Cost)

In the framework of the basic problem, consider the case where the cost has the multiplicative form

$$\mathop{E}_{\substack{w_k \\ k=0,1,\ldots,N-1}} \{g_N(x_N) \cdot g_{N-1}(x_{N-1}, u_{N-1}, w_{N-1}) \cdots g_0(x_0, u_0, w_0)\}.$$

Develop a DP-like algorithm for this problem assuming that $g_k(x_k, u_k, w_k) \geq 0$ for all x_k, u_k, w_k, and k.

1.10

Assume that we have a vessel whose maximum weight capacity is z and whose cargo is to consist of different quantities of N different items. Let v_i denote the value of the ith type of item, w_i the weight of ith type of item, and x_i the number of items of type i that are loaded in the vessel. The problem is to find the most valuable cargo, i.e., to maximize $\sum_{i=1}^{N} x_i v_i$ subject to the constraints $\sum_{i=1}^{N} x_i w_i \leq z$ and $x_i = 0, 1, 2, \ldots$ Formulate this problem in terms of DP.

1.11

Consider a device consisting of N stages connected in series, where each stage consists of a particular component. The components are subject to failure, and to increase the reliability of the device duplicate components are provided. For $j = 1, 2, \ldots, N$, let $(1 + m_j)$ be the number of components for the jth stage, let $p_j(m_j)$ be the probability of successful operation of the jth stage when $(1 + m_j)$ components are used, and let c_j denote the cost of a single component at the jth stage. Formulate in terms of DP the problem of finding the number of components at each stage that maximize the reliability of the device expressed by

$$p_1(m_1) \cdot p_2(m_2) \cdots p_N(m_N),$$

subject to the cost constraint $\sum_{j=1}^{N} c_j m_j \leq A$, where $A > 0$ is given.

1.12 (Minimization over a Subset of Policies)

This problem is primarily of theoretical interest (see the end of Section 1.5). Consider a variation of the basic problem whereby we seek

$$\min_{\pi \in \tilde{\Pi}} J_\pi(x_0),$$

where $\tilde{\Pi}$ is some given *subset* of the set of sequences $\{\mu_0, \mu_1, \ldots, \mu_{N-1}\}$ of functions $\mu_k : S_k \to C_k$ with $\mu_k(x_k) \in U_k(x_k)$ for all $x_k \in S_k$. Assume that

$$\pi^* = \{\mu_0^*, \mu_1^*, \ldots, \mu_{N-1}^*\}$$

belongs to $\tilde{\Pi}$ and attains the minimum in the DP algorithm; that is, for all $k = 0, 1, \ldots, N-1$ and $x_k \in S_k$,

$$\begin{aligned} J_k(x_k) &= \mathop{E}_{w_k} \Big\{ g_k\big(x_k, \mu_k^*(x_k), w_k\big) + J_{k+1}\big(f_k(x_k, \mu_k^*(x_k), w_k)\big) \Big\} \\ &= \min_{u_k \in U_k(x_k)} \mathop{E}_{w_k} \Big\{ g_k(x_k, u_k, w_k) + J_{k+1}\big(f_k(x_k, u_k, w_k)\big) \Big\}, \end{aligned}$$

with $J_N(x_N) = g_N(x_N)$. Assume further that the functions J_k are real-valued and that the preceding expectations are well defined and finite. Show that π^* is optimal within $\tilde{\Pi}$ and that the corresponding optimal cost is equal to $J_0(x_0)$.

1.13 (Semilinear Systems)

Consider a problem involving the system

$$x_{k+1} = A_k x_k + f_k(u_k) + w_k,$$

where $x_k \in \Re^n$, f_k are given functions, and A_k and w_k are random $n \times n$ matrices and n-vectors, respectively, with given probability distributions that do not depend on x_k, u_k or prior values of A_k and w_k. Assume that the cost is of the form

$$\mathop{E}_{\substack{A_k, w_k \\ k=0,1,\ldots,N-1}} \left\{ c_N' x_N + \sum_{k=0}^{N-1} \Big(c_k' x_k + g_k\big(\mu_k(x_k)\big) \Big) \right\},$$

where c_k are given vectors and g_k are given functions. Show that if the optimal cost for this problem is finite and the control constraint sets $U_k(x_k)$ are independent of x_k, then the cost-to-go functions of the DP algorithm are affine (linear plus constant). Assuming that there is at least one optimal policy, show that there exists an optimal policy that consists of constant functions μ_k^*; that is, $\mu_k^*(x_k) =$ constant for all $x_k \in \Re^n$.

1.14

A farmer annually producing x_k units of a certain crop stores $(1-u_k)x_k$ units of his production, where $0 \leq u_k \leq 1$, and invests the remaining $u_k x_k$ units, thus increasing the next year's production to a level x_{k+1} given by

$$x_{k+1} = x_k + w_k u_k x_k, \qquad k = 0, 1, \ldots, N-1.$$

The scalars w_k are independent random variables with identical probability distributions that do not depend either on x_k or u_k. Furthermore, $E\{w_k\} = \overline{w} > 0$. The problem is to find the optimal investment policy that maximizes the total expected product stored over N years

$$\mathop{E}_{\substack{w_k \\ k=0,1,\ldots,N-1}} \left\{ x_N + \sum_{k=0}^{N-1} (1-u_k)x_k \right\}.$$

Show the optimality of the following policy that consists of constant functions:

(a) If $\overline{w} > 1$, $\mu_0^*(x_0) = \cdots = \mu_{N-1}^*(x_{N-1}) = 1$.

(b) If $0 < \overline{w} < 1/N$, $\mu_0^*(x_0) = \cdots = \mu_{N-1}^*(x_{N-1}) = 0$.

(c) If $1/N \leq \overline{w} \leq 1$,

$$\mu_0^*(x_0) = \cdots = \mu_{N-\overline{k}-1}(x_{N-\overline{k}-1}) = 1,$$

$$\mu_{N-\overline{k}}^*(x_{N-\overline{k}}) = \cdots = \mu_{N-1}^*(x_{N-1}) = 0,$$

where $\overline{k}$ is such that $1/(\overline{k}+1) < \overline{w} \leq 1/\overline{k}$.

1.15

Let x_k denote the number of educators in a certain country at time k and let y_k denote the number of research scientists at time k. New scientists (potential educators or research scientists) are produced during the kth period by educators at a rate γ_k per educator, while educators and research scientists leave the field due to death, retirement, and transfer at a rate δ_k. The scalars γ_k, $k = 0, 1, \ldots, N-1$, are independent identically distributed random variables taking values within a closed and bounded interval of positive numbers. Similarly δ_k, $k = 0, 1, \ldots, N-1$, are independent identically distributed and take values in an interval $[\delta, \delta']$ with $0 < \delta \leq \delta' < 1$. By means of incentives, a science policy maker can determine the proportion u_k of new scientists produced at time k who become educators. Thus, the number of research scientists and educators evolves according to the equations

$$x_{k+1} = (1-\delta_k)x_k + u_k \gamma_k x_k,$$

$$y_{k+1} = (1-\delta_k)y_k + (1-u_k)\gamma_k x_k.$$

The initial numbers x_0, y_0 are known, and it is required to find a policy

$$\{\mu_0^*(x_0, y_0), \ldots, \mu_{N-1}^*(x_{N-1}, y_{N-1})\}$$

with

$$0 < \alpha \le \mu_k^*(x_k, y_k) \le \beta < 1, \qquad \text{for all } x_k, y_k, \text{and } k,$$

which maximizes $E_{\gamma_k, \delta_k}\{y_N\}$ (i.e., the expected final number of research scientists after N periods). The scalars α and β are given.

(a) Show that the cost-to-go functions $J_k(x_k, y_k)$ are linear; that is, for some scalars ξ_k, ζ_k,

$$J_k(x_k, y_k) = \xi_k x_k + \zeta_k y_k.$$

(b) Derive an optimal policy $\{\mu_0^*, \ldots, \mu_{N-1}^*\}$ under the assumption $E\{\gamma_k\} > E\{\delta_k\}$ and show that this optimal policy can consist of constant functions.

(c) Assume that the proportion of new scientists who become educators at time k is $u_k + \epsilon_k$ (rather than u_k), where ϵ_k are identically distributed independent random variables that are also independent of γ_k, δ_k and take values in the interval $[-\alpha, 1-\beta]$. Derive the form of the cost-to-go functions and the optimal policy.

W **1.16**

Given a sequence of matrix multiplications

$$M_1 M_2 \cdots M_k M_{k+1} \cdots M_N,$$

where each M_k is a matrix of dimension $n_k \times n_{k+1}$, the order in which multiplications are carried out can make a difference. For example, if $n_1 = 1$, $n_2 = 10$, $n_3 = 1$, and $n_4 = 10$, the calculation $\big((M_1 M_2) M_3\big)$ requires 20 scalar multiplications, but the calculation $\big(M_1 (M_2 M_3)\big)$ requires 200 scalar multiplications (multiplying an $m \times n$ matrix with an $n \times k$ matrix requires mnk scalar multiplications). Derive a DP algorithm for finding the optimal multiplication order. Solve the problem for $N = 3$, $n_1 = 2$, $n_2 = 10$, $n_3 = 5$, and $n_4 = 1$.

1.17

The paragraphing problem deals with breaking up a sequence of N words of given lengths into lines of length A. Let $w_1, \ldots, w_N$ be the words and let $L_1, \ldots, L_N$ be their lengths. In a simple version of the problem, words are separated by blanks whose ideal width is b, but blanks can stretch or shrink if necessary, so that a line $w_i, w_{i+1}, \ldots, w_{i+k}$ has length exactly A. The cost associated with the line is $(k+1)|b' - b|$, where $b' = (A - L_i - \cdots - L_{i+k})/(k+1)$ is the actual average width of the blanks, except if we have the last line ($N = i + k$), in which case the cost is zero when $b' \ge b$. Formulate a DP algorithm for finding the minimum cost separation. *Hint*: Consider the subproblems of optimally separating $w_i, \ldots, w_N$ for $i = 1, \ldots, N$.

1.18 (Computer Assignment)

In the classical game of blackjack the player draws cards knowing only one card of the dealer. The player loses upon reaching a sum of cards exceeding 21. If the player stops before exceeding 21, the dealer draws cards until reaching 17 or higher. The dealer loses upon reaching a sum exceeding 21 or stopping at a lower sum than the player's. If player and dealer end up with an equal sum no one wins. In all other cases the dealer wins. An ace for the player may be counted as a 1 or an 11 as the player chooses. An ace for the dealer is counted as an 11 if this results in a sum from 17 to 21 and as a 1 otherwise. Jacks, queens, and kings count as 10 for both dealer and player. We assume an infinite card deck so the probability of a particular card showing up is independent of earlier cards.

(a) For every possible initial dealer card, calculate the probability that the dealer will reach a sum of 17, 18, 19, 20, 21, or over 21.

(b) Calculate the optimal choice of the player (draw or stop) for each of the possible combinations of dealer's card and player's sum of 12 to 20. Assume that the player's cards do not include an ace.

(c) Repeat part (b) for the case where the player's cards include an ace.

1.19 [Shr81]

A decision maker must continually choose between two activities over a time interval $[0, T]$. Choosing activity i at time t, where $i = 1, 2$, earns reward at a rate $g_i(t)$, and every switch between the two activities costs $c > 0$. Thus, for example, the reward for starting with activity 1, switching to 2 at time t_1, and switching back to 1 at time $t_2 > t_1$ earns total reward

$$\int_0^{t_1} g_1(t)\,dt + \int_{t_1}^{t_2} g_2(t)\,dt + \int_{t_2}^{T} g_1(t)\,dt - 2c.$$

We want to find a set of switching times that maximize the total reward. Assume that the function $g_1(t) - g_2(t)$ changes sign a finite number of times in the interval $[0, T]$. Formulate the problem as a finite horizon problem and write the corresponding DP algorithm.

1.20 (Games)

(a) Consider a smaller version of a popular puzzle game. Three square tiles numbered 1, 2, and 3 are placed in a 2×2 grid with one space left empty. The two tiles adjacent to the empty space can be moved into that space, thereby creating new configurations. Use a DP argument to answer the question whether it is possible to generate a given configuration starting from any other configuration.

(b) From a pile of eleven matchsticks, two players take turns removing one or four sticks. The player who removes the last stick wins. Use a DP argument to show that there is a winning strategy for the player who plays first.

1.21 (The Counterfeit Coin Problem)

We are given six coins, one of which is counterfeit and is known to have different weight than the rest. Construct a strategy to find the counterfeit coin using a two-pan scale in a minimum average number of tries. *Hint*: There are two initial decisions that make sense: (1) test two of the coins against two others, and (2) test one of the coins against one other.

1.22 (DP for Uncontrollable State Components)

Consider the case where the state is a composite (x_k, y_k) of two components x_k and y_k. The evolution of the main component is affected by the control u_k according to the equation

$$x_{k+1} = f_k(x_k, y_k, u_k, w_k),$$

where the probability distribution $P_k(w_k \mid x_k, y_k, u_k)$ is given. The evolution of the other component is governed by a given conditional distribution $P_k(y_k \mid x_k)$ and cannot be affected by the control, except indirectly through x_k. Define

$$\hat{J}_k(x_k) = \underset{y_k}{E}\big\{J_k(x_k, y_k) \mid x_k\big\}.$$

Show that an equivalent DP algorithm is given by

$$\hat{J}_k(x_k) = \underset{y_k}{E}\Big\{\min_{u_k \in U_k(x_k, y_k)} \underset{w_k}{E}\big\{g_k(x_k, y_k, u_k, w_k) + \hat{J}_{k+1}\big(f_k(x_k, y_k, u_k, w_k)\big)\big\}\Big| \, x_k\Big\}.$$

Discuss the advantages of this algorithm.

1.23 (Monotonicity Property of DP)

An evident, yet very important property of the DP algorithm is that if the terminal cost g_N is changed to a uniformly larger cost $\overline{g}_N$ [i.e., $g_N(x_N) \leq \overline{g}_N(x_N)$ for all x_N], then the last stage cost-to-go $J_{N-1}(x_{N-1})$ will be uniformly increased. More generally, given two functions J_{k+1} and $\bar{J}_{k+1}$ with $J_{k+1}(x_{k+1}) \leq \bar{J}_{k+1}(x_{k+1})$ for all x_{k+1}, we have, for all x_k and $u_k \in U_k(x_k)$,

$$\underset{w_k}{E}\big\{g_k(x_k, u_k, w_k) + J_{k+1}\big(f_k(x_k, u_k, w_k)\big)\big\} \leq \underset{w_k}{E}\big\{g_k(x_k, u_k, w_k) + \bar{J}_{k+1}\big(f_k(x_k, u_k, w_k)\big)\big\}.$$

Suppose now that in the basic problem the system and cost are time invariant; that is, $S_k \equiv S$, $C_k \equiv C$, $D_k \equiv D$, $f_k \equiv f$, $U_k(x_k) \equiv U(x_k)$, and $g_k \equiv g$ for some S, C, D, f, $U(\cdot)$, and g. Show that if in the DP algorithm we have $J_{N-1}(x) \leq J_N(x)$ for all $x \in S$, then

$$J_k(x) \leq J_{k+1}(x), \qquad \text{for all } x \in S \text{ and } k.$$

Similarly, if we have $J_{N-1}(x) \geq J_N(x)$ for all $x \in S$, then

$$J_k(x) \geq J_{k+1}(x), \qquad \text{for all } x \in S \text{ and } k.$$

1.24

An unscrupulous innkeeper charges a different rate for a room as the day progresses, depending on whether he has many or few vacancies. His objective is to maximize his expected total income during the day. Let x be the number of empty rooms at the start of the day, and let y be the number of customers that will ask for a room in the course of the day. We assume (somewhat unrealistically) that the innkeeper knows y with certainty, and upon arrival of a customer, quotes one of m prices r_i, $i = 1, \ldots, m$, where $0 < r_1 \leq r_2 \leq \cdots \leq r_m$. A quote of a rate r_i is accepted with probability p_i and is rejected with probability $1 - p_i$, in which case the customer departs, never to return during that day.

(a) Formulate this as a problem with y stages and show that the maximal expected income as a function of x and y satisfies the recursion

$$J(x, y) = \max_{i=1,\ldots,m} \left[p_i\big(r_i + J(x-1, y-1)\big) + (1-p_i)J(x, y-1)\right],$$

for all $x \geq 1$ and $y \geq 1$, with initial conditions

$$J(x, 0) = J(0, y) = 0, \qquad \text{for all } x \text{ and } y.$$

Assuming that the product $p_i r_i$ is monotonically nondecreasing with i, show that the innkeeper should always charge the highest rate r_m.

(b) Consider a variant of the problem where each arriving customer offers a price r_i for a room, which the innkeeper may accept or reject, in which case the customer departs, never to return during that day. Show that an appropriate DP algorithm is

$$J(x, y) = \sum_{i=1}^{m} p_i \max\left[r_i + J(x-1, y-1),\ J(x, y-1)\right],$$

with initial conditions

$$J(x, 0) = J(0, y) = 0, \qquad \text{for all } x \text{ and } y.$$

Show also that for given x and y it is optimal to accept a customer's offer if it is larger than some threshold $\overline{r}(x, y)$. *Hint*: This part is related to DP for uncontrollable state components (cf. Exercise 1.22).

1.25 (Investing in a Stock)

An investor observes at the beginning of each period k the price x_k of a stock and decides whether to buy 1 unit, sell 1 unit, or do nothing. There is a transaction cost c for buying or selling. The stock price can take one of n different values $v^1, \ldots, v^n$ and the transition probabilities $p_{ij}^k = P\{x_{k+1} = v^j \mid x_k = v^i\}$ are known. The investor wants to maximize the total worth of his stock at a fixed final period N minus his investment costs from period 0 to period $N-1$ (revenue from a sale is viewed as negative cost). We assume that $E\{x_N \mid x_k = x\} - x$ is monotonically nonincreasing as a function of x; that is, the expected profit from a purchase is a nonincreasing function of the purchase price.

(a) Assume that the investor starts with N or more units of stock and an unlimited amount of cash, so that a purchase or sale decision is possible at each period regardless of the past decisions and the current price. Show that for every period k, there is a threshold $\overline{x}_k$ such that it is optimal to buy if $x_k \leq \overline{x}_k - c$, sell if $x_k \geq \overline{x}_k + c$, and do nothing otherwise. Assuming that p_{ij}^k is independent of k, under what conditions can you argue that as $k \to \infty$, $\overline{x}_k$ tends to a constant value, which can be viewed as the long term average price of the stock? *Hint*: Formulate the problem as one of maximizing $E\Big\{\sum_{k=0}^{N-1}\big(u_k g_k(x_k) - c|u_k|\big)\Big\}$, where $u_k \in \{-1, 0, 1\}$ and $g_k(x_k) = E\{x_N \mid x_k\} - x_k$.

(b) Formulate an efficient DP algorithm for the case where the investor starts with less than N units of stock and an unlimited amount of cash. Show that it is still optimal to buy if $x_k \leq \overline{x}_k - c$ and it is still not optimal to sell if $x_k < \overline{x}_k + c$. Could it be optimal to buy at any prices x_k greater than $\overline{x}_k - c$?

(c) Consider the situation where the investor initially has N or more units of stock but a limited amount of cash. We model this situation by imposing the constraint that for any time k the number of purchase decisions up to k should not exceed the number of sale decisions up to k by more than a given fixed number m. Formulate an efficient DP algorithm for this case. Show that it is still optimal to sell if $x_k \geq \overline{x}_k + c$ and it is still not optimal to buy if $x_k > \overline{x}_k - c$.

(d) Consider the situation where the investor initially has less than N units of stock as well as a limited amount of cash. Derive a DP algorithm for this problem.

(e) How would the analysis of (a)-(d) be affected if cash is invested at a given fixed interest rate?

(f) Formulate problems similar to the ones of parts (a)-(c), where the investor can buy or sell two or more types of stock. Derive the corresponding DP algorithms.

2

Deterministic Systems and the Shortest Path Problem

Contents

In this chapter, we focus on deterministic problems, that is, problems where each disturbance w_k can take only one value. Such problems arise in many important contexts and they also arise in cases where the problem is really stochastic but, as an approximation, the disturbance is fixed at some typical value; see Chapter 6.

An important property of deterministic problems is that, in contrast with stochastic problems, *using feedback results in no advantage in terms of cost reduction.* In other words, minimizing the cost over admissible policies $\{\mu_0, \ldots, \mu_{N-1}\}$ results in the same optimal cost as minimizing over *sequences of control vectors* $\{u_0, \ldots, u_{N-1}\}$. This is true because given a policy and the initial state, the future evolution of the system's state and the corresponding future controls are perfectly predictable. In particular, the cost achieved by an admissible policy $\{\mu_0, \ldots, \mu_{N-1}\}$ for a deterministic problem is also achieved by the control sequence $\{u_0, \ldots, u_{N-1}\}$ defined by

$$u_k = \mu_k(x_k), \qquad k = 0, 1, \ldots, N-1,$$

where the states $x_1, \ldots, x_{N-1}$ are determined from x_0 and $\{\mu_0, \ldots, \mu_{N-1}\}$ by

$$x_{k+1} = f_k\big(x_k, \mu_k(x_k)\big), \qquad k = 0, 1, \ldots, N-1.$$

As a result, we may restrict attention to sequences of controls without loss of optimality.

The difference just discussed between deterministic and stochastic problems often has important computational implications. In particular, in a deterministic problem with a "continuous space" character (states and controls are Euclidean vectors), optimal control sequences may be found by variational techniques to be discussed in Chapter 3, and by widely used optimal control algorithms such as steepest descent, conjugate gradient, and Newton's method (see e.g., [Ber95], [Lue84], and [Pol71]). These algorithms, when applicable, are usually more efficient than DP. On the other hand, DP *has a wider scope of applicability since it can handle difficult constraint sets such as integer or discrete sets.* Furthermore, DP *leads to a globally optimal solution* as opposed to variational techniques, for which this cannot be guaranteed in general.

In this chapter, we consider deterministic problems with a discrete character for which variational optimal control techniques are inapplicable, and specialized forms of DP are the principal solution methods.

2.1 FINITE-STATE SYSTEMS AND SHORTEST PATHS

Consider a deterministic problem where the state space S_k is a finite set for each k. Then at any state x_k, a control u_k can be associated with a transition from the state x_k to the state $f_k(x_k, u_k)$, at a cost $g_k(x_k, u_k)$. Thus a finite-state deterministic problem can be equivalently represented by a graph such as the one of Fig. 2.1.1, where the arcs correspond to transitions between states at successive stages and each arc has a cost associated with it. To handle the final stage, we have also added an artificial terminal node t. Each state x_N at stage N is connected to the terminal node t with an arc having cost $g_N(x_N)$. Control sequences correspond to paths originating at the initial state (node s at stage 0) and terminating at one of the nodes corresponding to the final stage N. If we view the cost of an arc as its length, we see that *a deterministic finite-state problem is equivalent to finding a minimum-length (or shortest) path from the initial node s of the graph to the terminal node t.* Here, by a path we mean a sequence of arcs of the form $(j_1, j_2), (j_2, j_3), \dots, (j_{k-1}, j_k)$, and by the length of a path we mean the sum of the lengths of its arcs.

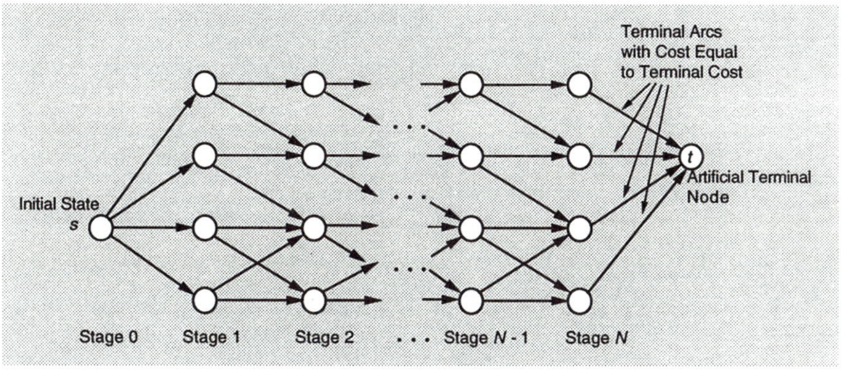

Figure 2.1.1 Transition graph for a deterministic finite-state system. Nodes correspond to states. An arc with start and end nodes x_k and x_{k+1}, respectively, corresponds to a transition of the form $x_{k+1} = f_k(x_k, u_k)$. The length of this arc is equal to the cost of the corresponding transition $g_k(x_k, u_k)$. The problem is equivalent to finding a shortest path from the initial node s to the terminal node t.

Let us denote

$$a_{ij}^k = \text{ Cost of transition at time } k \text{ from state } i \in S_k \text{ to state } j \in S_{k+1},$$

$$a_{it}^N = \text{ Terminal cost of state } i \in S_N \text{ [which is } g_N(i)],$$

where we adopt the convention $a_{ij}^k = \infty$ if there is no control that moves the state from i to j at time k. The DP algorithm takes the form

$$J_N(i) = a_{it}^N, \qquad i \in S_N, \tag{1.1}$$

$$J_k(i) = \min_{j \in S_{k+1}} \left[a_{ij}^k + J_{k+1}(j)\right], \qquad i \in S_k, \qquad k = 0, 1, \dots, N-1. \tag{1.2}$$

The optimal cost is $J_0(s)$ and is equal to the length of the shortest path from s to t.

A Forward DP Algorithm

The preceding algorithm proceeds *backward* in time. It is possible to derive an equivalent algorithm that proceeds *forward* in time by means of the following simple observation. An optimal path from s to t is also an optimal path from t to s in a "reverse" shortest path problem where the direction of each arc is reversed and its length is left unchanged. The DP algorithm corresponding to this "reverse" problem starts from the states $x_1 \in S_1$ of stage 1, proceeds to states $x_2 \in S_2$ of stage 2, and continues all the way to states $x_N \in S_N$ of stage N. It is given by

$$\tilde{J}_N(j) = a_{sj}^0, \qquad j \in S_1, \tag{1.3}$$

$$\tilde{J}_k(j) = \min_{i \in S_{N-k}} \left[a_{ij}^{N-k} + \tilde{J}_{k+1}(i)\right], \qquad j \in S_{N-k+1}, \qquad k = 1, 2, \ldots, N-1. \tag{1.4}$$

The optimal cost is

$$\tilde{J}_0(t) = \min_{i \in S_N} \left[a_{it}^N + \tilde{J}_1(i)\right]. \tag{1.5}$$

The backward algorithm (1.1)-(1.2) and the forward algorithm (1.3)-(1.4) yield the same result in the sense that

$$J_0(s) = \tilde{J}_0(t),$$

and an optimal control sequence (or shortest path) obtained from any one of the two is optimal for the original problem. We may view $\tilde{J}_k(j)$ in Eq. (1.4) as an *optimal cost-to-arrive* to state j from the initial state s. This should be contrasted with $J_k(i)$ in Eq. (1.2), which represents the optimal cost-to-go from state i to the terminal state t.

An important use of the forward DP algorithm arises in real-time applications where the stage k problem data are unknown prior to stage k. An example will be given in connection with the state estimation of Hidden Markov Models in Section 2.2.2. Note that to derive the forward DP algorithm, we used the shortest path formulation, which is available only for deterministic problems. Indeed, for stochastic problems, there is no analog of the forward DP algorithm.

In conclusion, *a deterministic finite-state problem is equivalent to a special type of shortest path problem and can be solved by either the ordinary (backward) DP algorithm or by an alternative forward DP algorithm.* It is also interesting to note that *any shortest path problem can be posed as a deterministic finite-state DP problem*, as we now show.

Converting a Shortest Path Problem to a Deterministic Finite-State Problem

Let $\{1, 2, \ldots, N, t\}$ be the set of nodes of a graph, and let a_{ij} be the cost of moving from node i to node j (also referred to as the *length* of the arc joining i and j). Node t is a special node, which we call the *destination.* We allow the possibility $a_{ij} = \infty$ to account for the case where there is no arc joining nodes i and j. We want to find a shortest path from each node i to node t, that is, a sequence of moves that minimizes total cost to get to t from each of the nodes $1, 2, \ldots, N$.

For the problem to have a solution, it is necessary to make an assumption relating to cycles, that is, paths of the form $(i, j_1), (j_1, j_2), \ldots, (j_k, i)$ that start and end at the same node. We must exclude the possibility that a cycle has negative total length. Otherwise, it would be possible to decrease the length of some paths to arbitrarily small values simply by adding more and more negative-length cycles. We thus assume that *all cycles have nonnegative length.* With this assumption, it is clear that an optimal path need not take more than N moves, so we may limit the number of moves to N. We formulate the problem as one where *we require exactly N moves but allow degenerate moves from a node i to itself with cost $a_{ii} = 0$.* We denote for $i = 1, \ldots, N$, $k = 0, \ldots, N-1$,

$$J_k(i) = \text{optimal cost of getting from } i \text{ to } t \text{ in } N-k \text{ moves.}$$

Then the cost of the optimal path from i to t is $J_0(i)$.

It is possible to formulate this problem within the framework of the basic problem and subsequently apply the DP algorithm. For simplicity, however, we write directly the DP equation, which takes the intuitively clear form

$$\text{optimal cost from } i \text{ to } t \text{ in } N-k \text{ moves}$$
$$= \min_{j=1,\ldots,N} \big[a_{ij} + (\text{optimal cost from } j \text{ to } t \text{ in } N-k-1 \text{ moves})\big],$$

or

$$J_k(i) = \min_{j=1,\ldots,N} \big[a_{ij} + J_{k+1}(j)\big], \qquad k = 0, 1, \ldots, N-2,$$

with

$$J_{N-1}(i) = a_{it}, \qquad i = 1, 2, \ldots, N.$$

The optimal policy when at node i after k moves is to move to a node j^* that minimizes $a_{ij} + J_{k+1}(j)$ over all $j = 1, \ldots, N$. If the optimal path obtained from the algorithm contains degenerate moves from a node to itself, this simply means that the path involves in reality less than N moves.

To demonstrate the algorithm, consider the problem shown in Fig. 2.1.2(a) where the costs a_{ij} with $i \neq j$ are shown along the connecting line segments (we assume $a_{ij} = a_{ji}$). Figure 2.1.2(b) shows the cost-to-go $J_k(i)$ at each node i and index k together with the optimal path. The optimal paths are

$$1 \to 5, \ 2 \to 3 \to 4 \to 5, \ 3 \to 4 \to 5, \ 4 \to 5.$$

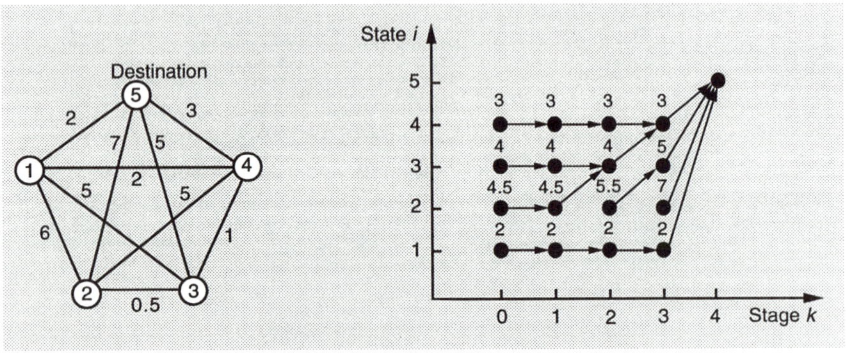

Figure 2.1.2 (a) Shortest path problem data. The destination is 5. Arc lengths are equal in both directions and are shown along the line segments connecting nodes. (b) Costs-to-go generated by the DP algorithm. The number along stage k and state i is $J_k(i)$. Arrows indicate the optimal moves at each stage and node.

2.2 SOME SHORTEST PATH APPLICATIONS

The shortest path problem appears in many diverse contexts. We provide some examples.

2.2.1 Critical Path Analysis

Consider the planning of a project involving several activities, some of which must be completed before others can begin. The duration of each activity is known in advance. We want to find the time required to complete the project, as well as the *critical* activities, those that even if slightly delayed will result in a corresponding delay of completion of the overall project.

The problem can be represented by a graph with nodes $1, \ldots, N$ such as the one shown in Fig. 2.2.1. Here nodes represent completion of some phase of the project. An arc (i, j) represents an activity that starts once phase i is completed and has known duration $t_{ij} > 0$. A phase (node) j is completed when all activities or arcs (i, j) that are incoming to j are completed. The special nodes 1 and N represent the start and end of the project. Node 1 has no incoming arcs, while node N has no outgoing arcs. Furthermore, there is at least one path from node 1 to every other node.

An important characteristic of an activity network is that it is *acyclic*; that is, it has no cycles. This is inherent in the problem formulation and the interpretation of nodes as phase completions.

For any path $p = \{(1, j_1), (j_1, j_2,), \ldots, (j_k, i)\}$ from node 1 to a node i, let D_p be the duration of the path defined as the sum of durations of its

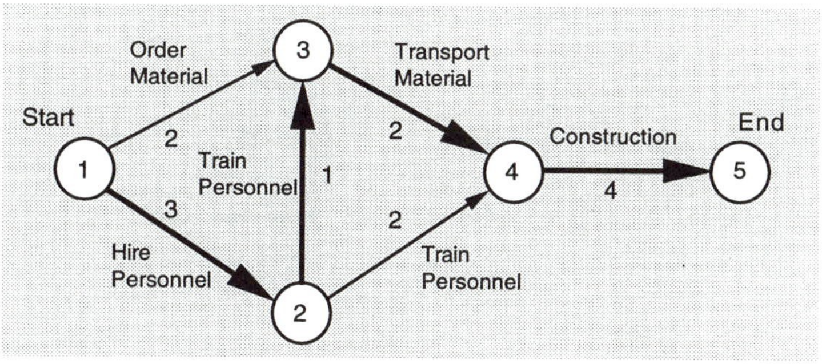

Figure 2.2.1 Graph of an activity network. Arcs represent activities and are labeled by the corresponding duration. Nodes represent completion of some phase of the project. A phase is completed if all activities associated with incoming arcs at the corresponding node are completed. The project is completed when all phases are completed. The project duration time is the length of the longest path from node 1 to node 5, which is shown with thick line.

activities; that is,

$$D_p = t_{ij_1} + t_{j_1 j_2} + \cdots + t_{j_k i}.$$

Then the time T_i required to complete phase i is

$$T_i = \max_{\substack{\text{paths } p \\ \text{from 1 to } i}} D_p.$$

The maximum above is attained by some path because there can be only a finite number of paths from 1 to i, since the network is acyclic. Thus to find T_i, we should find the *longest* path from 1 to i. Because the graph is acyclic, this problem may also be viewed as a shortest path problem with the length of each arc (i, j) being $-t_{ij}$. In particular, finding the duration of the project is equivalent to finding the shortest path from 1 to N.

Let us denote by S_1 the set of phases that do not depend on completion of any other phase, and more generally, for $k = 1, 2, \ldots$, let S_k be the set

$$S_k = \{i \mid \text{all paths from 1 to } i \text{ have } k \text{ arcs or less}\},$$

with $S_0 = \{1\}$. Using the acyclicity of the network, it can be shown that there is some integer $\overline{k}$ such that all the sets S_k with $k \leq \overline{k}$ are nonempty and all the sets S_k with $k > \overline{k}$ are empty. The sets S_k can be viewed as the state spaces for the equivalent DP problem. Using maximization in place of minimization while changing the sign of the arc lengths, the DP algorithm for the shortest path problem is

$$T_i = \max_{\substack{(j,i) \text{ such that} \\ j \in S_{k-1}}} \big[t_{ji} + T_j\big], \qquad \text{for all } i \in S_k \text{ with } i \notin S_{k-1}.$$

Note that this is a forward algorithm; that is, it starts at the origin 1 and proceeds towards the destination N. An alternative backward algorithm, which starts at N and proceeds towards 1, is also possible, as discussed in the preceding section. A shortest path is also called a *critical* path. A delay by a given amount in the completion of one of the activities on the critical path will delay the completion of the overall project by the same amount.

As an example, for the activity network of Fig. 2.2.1, we have

$$S_0 = \{1\}, \quad S_1 = \{1,2\}, \quad S_2 = \{1,2,3\},$$

$$S_3 = \{1,2,3,4\}, \quad S_4 = \{1,2,3,4,5\}.$$

A calculation using the preceding formula yields

$$T_1 = 0,\ T_2 = 3,\ T_3 = 4,\ T_4 = 6,\ T_5 = 10.$$

The critical path is $1 \to 2 \to 3 \to 4 \to 5$.

2.2.2 Hidden Markov Models and the Viterbi Algorithm

Consider a Markov chain with a finite number of states and given state transition probabilities p_{ij}. Suppose that when a transition occurs, we obtain an observation that relates to that transition. Given a sequence of observations, we want to estimate in some optimal sense the sequence of corresponding transitions. We are given the probability $r(z;i,j)$ of an observation taking value z when the state transition is from i to j. We assume independent observations; that is, an observation depends only on its corresponding transition and not on other transitions. We are also given the probability π_i that the initial state takes value i. The probabilities p_{ij} and $r(z;i,j)$ are assumed to be independent of time for notational convenience. The methodology to be described admits a straightforward extension to the case of time-varying system and observation statistics.

Markov chains whose state transitions are imperfectly observed according to the probabilistic mechanism just described are called *Hidden Markov Models (HMMs for short)* or *partially observable Markov chains*. In Chapter 5 we will discuss the control of such Markov chains in the context of stochastic optimal control problems with imperfect state information. In the present section, we will focus on the problem of estimating the state sequence given a corresponding observation sequence. This is an important problem, which arises in a broad variety of practical contexts.

We use a "most likely state" estimation criterion, whereby given the observation sequence $Z_N = \{z_1, z_2, \ldots, z_N\}$, we adopt as our estimate the state transition sequence $\hat{X}_N = \{\hat{x}_0, \hat{x}_1, \ldots, \hat{x}_N\}$ that maximizes over all $X_N = \{x_0, x_1, \ldots, x_N\}$ the conditional probability $p(X_N \mid Z_N)$. It turns out that $\hat{X}_N$ can be found by solving a shortest path problem, as we now describe.

We have

$$p(X_N \mid Z_N) = \frac{p(X_N, Z_N)}{p(Z_N)},$$

where $p(X_N, Z_N)$ and $p(Z_N)$ are the unconditional probabilities of occurrence of (X_N, Z_N) and Z_N, respectively. Since $p(Z_N)$ is a positive constant once Z_N is known, we can maximize $p(X_N, Z_N)$ in place of $p(X_N \mid Z_N)$. The probability $p(X_N, Z_N)$ can be written as

$$\begin{aligned} p(X_N, Z_N) &= p(x_0, x_1, \ldots, x_N, z_1, z_2, \ldots, z_N) \\ &= \pi_{x_0} p(x_1, \ldots, x_N, z_1, z_2, \ldots, z_N \mid x_0) \\ &= \pi_{x_0} p(x_1, z_1 \mid x_0) p(x_2, \ldots, x_N, z_2, \ldots, z_N \mid x_0, x_1, z_1) \\ &= \pi_{x_0} p_{x_0 x_1} r(z_1; x_0, x_1) p(x_2, \ldots, x_N, z_2, \ldots, z_N \mid x_0, x_1, z_1). \end{aligned}$$

This calculation can be continued by writing

$$\begin{aligned} &p(x_2, \ldots, x_N, z_2, \ldots, z_N \mid x_0, x_1, z_1) \\ &\quad = p(x_2, z_2 \mid x_0, x_1, z_1) p(x_3, \ldots, x_N, z_3, \ldots, z_N \mid x_0, x_1, z_1, x_2, z_2) \\ &\quad = p_{x_1 x_2} r(z_2; x_1, x_2) p(x_3, \ldots, x_N, z_3, \ldots, z_N \mid x_0, x_1, z_1, x_2, z_2), \end{aligned}$$

where to obtain the last equation, we used the assumed independence of the observations. Combining the above two relations, we have

$$\begin{aligned} p(X_N, Z_N) = \pi_{x_0} p_{x_0 x_1} &r(z_1; x_0, x_1) p_{x_1 x_2} r(z_2; x_1, x_2) \\ &\cdot p(x_3, \ldots, x_N, z_3, \ldots, z_N \mid x_0, x_1, z_1, x_2, z_2), \end{aligned}$$

and continuing in the same manner, we obtain

$$p(X_N, Z_N) = \pi_{x_0} \prod_{k=1}^{N} p_{x_{k-1} x_k} r(z_k; x_{k-1}, x_k). \tag{2.1}$$

We now show how the maximization of the above expression can be viewed as a shortest path problem. In particular, we construct a graph of state-time pairs, called the *trellis diagram*, by concatenating $N+1$ copies of the state space, and by preceding and following them with dummy nodes s and t, respectively, as shown in Fig. 2.2.2. The nodes of the kth copy correspond to the states x_{k-1} at time $k-1$. An arc connects a node x_{k-1} of the kth copy with a node x_k of the $(k+1)$st copy if the corresponding transition probability $p_{x_{k-1} x_k}$ is positive. Since maximizing a positive cost function is equivalent to maximizing its logarithm, we see from Eq. (2.1) that, given the observation sequence $Z_N = \{z_1, z_2, \ldots, z_N\}$, the problem of maximizing $p(X_N, Z_N)$ is equivalent to the problem

$$\begin{aligned} &\text{minimize } -\ln(\pi_{x_0}) - \sum_{k=1}^{N} \ln\big(p_{x_{k-1} x_k} r(z_k; x_{k-1}, x_k)\big) \\ &\text{over all possible sequences } \{x_0, x_1, \ldots, x_N\}. \end{aligned} \tag{2.2}$$

By assigning to an arc (s, x_0) the length $-\ln(\pi_{x_0})$, to an arc (x_N, t) the length 0, and to an arc (x_{k-1}, x_k) the length $-\ln\big(p_{x_{k-1}x_k} r(z_k; x_{k-1}, x_k)\big)$, we see that the above minimization problem is equivalent to the problem of finding the shortest path from s to t in the trellis diagram. This shortest path defines the estimated state sequence $\{\hat{x}_0, \hat{x}_1, \ldots, \hat{x}_N\}$.

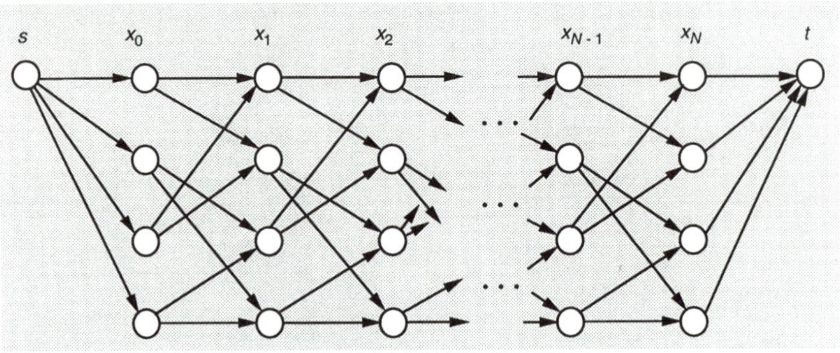

Figure 2.2.2 State estimation of an HMM viewed as a problem of finding a shortest path from s to t. Length of arcs from s to states x_0 is $-\ln(\pi_{x_0})$, and length of arcs from states x_N to t is zero. Length of an arc from a state x_{k-1} to x_k is $-\ln\big(p_{x_{k-1}x_k} r(z_k; x_{k-1}, x_k)\big)$, where z_k is the kth observation.

In practice, the shortest path is most conveniently constructed sequentially by forward DP, that is, by first calculating the shortest distance from s to each node x_1, then using these distances to calculate the shortest distance from s to each node x_2, etc. In particular, suppose that we have computed the shortest distances $D_k(x_k)$ from s to all states x_k on the basis of the observation sequence Z_k, and suppose that the new observation z_{k+1} is obtained. Then the shortest distances $D_{k+1}(x_{k+1})$ from s to any state x_{k+1} can be computed by the DP recursion

$$D_{k+1}(x_{k+1}) = \min_{\substack{\text{all } x_k \text{ such that} \\ p_{x_k x_{k+1}} > 0}} \Big[D_k(x_k) - \ln\big(p_{x_k x_{k+1}} r(z_{k+1}; x_k, x_{k+1})\big)\Big].$$

The initial condition is $D_0(x_0) = -\ln(\pi_{x_0})$. The final estimated state sequence $\hat{X}_N$ corresponds to the shortest path from s to the final state $\hat{x}_N$ that minimizes $D_N(x_N)$ over the finite set of possible states x_N. An advantage of this procedure is that it can be executed in real time, as soon as each new observation is obtained.

There are a number of practical schemes for estimating a portion of the state sequence without waiting to receive the entire observation sequence Z_N, and this is useful if Z_N is a long sequence. For example, one can check fairly easily whether for some k, all shortest paths from s to states x_k pass through a single node in the subgraph of states $x_0, \ldots, x_{k-1}$.

If so, it can be seen from Fig. 2.2.3 that the shortest path from s to that node will not be affected by reception of additional observations, and therefore the subsequence of state estimates up to that node can be determined without waiting for the remaining observations.

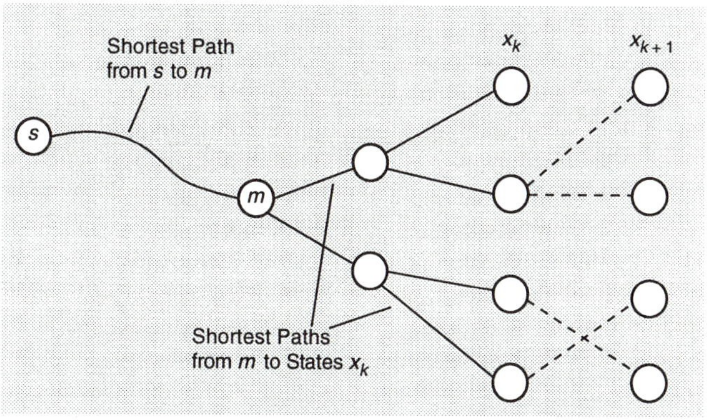

Figure 2.2.3 Estimating a portion of the state sequence prior to receiving the entire observation sequence. Suppose that the shortest paths from s to all states x_k pass through a single node m. If an additional observation is received, the shortest paths from s to all states x_{k+1} will continue to pass through m. Therefore, the portion of the state sequence up to node m can be safely estimated because additional observations will not change the initial portion of the shortest paths from s up to m.

The shortest path-based estimation procedure just described is known as the *Viterbi algorithm*, and finds numerous applications in a variety of contexts. An example is *speech recognition*, where the basic goal is to transcribe a spoken word sequence in terms of elementary speech units called *phonemes*. One possibility is to associate the states of the HHM with phonemes, and given a sequence of recorded phonemes $Z_N = \{z_1, \ldots, z_N\}$, to find a phonemic sequence $\hat{X}_N = \{\hat{x}_1, \ldots, \hat{x}_N\}$ that maximizes over all $X_N = \{x_1, \ldots, x_N\}$ the conditional probability $p(X_N \mid Z_N)$. The probabilities $r(z_k; x_{k-1}, x_k)$ and $p_{x_{k-1}x_k}$ can be experimentally obtained, if necessary by specialized "training" for each speaker that uses the speech recognition system. The Viterbi algorithm can then be used to find the most likely phonemic sequence. There are also other HMMs used for word and sentence recognition, where only phonemic sequences that constitute words from a given dictionary are considered. We refer the reader to [Pic90] and [Rab89] for a general review of HMMs applied to speech recognition and for further references to related work. It is also possible to use similar models for computerized recognition of handwriting.

The Viterbi algorithm was originally developed as a scheme for decoding data after transmission over a noisy communication channel. The following example describes this context in some detail.

Example 2.1 (Convolutional Coding and Decoding)

When binary data are transmitted over a noisy communication channel, it is often essential to use coding as a means of enhancing reliability of communication. A common type of coding method, called *convolutional coding*, converts a source-generated binary data sequence

$$\{w_1, w_2, \ldots\}, \qquad w_k \in \{0,1\}, \qquad k = 1,2,\ldots,$$

into a coded sequence $\{y_1, y_2, \ldots\}$, where each y_k is an n-dimensional vector with binary coordinates, called *codeword*,

$$y_k = \begin{pmatrix} y_k^1 \\ \vdots \\ y_k^n \end{pmatrix}, \qquad y_k^i \in \{0,1\}, \qquad i = 1,\ldots,n, \ k = 1,2,\ldots.$$

The sequence $\{y_1, y_2, \ldots\}$ is then transmitted over a noisy channel and gets transformed into a sequence $\{z_1, z_2, \ldots,\}$, which is then decoded to yield the decoded data sequence $\{\hat{w}_1, \hat{w}_2, \ldots\}$; see Fig. 2.2.4. The objective is to design the encoder/decoder scheme so that the decoded sequence is as close to the original as possible.

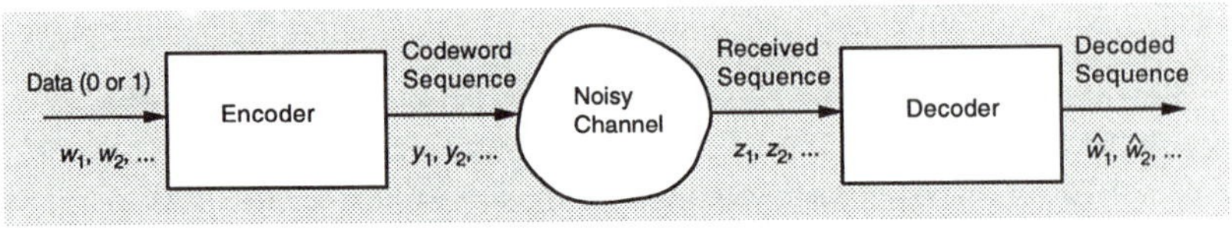

Figure 2.2.4 Encoder/decoder scheme.

The problem just discussed is central in information theory and can be approached in several different ways. In a particularly popular and effective technique called *convolutional coding*, the vectors y_k are related to w_k via equations of the form

$$y_k = Cx_{k-1} + dw_k, \qquad k = 1,2,\ldots, \tag{2.3}$$

$$x_k = Ax_{k-1} + bw_k, \qquad k = 1,2,\ldots, \qquad x_0: \text{ given}, \tag{2.4}$$

where x_k is an m-dimensional vector with binary coordinates, which we view as state, and C, d, A, and b are $n \times m$, $n \times 1$, $m \times m$, and $m \times 1$ matrices, respectively, with binary coordinates. The products and the sums involved in the expressions $Cx_{k-1} + dw_k$ and $Ax_{k-1} + bw_k$ are calculated using modulo 2 arithmetic.

As an example, let $m = 2$, $n = 3$, and

$$C = \begin{pmatrix} 1 & 0 \\ 0 & 1 \\ 0 & 1 \end{pmatrix}, \qquad A = \begin{pmatrix} 0 & 1 \\ 1 & 1 \end{pmatrix},$$

$$d = \begin{pmatrix} 1 \\ 1 \\ 1 \end{pmatrix}, \qquad b = \begin{pmatrix} 0 \\ 1 \end{pmatrix}.$$

Then the evolution of the system (2.3)-(2.4) can be represented by the diagram shown in Fig. 2.2.5. Given the initial x_0, this diagram can be used to generate the codeword sequence $\{y_1, y_2, \ldots\}$ corresponding to a data sequence $\{w_1, w_2, \ldots\}$. For example, when the initial state is $x_0 = 00$, the data sequence

$$\{w_1, w_2, w_3, w_4\} = \{1, 0, 0, 1\}$$

generates the state sequence

$$\{x_0, x_1, x_2, x_3, x_4\} = \{00, 01, 11, 10, 00\},$$

and the codeword sequence

$$\{y_1, y_2, y_3, y_4\} = \{111, 011, 111, 011\}.$$

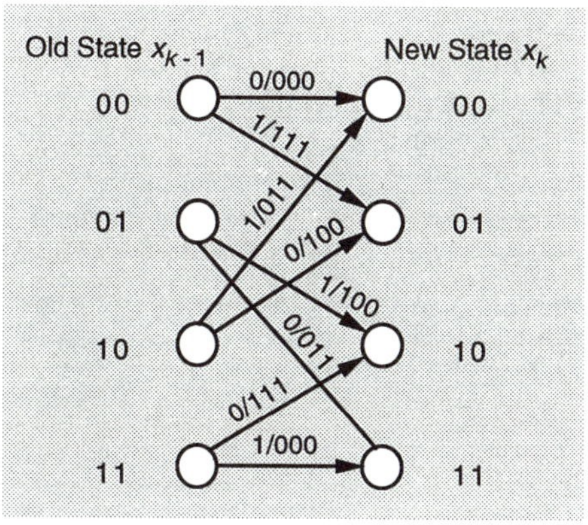

Figure 2.2.5 State transition diagram for convolutional coding. The binary number pair on each arc is the data/codeword pair w_k/y_k for the corresponding transition. So for example, when $x_{k-1} = 01$, a zero data bit ($w_k = 0$) effects a transition to $x_k = 11$ and generates the codeword 001.

Assume now that the characteristics of the noisy transmission channel are such that a codeword y is actually received as z with known probability $p(z \mid y)$, where z is any n-bit binary number. We assume independent errors so that

$$p(Z_N \mid Y_N) = \prod_{k=1}^{N} p(z_k \mid y_k), \tag{2.5}$$

where $Z_N = \{z_1, \ldots, z_N\}$ is the received sequence and $Y_N = \{y_1, \ldots, y_N\}$ is the transmitted sequence. By associating the codewords y with state transitions, we formulate a maximum likelihood estimation problem, whereby we want to find a sequence $\hat{Y}_N = \{\hat{y}_1, \hat{y}_2, \ldots, \hat{y}_N\}$ such that

$$p(Z_N \mid \hat{Y}_N) = \max_{Y_N} p(Z_N \mid Y_N).$$

The constraint on Y_N is that it must be a feasible codeword sequence (i.e., it must correspond to some initial state and data sequence, or equivalently, to a sequence of arcs of the trellis diagram).

Let us construct a trellis diagram by concatenating N state transition diagrams and appending dummy nodes s and t on its left and right sides, which are connected with zero-length arcs to the states x_0 and x_N, respectively. By using Eq. (2.5), we see that, given the received sequence $Z_N = \{z_1, z_2, \ldots, z_N\}$, the problem of maximizing $p(Z_N \mid Y_N)$ is equivalent to the problem

$$\begin{aligned} &\text{minimize} \sum_{k=1}^{N} -\ln\big(p(z_k \mid y_k)\big) \\ &\text{over all binary sequences } \{y_1, y_2, \ldots, y_N\}. \end{aligned} \tag{2.6}$$

This is equivalent to a problem of finding a shortest path in the trellis diagram from s to t, where the length of the arc associated with the codeword y_k is $-\ln\big(p(z_k \mid y_k)\big)$, and the lengths of each arc incident to a dummy node is zero. From the shortest path and the trellis diagram, we can then obtain the corresponding data sequence $\{\hat{w}_1, \ldots, \hat{w}_N\}$, which is accepted as the decoded data.

The maximum likelihood estimate $\hat{Y}_N$ can be found by solving the corresponding shortest path problem using the Viterbi algorithm. In particular, the shortest distances $D_{k+1}(x_{k+1})$ from s to any state x_{k+1} are computed by the DP recursion

$$D_{k+1}(x_{k+1}) = \min_{\substack{\text{all } x_k \text{ such that} \\ (x_k, x_{k+1}) \text{ is an arc}}} \Big[D_k(x_k) - \ln\big(p(z_{k+1} \mid y_{k+1})\big)\Big],$$

where y_{k+1} is the codeword corresponding to the arc (x_k, x_{k+1}). The final state $\hat{x}_N$ on the shortest path is the one that minimizes $D_N(x_N)$ over x_N.

2.3 SHORTEST PATH ALGORITHMS

We have seen that shortest path problems and deterministic finite-state optimal control problems are equivalent. The computational implications of this are twofold.

(a) One can use DP to solve general shortest path problems. Note that there are several other shortest path methods, some of which have superior theoretical worst-case performance to DP. However, DP is often preferred in practice, particularly for problems with an acyclic graph structure and also when a parallel computer is available.

(b) One can use general shortest path methods (other than DP) for deterministic finite-state optimal control problems. Even though DP is preferable in most cases to other shortest path methods because it is tailored to the sequential nature of optimal control problems, there are important cases where other shortest path methods are preferable.

In this section we discuss several alternative shortest path methods. We motivate these methods by focusing on shortest path problems with a very large number of nodes. Suppose that there is only one origin and only one destination, as in shortest path problems arising from deterministic optimal control (cf. Fig. 2.1.1). Then it is often true that most of the nodes are not relevant to the shortest path problem in the sense that they are unlikely candidates for inclusion in a shortest path between the given origin and destination. Unfortunately, however, in the DP algorithm every node and arc will participate in the computation, so there arises the possibility of other more efficient methods.

A similar situation arises in some search problems that are common in artificial intelligence. We provide some examples.

Example 3.1 (The Four Queens Problem)

Four queens must be placed on a 4×4 portion of a chessboard so that no queen can attack another. In other words, the placement must be such that every row, column, or diagonal of the 4×4 board contains at most one queen. Equivalently, we can view the problem as a sequence of problems; first, placing a queen in one of the first two squares in the top row, then placing another queen in the second row so that it is not attacked by the first, and similarly placing the third and fourth queens. (It is sufficient to consider only the first two squares of the top row, since the other two squares lead to symmetric positions.) We can associate positions with nodes of an acyclic graph where the root node s corresponds to the position with no queens and the terminal nodes correspond to the positions where no additional queens can be placed without some queen attacking another. Let us connect each terminal position with an artificial node t by means of an arc. Let us also assign to all arcs length zero except for the artificial arcs connecting terminal positions with less than four queens with the artificial node t. These latter arcs are assigned the length ∞ (see Fig. 2.3.1) to express the fact that they correspond to dead-end positions that cannot lead to a solution. Then, the four queens problem reduces to finding a shortest path from node s to node t. Note that once the nodes of the graph are enumerated the problem is essentially solved. Here the number of nodes is small. However, we can think of similar problems with much larger memory requirements. For example, there is an eight queens problem where the board is 8×8 instead of 4×4.

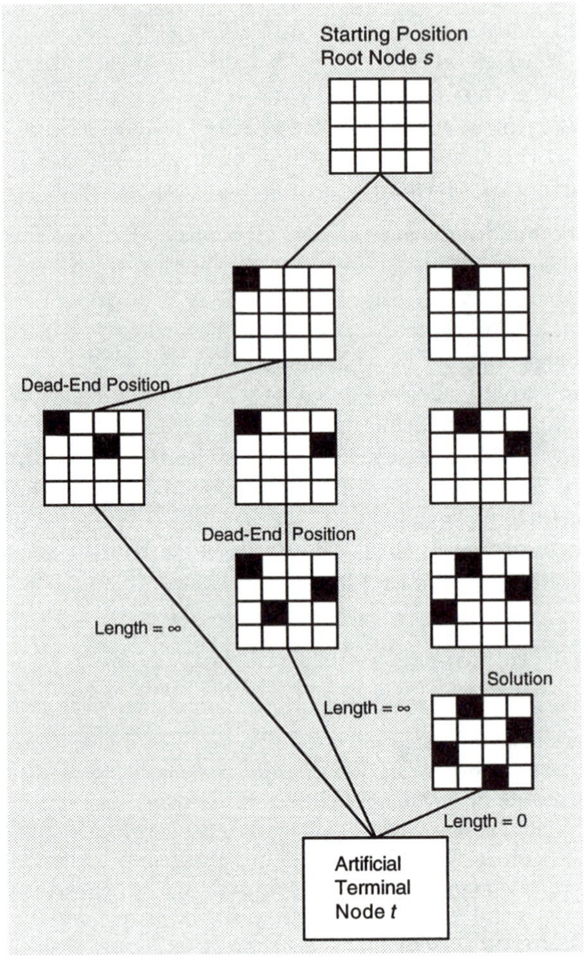

Figure 2.3.1 Shortest path formulation of the four queens problem. Symmetric positions resulting from placing a queen in one of the rightmost squares in the top row have been ignored. Squares containing a queen have been darkened. All arcs have length zero except for those connecting dead-end positions to the artificial terminal node.

Example 3.2 (The Traveling Salesman Problem)

An important model for scheduling a sequence of operations is the classical traveling salesman problem. Here we are given N cities and the mileage between each pair of cities. We wish to find a minimum-mileage trip that visits each of the cities exactly once and returns to the origin node. To convert this problem to a shortest path problem, we associate a node with every sequence of n distinct cities, where $n \leq N$. The construction and arc lengths of the corresponding graph are explained by means of an example in Fig. 2.3.2. The origin node s consists of city A, taken as the start. A sequence of n cities ($n < N$) yields a sequence of $(n + 1)$ cities by adding a new city. Two such sequences are connected by an arc with length equal to the mileage between the last two of the $n+1$ cities. Each sequence of N cities is connected to an artificial terminal node t with an arc having length equal to the distance from the last city of the sequence to the starting city A. Note that the number of nodes grows exponentially with the number of cities.

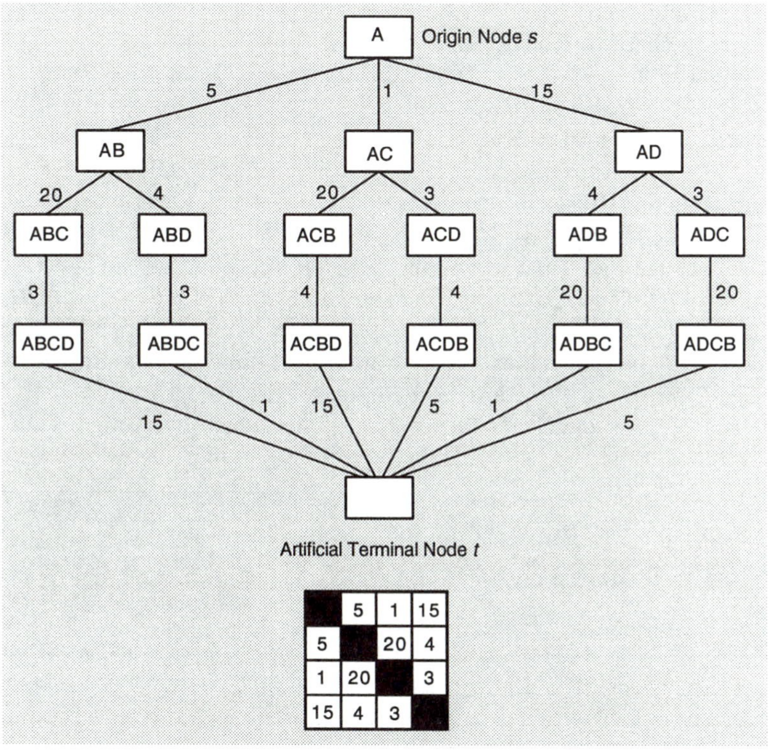

	5	1	15
5		20	4
1	20		3
15	4	3	

Figure 2.3.2 Example of a shortest path formulation of the traveling salesman problem. The distance between the four cities A, B, C, and D are shown in the table. The arc lengths are shown next to the arcs.

In the shortest path problem that we will consider in this section, there is a special node s, called the *origin*, and a special node t, called the *destination*. Exercise 2.6 deals with the case where there are multiple destinations. A node j is called a *child* of node i if there is an arc (i, j) connecting i with j. The length of arc (i, j) is denoted by a_{ij} and we assume that *all arcs have nonnegative length*. Exercise 2.7 deals with the case where all cycle lengths (rather than arc lengths) are assumed nonnegative. We wish to find a shortest path from origin to destination.

2.3.1 Label Correcting Methods

We now discuss a general type of shortest path algorithm. The idea is to progressively discover shorter paths from the origin s to every other node i, and to maintain the length of the shortest path found so far in a variable d_i called the *label of* i. Each time d_i is reduced following the discovery of a shorter path to i, the algorithm checks to see if the labels d_j of the children j of i can be "corrected," that is, they can be reduced by setting them to $d_i + a_{ij}$ [the length of the shortest path to i found thus far followed by arc (i, j)]. The label d_t of the destination is maintained in a variable called UPPER, which plays a special role in the algorithm. Initially, $d_s = 0$ and $d_i = \infty$ for all other nodes i.

The algorithm also makes use of a list of nodes called OPEN (another name frequently used is *candidate list*). The list OPEN contains nodes that are currently active in the sense that they are candidates for further examination by the algorithm and possible inclusion in the shortest path. Initially, OPEN contains just the origin node s. Each node that has entered OPEN at least once, except s, is assigned a "parent," which is some other node. The parent nodes are not necessary for the computation of the shortest distance; they are needed for tracing the shortest path to the origin after the algorithm terminates. The steps of the algorithm are as follows:

Label Correcting Algorithm

Step 1: Remove a node i from OPEN and for each child j of i, execute step 2.

Step 2: If $d_i + a_{ij} < \min\{d_j, \text{UPPER}\}$, set $d_j = d_i + a_{ij}$ and set i to be the parent of j. In addition, if $j \neq t$, place j in OPEN if it is not already in OPEN, while if $j = t$, set UPPER to the new value $d_i + a_{it}$ of d_t.

Step 3: If OPEN is empty, terminate; else go to step 1.

It can be seen by induction that, throughout the algorithm, d_j is either ∞ (if node j has not yet entered the OPEN list), or else it is the length of some path from s to j consisting of nodes that have entered the OPEN list at least once. In the latter case, this path can be constructed by tracing backward the parent nodes starting with the parent of node j.

Furthermore, UPPER is either ∞, or else it is the length of some path from s to t, and consequently it is an upper bound of the shortest distance from s to t. The idea in the algorithm is that when a path from s to j is discovered that is shorter than those considered earlier ($d_i + a_{ij} < d_j$ in step 2), the value of d_j is accordingly reduced, and node j enters the OPEN list so that paths passing through j and reaching the children of j can be taken into account. It makes sense to do so, however, only when the path considered has a chance of leading to a path from s to t with length smaller than the upper bound UPPER of the shortest distance from s to t. In view of the nonnegativity of the arc lengths, this is possible only if the path length $d_i + a_{ij}$ is smaller than UPPER. This provides the rationale for entering j into OPEN in step 2 only if $d_i + a_{ij}$ is less than UPPER.

Tracing the steps of the algorithm, we see that it will first remove node s from OPEN and sequentially examine its children. If t is not a child of s, the algorithm will place all children j of s in OPEN after setting $d_j = a_{sj}$. If t is a child of s, then the algorithm will place all children j of s examined before t in OPEN and will set their labels to a_{sj}; then it will examine t and set UPPER to a_{st}; finally, it will place each of the remaining children j of s in OPEN only if a_{sj} is less than the current value of UPPER, which is a_{st}. The algorithm will subsequently take a child $i \neq t$ of s from OPEN, and sequentially place in OPEN those of its children $j \neq t$ that satisfy the criterion of step 2, etc. Note that the origin s can never reenter OPEN because d_s cannot be reduced from its initial value of zero. Furthermore, by the rules of the algorithm, the destination can never enter OPEN. When the algorithm terminates, we will show shortly that a shortest path can be obtained by tracing backwards the parent nodes starting from t and going towards s. Figure 2.3.3 illustrates the use of the algorithm to solve the traveling salesman problem of Fig. 2.3.2.

The following proposition establishes the validity of the algorithm.

Proposition 3.1: If there exists at least one path from the origin to the destination, the label correcting algorithm terminates with UPPER equal to the shortest distance from the origin to the destination. Otherwise the algorithm terminates with UPPER $= \infty$.

Proof: We first show that the algorithm will terminate. Indeed, each time a node j enters the OPEN list, its label is decreased and becomes equal to the length of some path from s to j. On the other hand, the number of distinct lengths of paths from s to j that are smaller than any given number is finite. The reason is that each path can be decomposed into a path with no repeated nodes (there is a finite number of distinct such paths), plus a (possibly empty) set of cycles, each having a nonnegative length. Therefore, there can be only a finite number of label reductions,

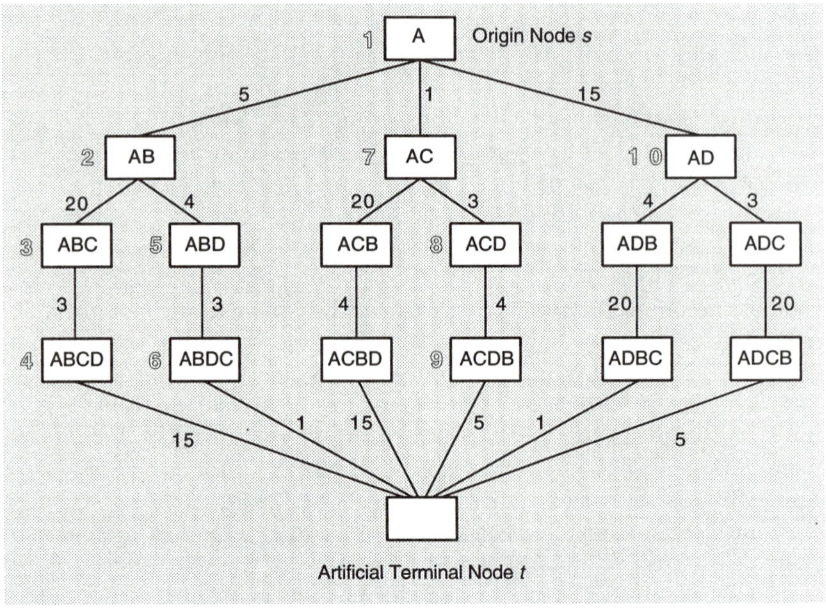

Iter. No.	Node Exiting OPEN	OPEN at the End of Iteration	UPPER
0	-	1	∞
1	1	2, 7,10	∞
2	2	3, 5, 7, 10	∞
3	3	4, 5, 7, 10	∞
4	4	5, 7, 10	43
5	5	6, 7, 10	43
6	6	7, 10	13
7	7	8, 10	13
8	8	9, 10	13
9	9	10	13
10	10	Empty	13

Figure 2.3.3 The algorithm applied to the traveling salesman problem of Fig. 2.3.2. The optimal solution ABDC is found after examining nodes 1 through 10 in the figure in that order. The table shows the successive contents of the OPEN list.

implying that the algorithm will terminate.

Suppose that there is no path from s to t. Then a node i such that (i, t) is an arc cannot enter the OPEN list, because as argued earlier, this would establish that there is a path from s to i, and therefore also a path from s to t. Thus, based on the rules of the algorithm, UPPER can never be reduced from its initial value of ∞.

Suppose now that there is a path from s to t. Then, since there is a finite number of distinct lengths of paths from s to t that are smaller than any given number, there is also a shortest path. Let $(s, j_1, j_2, \ldots, j_k, t)$ be a shortest path and let d^* be the corresponding shortest distance. We will show that the value of UPPER upon termination must be equal to d^*. Indeed, each subpath $(s, j_1, \ldots, j_m)$, $m = 1, \ldots, k$, of the shortest path $(s, j_1, j_2, \ldots, j_k, t)$ must be a shortest path from s to j_m. If the value of UPPER is larger than d^* at termination, the same must be true throughout the algorithm, and therefore UPPER will also be larger than the length of all the paths $(s, j_1, \ldots, j_m)$, $m = 1, \ldots, k$, throughout the algorithm, in view of the nonnegative arc length assumption. It follows that node j_k will never enter the OPEN list with d_{j_k} equal to the shortest distance from s to j_k, since in this case UPPER would be set to d^* in step 2 immediately following the next time node j_k is examined by the algorithm in step 2. Similarly, and using also the nonnegative length assumption, this means that node j_{k-1} will never enter the OPEN list with $d_{j_{k-1}}$ equal to the shortest distance from s to j_{k-1}. Proceeding backward, we conclude that j_1 never enters the OPEN list with d_{j_1} equal to the shortest distance from s to j_1 [which is equal to the length of the arc (s, j_1)]. This happens, however, at the first iteration of the algorithm, obtaining a contradiction. It follows that at termination, UPPER will be equal to the shortest distance from s to t. **Q.E.D.**

From the preceding proof, it can also be seen that, upon termination of the algorithm, the path constructed by tracing the parent nodes backward from t to s has length equal to UPPER, so it is a shortest path from s to t. Thus the algorithm yields not just the shortest distance but also a shortest path, provided that we keep track of the parent of each node that enters OPEN.

An important property of the algorithm is that nodes j for which $d_i + a_{ij} \geq$ UPPER in step 2 need not enter OPEN and be examined later. As a result the number of nodes that enter OPEN may be much smaller than the total number of nodes. Furthermore, if we know a good lower bound to the shortest distance from s to t (or the shortest distance itself), we can terminate the computation once UPPER reaches that bound within an acceptable tolerance. This is useful, for example, in the four queens problem, where the shortest distance is known to be zero or infinity. Then the algorithm will terminate once a solution is found.

Specific Label Correcting Methods

There is considerable freedom in selecting the node to be removed from OPEN at each iteration. This gives rise to several different methods. The following are some of the most important.

(a) *Breadth-first search*, also known as the *Bellman-Ford* method, which adopts a first-in/first-out policy; that is, the node is always removed from the top of OPEN and each node entering OPEN is placed at the bottom of OPEN. [Here and in the following methods except (c), we assume that OPEN is structured as a queue.]

(b) *Depth-first search*, which adopts a last-in/first-out policy; that is, the node is always removed from the top of OPEN and each node entering OPEN is placed at the top of OPEN. One motivation for this method is that it often requires relatively little memory. For example, suppose that the graph has a tree-like structure whereby there is a unique path from the origin node to every node other than the destination as shown in Fig. 2.3.4. Then the nodes will enter OPEN only once and in the order shown in Fig. 2.3.4. At any one time, it is only necessary to store a small portion of the graph as shown in Fig. 2.3.5.

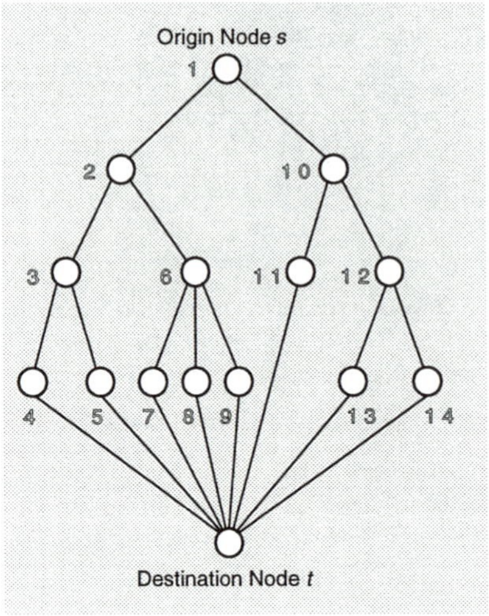

Figure 2.3.4 Searching a tree in depth-first fashion. The numbers next to the nodes indicate the order in which nodes exit the OPEN list.

(c) *Best-first search*, which at each iteration removes from OPEN a node with minimum value of label, i.e., a node i with

$$d_i = \min_{j \text{ in OPEN}} d_j.$$

This method, also known as *Dijkstra's method* or *label setting method*, has a particularly interesting property. It can be shown that in this method, a node will enter the OPEN list at most once (see Exercise 2.4). The drawback of this method is the overhead required to find at

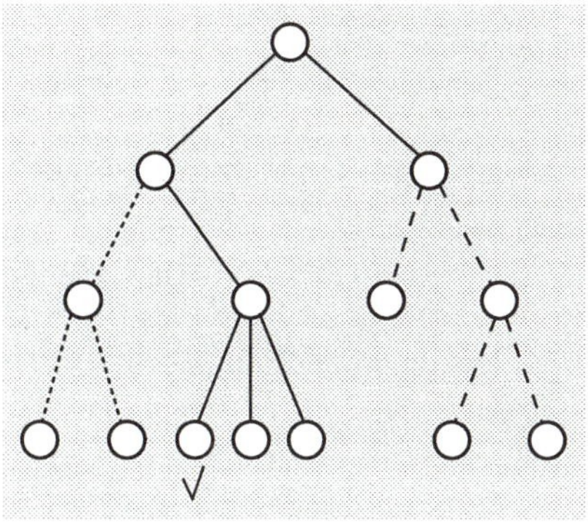

Figure 2.3.5 Memory requirements of depth-first search for the graph of Fig. 2.3.4. At the time the node marked by the checkmark exits the OPEN list, only the solid-line portion of the tree is needed in memory. The dotted-line portion has been generated and purged from memory based on the rule that for a graph where there is only one path from the origin to every node other than the destination, it is unnecessary to store a node once all of its successors are out of the OPEN list. The broken-line portion of the tree has not yet been generated.

each iteration the node of OPEN that has minimum label. Several sophisticated methods have been developed to carry out this operation efficiently (see e.g., [Ber91a]).

(d) *D'Esopo-Pape method*, which at each iteration removes the node that is at the top of OPEN, but inserts a node at the top of OPEN if it has already been in OPEN earlier, and inserts the node at the bottom of OPEN otherwise.

(e) *Small-Label-First (SLF) method*, which at each iteration removes the node that is at the top of OPEN, but inserts a node i at the top of OPEN if its label d_i is less than or equal to the label d_j of the top node j of OPEN; otherwise it inserts i at the bottom of OPEN. This is a low-overhead approximation to the best-first search strategy. As a complement to the SLF strategy, one may also try to avoid removing nodes with relatively large labels from OPEN using the following device, known as the *Large-Label-Last (LLL) strategy*: at the start of an iteration, the top node of OPEN is compared with the average of the labels of the nodes in OPEN and if it is larger, the top node is placed at the bottom of OPEN and the new top node of OPEN is similarly tested against the average of the labels. In this way the removal of nodes with relatively large labels is postponed in favor of nodes with relatively small labels. The extra overhead required in this method is to maintain the sum of the labels and the number of nodes in OPEN. When starting a new iteration, the ratio of these numbers gives the desired average. There are also several other methods, which are based on the idea of examining nodes with small labels first (see [Ber93] and [BGM94] for detailed descriptions and computational studies).

Generally, it appears that for nonnegative arc lengths, the number of iterations is reduced as the method is more successful in removing from

OPEN nodes with a relatively small label. For a supporting heuristic argument, note that for a node j to reenter OPEN, some node i such that $d_i + a_{ij} < d_j$ must first exit OPEN. Thus, the smaller d_j was at the previous exit of j from OPEN, the less likely it is that $d_i + a_{ij}$ will subsequently become less than d_j for some node i in OPEN and arc (i, j). In particular, if $d_j \leq \min_{i \text{ in OPEN}} d_i$, it is impossible that subsequent to the exit of j from OPEN we will have $d_i + a_{ij} < d_j$ for some i in OPEN, since the arc lengths a_{ij} are nonnegative. The SLF and other related but more sophisticated methods, often require a number of iterations, which is close to the minimum (the one required by the best-first method). However, they can be much faster than the best-first method, because they require much less overhead for determining the node to be removed from OPEN.

Variations of Label Correcting Methods

We mentioned earlier that a key idea of the algorithm is to save computation by foregoing the examination of nodes j that cannot lie on a shortest path. This is based on the test $d_i + a_{ij} <$ UPPER that node j must pass before it can be placed in the OPEN list in step 2. We can strengthen this test if we can find a *positive underestimate* h_j of the shortest distance of node j to the destination. Such an estimate can be obtained from special knowledge about the problem at hand. We may then speed up the computation substantially by placing a node j in OPEN in step 2 only when $d_i + a_{ij} + h_j <$ UPPER (instead of $d_i + a_{ij} <$ UPPER). In this way, fewer nodes will potentially be placed in OPEN before termination. Using the fact that h_j is an underestimate of the true shortest distance from j to the destination, the argument given in the proof of Prop. 3.1 shows that the algorithm will terminate with a shortest path.

The idea just described is one way to sharpen the test $d_i + a_{ij} <$ UPPER for admission of node j into the OPEN list. An alternative idea is to try to reduce the value of UPPER by obtaining for the node j in step 2 an *upper bound* m_j of the shortest distance from j to the destination. Then if $d_j + m_j <$ UPPER after step 2, we can reduce UPPER to $d_j + m_j$, thereby making the test for future admissibility into OPEN more stringent. This idea is used in some versions of the branch-and-bound algorithm, one of which we now briefly describe (see also [PaS82] and [Pea84] for further discussion).

Example 3.3 (Branch-and-Bound Algorithm)

Consider a problem of minimizing a cost function $f(x)$ over a *finite* set of feasible solutions X. The branch-and-bound algorithm uses an acyclic graph with nodes that correspond on a one-to-one basis with a collection $\mathcal{X}$ of subsets X. We require the following:

1. $X \in \mathcal{X}$ (i.e., the set of all solutions is a node).

2. If x is a solution, then $\{x\} \in \mathcal{X}$ (i.e., each solution viewed as a singleton set is a node).
3. If a set $Y \in \mathcal{X}$ contains more than one solution $x \in X$, then there exist sets $Y_1, \ldots, Y_n \in \mathcal{X}$ such that $Y_i \neq Y$ for all i and

$$\bigcup_{i=1}^{n} Y_i = Y.$$

The set Y is called the *parent* of $Y_1, \ldots, Y_n$, and the sets $Y_1, \ldots, Y_n$ are called the *children* of Y.

4. Each set in $\mathcal{X}$ other than X has at least one parent.

The collection of sets $\mathcal{X}$ defines an acyclic graph with root node the set of all feasible solutions X and terminal nodes the singleton solutions $\{x\}$, $x \in X$ (see Fig. 2.3.6). The arcs of the graph are those that connect parents Y and their children Y_i. Suppose that for every nonterminal node Y there is an algorithm that calculates upper and lower bounds $\underline{f}_Y$ and $\overline{f}_Y$ for the minimum cost over Y:

$$\underline{f}_Y \leq \min_{x \in Y} f(x) \leq \overline{f}_Y.$$

Assume further that the upper and lower bounds are exact for each singleton solution node $\{x\}$:

$$\underline{f}_{\{x\}} = f(x) = \overline{f}_{\{x\}}, \qquad \text{for all } x \in X.$$

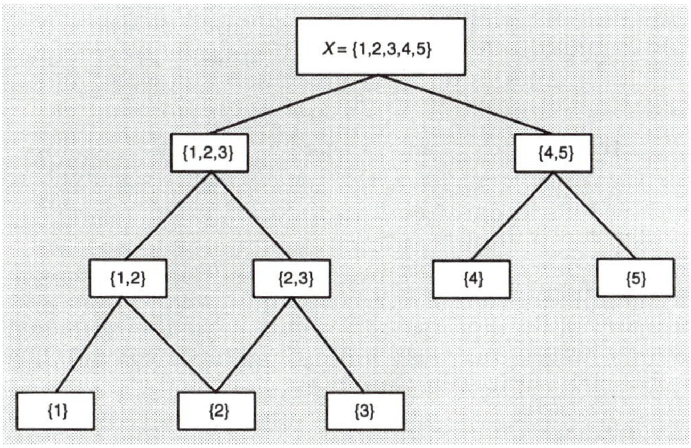

Figure 2.3.6 A tree corresponding to a branch-and-bound algorithm. Each node (subset) except those consisting of a single solution is partitioned into several other nodes (subsets).

Define now the length of an arc involving a parent Y and a child Y_i, to be the lower bound difference

$$\underline{f}_{Y_i} - \underline{f}_Y.$$

Then the length of any path from the origin node X to any node Y is $\underline{f}_Y$. Since $\underline{f}_{\{x\}} = f(x)$ for all feasible solutions $x \in X$, it is clear that minimizing $f(x)$ over $x \in X$ is equivalent to finding a shortest path from the origin node to one of the singleton nodes $\{x\}$.

Consider now a variation of the label correcting method where in addition we use our knowledge of the upper bounds $\overline{f}_Y$ to reduce the value of UPPER. Initially, OPEN contains just X, and UPPER equals $\overline{f}_X$.

Branch-and-Bound Algorithm

Step 1: Remove a node Y from OPEN. For each child Y_j of Y, do the following: If $\underline{f}_{Y_j}$ < UPPER, then place Y_j in OPEN. If in addition $\overline{f}_{Y_j}$ < UPPER, then set UPPER $= \overline{f}_{Y_j}$, and if Y_j consists of a single solution, mark that solution as being the best solution found so far.

Step 2: (Termination Test) If OPEN is nonempty, go to step 1. Otherwise, terminate; the best solution found so far is optimal.

An alternative termination step 2 for the preceding algorithm is to set a tolerance $\epsilon > 0$, and check whether UPPER and the minimum lower bound $\underline{f}_Y$ over all sets Y in the OPEN list differ by less than ϵ. If so, the algorithm is terminated, and some set in OPEN must contain a solution that is within ϵ of being optimal. There are a number of other variations of the algorithm. For example, if the upper bound $\overline{f}_Y$ at a node is actually the cost $f(x)$ of some element $x \in Y$, then this element can be taken as the best solution found so far whenever $\overline{f}_Y$ < UPPER in step 2. Other variations relate to the method of selecting a node from OPEN in step 1. For example, two strategies of the best-first type are to select the node with minimal lower or upper bound. In closing, we note that to apply branch-and-bound effectively it is important to have algorithms for generating as sharp as practically possible upper and lower bounds at each node. Then, fewer nodes will be admitted into OPEN, with attendant computational savings.

2.3.2 Auction Algorithms

We now discuss another class of algorithms for finding a shortest path from the origin s to the destination t. These are called *auction* algorithms because they can be shown to be closely related to an algorithm for optimally assigning persons to objects, which resembles a real-life auction; see [Ber91a]. The main algorithm is very simple. It maintains a single path starting at the origin. At each iteration, the path is either *extended* by adding a new node, or *contracted* by deleting its terminal node. When the destination becomes the terminal node of the path, the algorithm terminates.

To get an intuitive sense of the algorithm, think of a mouse moving in a graph-like maze, trying to reach a destination. The mouse criss-crosses the maze, either advancing or backtracking along its current path. Each time the mouse backtracks from a node, it records a measure of the desirability of revisiting and advancing from that node in the future (this will be represented by a suitable variable). The mouse revisits and proceeds forward from a node when the node's measure of desirability is judged superior to those of other nodes. The algorithm emulates efficiently this search process using simple data structures.

The algorithm maintains a path $P = \big((s, i_1), (i_1, i_2), \ldots, (i_{k-1}, i_k)\big)$ with no cycles, and modifies P using two operations, *extension* and *contraction*. If i_{k+1} is a node not on P and (i_k, i_{k+1}) is an arc, an *extension of P by i_{k+1}* replaces P by the path $\big((s, i_1), (i_1, i_2), \ldots, (i_{k-1}, i_k), (i_k, i_{k+1})\big)$. If P does not consist of just the origin node s, a *contraction of P* replaces P by the path $\big((s, i_1), (i_1, i_2), \ldots, (i_{k-2}, i_{k-1})\big)$.

We introduce a variable p_i for each node i, called the *price of node i*. We denote by p the price vector of all node prices. The algorithm maintains a price vector p satisfying together with P the following property

$$p_i \leq a_{ij} + p_j, \qquad \text{for all arcs } (i, j), \tag{3.1a}$$

$$p_i = a_{ij} + p_j, \qquad \text{for all pairs of successive nodes } i \text{ and } j \text{ of } P, \tag{3.1b}$$

referred to as *complementary slackness* (CS for short), because of its relation to a corresponding linear programming notion (see [Ber91a]). We assume that the initial pair (P, p) satisfies CS. This is not restrictive, since the default pair

$$P = (s), \qquad p_i = 0, \quad \text{for all } i$$

satisfies CS in view of the nonnegative arc length assumption. To define the algorithm we also need to assume that *all cycles have positive length*; Exercise 2.9 indicates how this assumption can be relaxed.

It can be shown that if a pair (P, p) satisfies the CS conditions, then the portion of P between node s and any node $i \in P$ is a shortest path from s to i, while $p_s - p_i$ is the corresponding shortest distance. To see this, note that by Eq. (3.1b), $p_i - p_k$ is the length of the portion of P between i and k, and that every path connecting i to k must have length at least equal to $p_i - p_k$ [add Eq. (3.1a) along the arcs of the path].

The algorithm proceeds in iterations, transforming a pair (P, p) satisfying CS into another pair satisfying CS. At each iteration, the path P either is extended by a new node or else is contracted by deleting its terminal node. In the latter case the price of the terminal node is increased strictly. A degenerate case occurs when the path consists of just the origin node s; in this case the path either is extended or is left unchanged with the price p_s being strictly increased. The iteration is as follows.

Typical Iteration of the Auction Algorithm

Let i be the terminal node of P. If

$$p_i < \min_{\{j \mid (i,j) \text{ is an arc}\}} \big[a_{ij} + p_j\big], \tag{3.2}$$

go to Step 1; else go to Step 2.

Step 1 (Contract path): Set

$$p_i := \min_{\{j \mid (i,j) \text{ is an arc}\}} \big[a_{ij} + p_j\big], \tag{3.3}$$

and if $i \neq s$, contract P. Go to the next iteration.

Step 2 (Extend path): Extend P by node j_i where

$$j_i = \arg \min_{\{j \mid (i,j) \text{ is an arc}\}} \big[a_{ij} + p_j\big]. \tag{3.4}$$

If j_i is the destination t, stop; P is the desired shortest path. Otherwise, go to the next iteration.

It is easily seen that the algorithm maintains CS. Furthermore, the addition of the node j_i to P following an extension does not create a cycle, since otherwise, in view of the condition $p_i \leq a_{ij} + p_j$, for every arc (i,j) of the cycle we would have $p_i = a_{ij} + p_j$. By adding this equality along the cycle, we see that the length of the cycle must be zero, which is not possible by our assumptions.

Figure 2.3.7 illustrates the algorithm. It can be seen from the example of this figure that the terminal node traces the tree of shortest paths from the origin to the nodes that are closer to the origin than the given destination. This behavior is typical when the initial prices are all zero (see Exercise 2.10).

There is an interesting interpretation of the CS conditions in terms of a mechanical model. Think of each node as a ball, and for every arc (i,j), connect i and j with a string of length a_{ij}. (This requires that $a_{ij} = a_{ji} > 0$, which we assume for the sake of the interpretation.) Let the resulting balls-and-strings model be at an arbitrary position in three-dimensional space, and let p_i be the vertical coordinate of node i. Then the CS condition $p_i - p_j \leq a_{ij}$ clearly holds for all arcs (i,j), as illustrated in Fig. 2.3.8(b). If the model is picked up and left to hang from the origin node (by gravity – strings that are tight are perfectly vertical), then for all the tight strings (i,j) we have $p_i - p_j = a_{ij}$, so any tight chain of strings corresponds to a shortest path between the end nodes of the chain, as illustrated in Fig. 2.3.8(c). In particular, the length of the tight chain

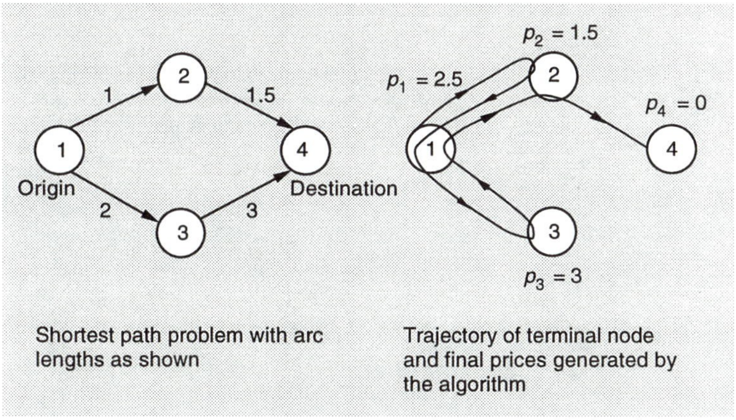

Iteration #	Path P prior to iteration	Price vector p prior to iteration	Type of action during iteration
1	(1)	$(0,0,0,0)$	contraction at 1
2	(1)	$(1,0,0,0)$	extension to 2
3	$(1,2)$	$(1,0,0,0)$	contraction at 2
4	(1)	$(1,1.5,0,0)$	contraction at 1
5	(1)	$(2,1.5,0,0)$	extension to 3
6	$(1,3)$	$(2,1.5,0,0)$	contraction at 3
7	(1)	$(2,1.5,3,0)$	contraction at 1
8	(1)	$(2.5,1.5,3,0)$	extension to 2
9	$(1,2)$	$(2.5,1.5,3,0)$	extension to 4
10	$(1,2,4)$	$(2.5,1.5,3,0)$	stop

Figure 2.3.7 An example illustrating the auction algorithm starting with $P = (1)$ and $p = 0$.

connecting the origin node s to any other node i is $p_s - p_i$ and is also equal to the shortest distance from s to i.

The algorithm can also be interpreted in terms of the balls-and-strings model; it can be viewed as a process whereby nodes are raised in stages as illustrated in Fig. 2.3.9. Initially all nodes are resting on a flat surface. At each stage, we raise the *last* node in a tight chain that starts at the origin to the level at which at least one more string becomes tight.

The following proposition establishes the validity of the auction algorithm.

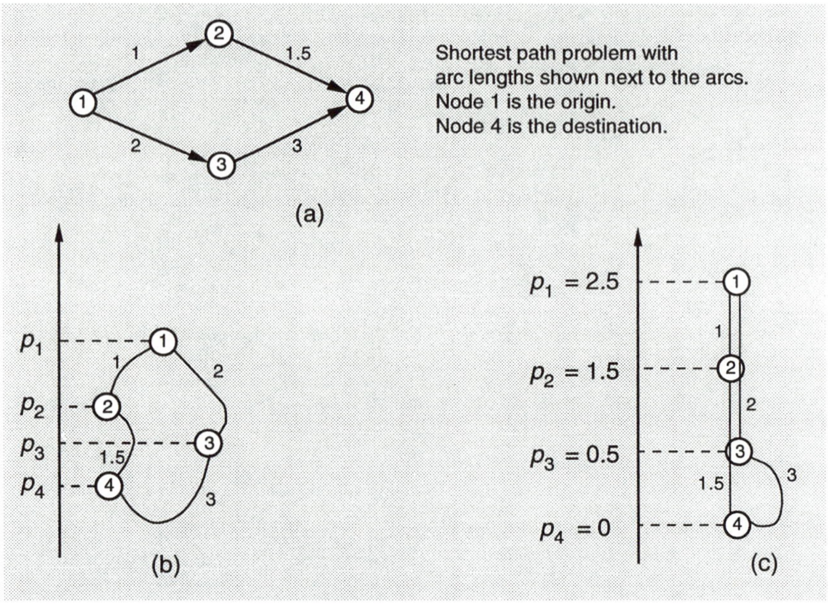

Figure 2.3.8 Illustration of the CS conditions for the shortest path problem. If each node is a ball, and for every arc (i, j), nodes i and j are connected with a string of length a_{ij}, the vertical coordinates p_i of the nodes satisfy $p_i - p_j \leq a_{ij}$, as shown in (b) for the problem given in (a). If the model is picked up and left to hang from the origin node s, then $p_s - p_i$ gives the shortest distance to each node i, as shown in (c).

Proposition 3.2: If there exists at least one path from the origin to the destination, the auction algorithm terminates with a shortest path from the origin to the destination. Otherwise the algorithm never terminates and $p_s \to \infty$.

Proof: We first show by induction that (P, p) satisfies CS throughout the algorithm. Indeed, the initial pair satisfies CS by assumption. Consider an iteration that starts with a pair (P, p) satisfying CS and produces a pair $(\overline{P}, \overline{p})$. Let i be the terminal node of P. If

$$p_i = \min_{\{j \mid (i,j) \text{ is an arc}\}} \big[a_{ij} + p_j\big], \tag{3.5}$$

then $\overline{P}$ is the extension of P by a node j_i and $\overline{p} = p$, implying that the CS condition (3.1b) holds for all arcs of P as well as arc (i, j_i) [since j_i attains the minimum in Eq. (3.5); cf. condition (3.4)].

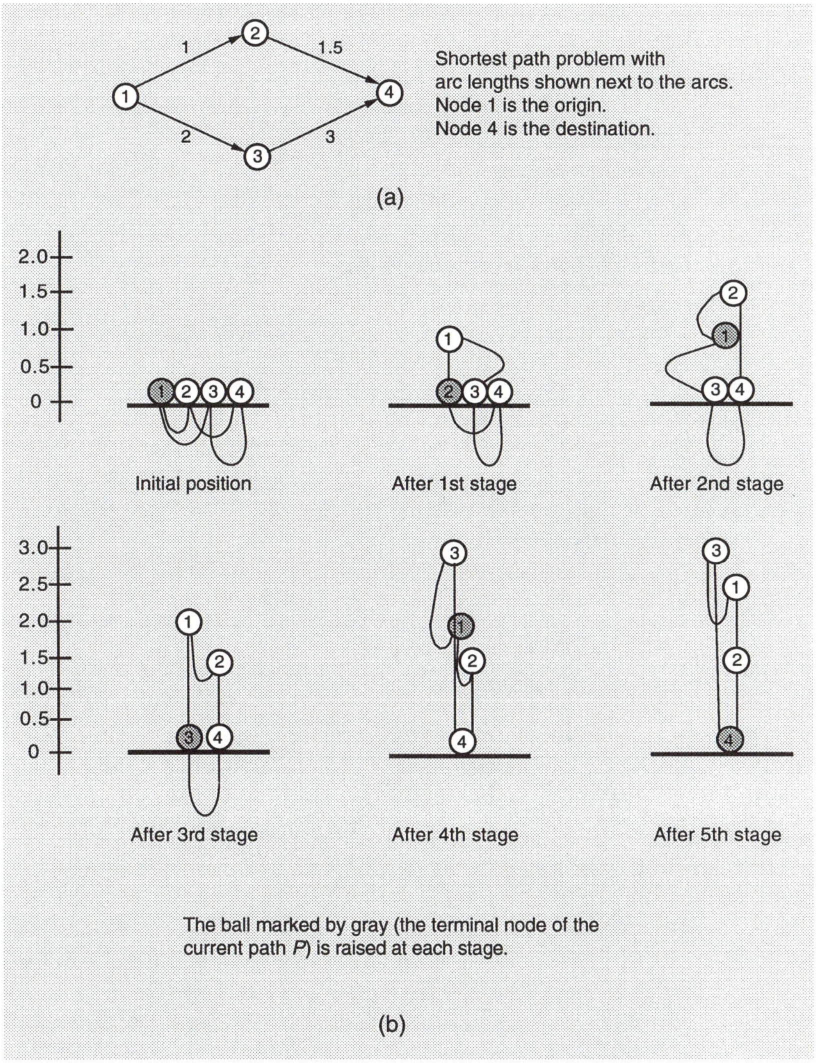

Figure 2.3.9 Illustration of the auction algorithm in terms of the balls-and-strings model for the problem shown in (a). The model initially rests on a flat surface, and various balls are then raised in stages. At each stage we raise a single ball $i \neq t$ (marked by gray), which is at a lower level than the origin s and can be reached from s through a sequence of tight strings; i should not have any tight string connecting it to another ball, which is at a lower level, that is, i should be the last ball in a tight chain hanging from s. (If s does not have any tight string connecting it to another ball, which is at a lower level, we use $i = s$.) We then raise i to the first level at which one of the strings connecting it to a ball at a lower level becomes tight. Each stage corresponds to a contraction plus all the extensions up to the next contraction. The ball i, which is being raised, corresponds to the terminal node of the current path P.

Suppose next that

$$p_i < \min_{\{j \mid (i,j) \text{ is an arc}\}} \left[a_{ij} + p_j\right].$$

Then if P is the degenerate path (s), the CS condition holds vacuously. Otherwise, $\overline{P}$ is obtained by contracting P, and for all nodes $j \in \overline{P}$, we have $\overline{p}_j = p_j$, implying conditions (3.1a) and (3.1b) for arcs outgoing from nodes of $\overline{P}$. Also, for the terminal node i, we have

$$\overline{p}_i = \min_{\{j \mid (i,j) \text{ is an arc}\}} \left[a_{ij} + p_j\right],$$

implying condition (3.1a) for arcs outgoing from that node as well. Finally, since $\overline{p}_i > p_i$ and $\overline{p}_k = p_k$ for all $k \neq i$, we have $\overline{p}_k \leq a_{kj} + \overline{p}_j$ for all arcs (k, j) outgoing from nodes $k \notin P$. This completes the induction proof that (P, p) satisfies CS throughout the algorithm.

Assume first that there is a path from node s to the destination t. By adding the CS condition (3.1a) along that path, we see that $p_s - p_t$ is an underestimate of the (finite) shortest distance from s to t. Since p_s is monotonically nondecreasing, and p_t is fixed throughout the algorithm, it follows that p_s must stay bounded.

We next claim that p_i must stay bounded for all i. Indeed, in order to have $p_i \to \infty$, node i must become the terminal node of P infinitely often. Each time this happens, $p_s - p_i$ is equal to the shortest distance from s to i, which is a contradiction since p_s is bounded.

We next show that the algorithm terminates. Indeed, it can be seen with a straightforward induction argument that for every node i, p_i is either equal to its initial value, or else it is the length of some path starting at i plus the initial price of the final node of the path; we call this the *modified length* of the path. Every path from s to i can be decomposed into a path with no cycles together with a finite number of cycles, each having positive length by assumption, so the number of distinct modified path lengths within any bounded interval is bounded. Now p_i was shown earlier to be bounded, and each time i becomes the terminal node by extension of the path P, p_i is strictly larger over the preceding time i became the terminal node of P, corresponding to a strictly larger modified path length. It follows that the number of times i can become a terminal node by extension of the path P is bounded. Since the number of path contractions between two consecutive path extensions is bounded by the number of nodes in the graph, the number of iterations of the algorithm is bounded, implying that the algorithm terminates.

Assume now that there is no path from node s to the destination. Then, the algorithm will never terminate, so by the preceding argument, some node i will become the terminal node by extension of the path P infinitely often and $p_i \to \infty$. At the end of iterations where this happens,

$p_s - p_i$ must be equal to the shortest distance from s to i, implying that $p_s \to \infty$. **Q.E.D.**

There are a number of variations of the basic algorithm for which we refer to the literature. The most significant of these relates to a forward/reverse version of the algorithm that maintains, in addition to the path P, another path R that *ends at the destination*. Such an algorithm can be interpreted in terms of the balls-and-strings model of Fig. 2.3.8 as follows. When the forward part of the algorithm is used, we raise nodes in stages as illustrated in Fig. 2.3.9. When the reverse part of the algorithm is used, we *lower* nodes in stages; at each stage, we lower the *top* node in a tight chain that ends at the destination to the level at which at least one more string becomes tight. The algorithm terminates when the origin and the destination are joined by a tight chain.

When implemented in its forward/reverse version, the auction algorithm has been found to perform very well for many types of problems, particularly those involving few destinations and a graph with relatively small diameter. However, the algorithm as presented here, can perform poorly in some specially constructed problems (in the language of computational complexity theory, it is nonpolynomial); for an example see Exercise 2.8. There are a number of improved versions of the basic algorithm [Ber91a], [Ber91b], [CDP92], including some polynomial variations developed in [BPS92]; see also Exercise 2.9. The auction algorithm is also well-suited for implementation in parallel computers; see [PoB94].

2.4 NOTES, SOURCES, AND EXERCISES

Work on the shortest path problem is very extensive. Literature surveys are given in [Dre69], [DeP84], and [GaP88]. For a detailed textbook treatment of shortest paths and a variety of associated computer codes, see [Ber91a]. For a discussion of application of shortest path methods in artificial intelligence, see [Pea84]. For a treatment of critical path analysis, see [Elm78]. A tutorial survey of Hidden Markov Models is given in [Rab89]. The Viterbi algorithm, first proposed in [Vit67], is also discussed in [For73]. For applications in communication systems, see [PrS94] and [Skl88]. For applications in speech recognition, see [Rab89] and [Pic90].

Label correcting methods draw their origin from the works of Bellman [Bel57] and Ford [For56]. The D'Esopo-Pape algorithm appeared in [Pap74] based on an earlier suggestion of D'Esopo. For a discussion of various implementations of Dijkstra's algorithm, see [Ber91a]. The SLF method and some variations were proposed by the author [Ber93]; see also [BGM94], where the LLL strategy as well as implementations on a parallel computer of various label correcting methods are discussed. The auction algorithm for shortest paths is due to the author [Ber91a], [Ber91b].

EXERCISES

2.1

Find a shortest path from each node to node 6 for the graph of Fig. 2.4.1 by using the DP algorithm.

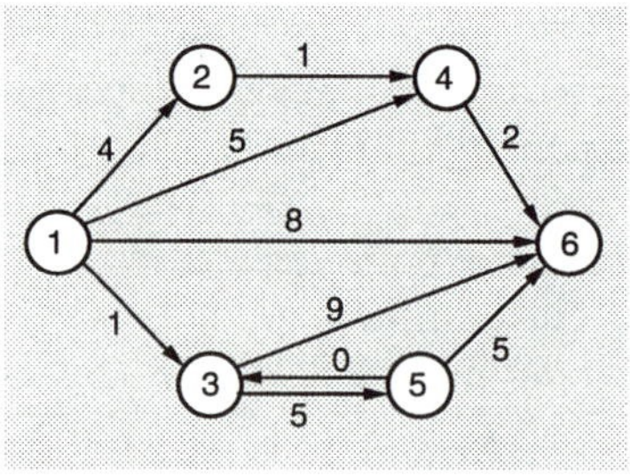

Figure 2.4.1 Graph for Exercise 2.1. The arc lengths are shown next to the arcs.

2.2

Find a shortest path from node 1 to node 5 for the graph of Fig. 2.4.2 by using the label correcting method of Section 2.3.1 and the auction algorithm of Section 2.3.2.

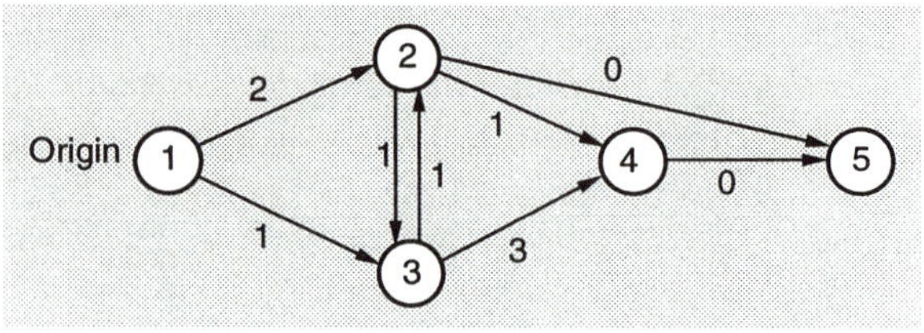

Figure 2.4.2 Graph for Exercise 2.2. The arc lengths are shown next to the arcs.

2.3

Air transportation is available between n cities, in some cases directly and in others through intermediate stops and change of carrier. The airfare between cities i and j is denoted by a_{ij}. We assume that $a_{ij} = a_{ji}$, and for notational convenience, we write $a_{ij} = \infty$ if there is no direct flight between i and j. The problem is to find the cheapest airfare for going between two cities perhaps through intermediate stops. Let $n = 6$ and $a_{12} = 30$, $a_{13} = 60$, $a_{14} = 25$, $a_{15} = a_{16} = \infty$, $a_{23} = a_{24} = a_{25} = \infty$, $a_{26} = 50$, $a_{34} = 35$, $a_{35} = a_{36} = \infty$, $a_{45} = 15$, $a_{46} = \infty$, $a_{56} = 15$. Find the cheapest airfare from every city to every other city by using the DP algorithm.

2.4 (Dijkstra's Algorithm for Shortest Paths)

Consider the best-first version of the label correcting algorithm of Section 2.3.1. Here at each iteration we remove from OPEN a node that has minimum label over all nodes in OPEN.

(a) Show that each node j will enter OPEN at most once, and show that at the time it exits OPEN, its label d_j is equal to the shortest distance from s to j. *Hint*: Use the nonnegative arc length assumption to argue that in the label correcting algorithm, in order for the node i that exits OPEN to reenter, there must exist another node k in OPEN with $d_k + a_{ki} < d_i$.

(b) Show that the number of arithmetic operations required for termination is bounded by cN^2 where N is the number of nodes and c is some constant.

2.5 (Label Correcting for Acyclic Graphs)

Consider a shortest path problem involving an acyclic graph. Let S_k be the set of nodes i such that all paths from the origin to i have k arcs or less and at least one such path has k arcs. Consider a label correcting algorithm that removes from OPEN a node of S_k only if there are no nodes of $S_1, \ldots, S_{k-1}$ in OPEN. Show that each node with enter OPEN at most once. How does this result relate to the type of shortest path problem arising from deterministic DP (cf. Fig. 2.1.1)?

2.6 (Label Correcting with Multiple Destinations)

Consider the problem of finding a shortest path from node s to each node in a subset T, assuming that all arc lengths are nonnegative. Show that the following modified version of the label correcting algorithm of Section 2.3.1 solves the problem. Initially, UPPER $= \infty$, $d_s = 0$, and $d_i = \infty$ for all $i \neq s$.

Step 1: Remove a node i from OPEN and for each child j of i, execute step 2.

Step 2: If $d_i + a_{ij} < \min\{d_j, \text{UPPER}\}$, set $d_j = d_i + a_{ij}$, set i to be the parent of j, and place j in OPEN if it is not already in OPEN. In addition, if $j \in T$, set UPPER $= \max_{t \in T} d_t$.

Step 3: If OPEN is empty, terminate; else go to step 1.

Prove a termination property such as the one of Prop. 3.1 for this algorithm.

2.7 (Label Correcting with Negative Arc Lengths)

Consider the problem of finding a shortest path from node s to node t, and assume that all cycle lengths are nonnegative (instead of all arc lengths being nonnegative). Suppose that a scalar u_j is known for each node j, which is an underestimate of the shortest distance from j to t (u_j can be taken $-\infty$ if no underestimate is known). Consider a modified version of the typical iteration of the label correcting algorithm of Section 2.3.1, where step 2 is replaced by the following:

Modified Step 2: If $d_i + a_{ij} < \min\{d_j, \text{UPPER} - u_j\}$, set $d_j = d_i + a_{ij}$ and set i to be the parent of j. In addition, if $j \neq t$, place j in OPEN if it is not already in OPEN, while if $j = t$, set UPPER to the new value $d_i + a_{it}$ of d_t.

(a) Show that the algorithm terminates.

(b) Why is the algorithm of Section 2.3.1 a special case of the one of this exercise?

2.8

Consider the problem of finding a shortest path from node 1 to node 5 in a graph consisting of arcs (1,2), (2,3), (3,4), (4,2), and (3,5) with corresponding lengths $a_{12} = 1$, $a_{23} = 1$, $a_{34} = 1$, $a_{42} = 1$, $a_{35} = L$, where L is a large integer. Show that the auction algorithm requires a number of iterations that is proportional to L.

2.9 (An Auction Algorithm with Graph Reduction [BPS92])

Consider the variant of the auction algorithm that has the following added feature: each time that a node j becomes the terminal node of the path P through an extension using arc (i, j), all incoming arcs (k, j) of j with $k \neq i$ are deleted from the graph. Also, each time that a node j with no outgoing arcs becomes the terminal node of P, the path P is contracted and the node j is deleted from the graph.

(a) Show that the arc deletion process leaves the shortest distance from s to t unaffected, and that the algorithm terminates either by finding a shortest path from s to t or by deleting s, depending on whether there exists at least one path from s to t or not.

(b) Show that the conclusion of part (a) holds even if there are cycles of zero length.

(c) Apply the algorithm to the problem of Exercise 2.8 and verify that it performs much better than the auction algorithm given in Section 2.3.2.

2.10

Consider the auction algorithm with all initial prices equal to 0. Show that if node j becomes for the first time the terminal node of P after node i does, then the shortest distance from 1 to j is not less than the shortest distance from 1 to i.

2.11 (DP on Two Parallel Processors [Las85])

Formulate a DP algorithm to solve the deterministic problem of Section 2.1 on a parallel computer with two processors. One processor should execute a forward algorithm and the other a backward algorithm.

2.12 (Doubling Algorithms)

Consider a deterministic finite-state problem that is time-invariant in the sense that the state and control spaces, the cost per stage, and the system equation are the same for each time period. Let $J_k(x, y)$ be the optimal cost to reach state y at time k starting from state x at time 0. Show that for all k

$$J_{2k}(x, y) = \min_z \bigl\{ J_k(x, z) + J_k(z, y) \bigr\}.$$

Discuss how this equation may be used with advantage to solve problems with a large number of stages.

2.13 (Distributed Shortest Path Computation [Ber82a])

Consider the problem of finding a shortest path from nodes $1, 2, \ldots, N$ to node t, and assume that all arc lengths are nonnegative and all cycle lengths are positive. Consider the iteration

$$d_i^{k+1} = \min_j \bigl[a_{ij} + d_j^k\bigr], \qquad i = 1, 2, \ldots, N, \tag{4.1}$$

$$d_t^{k+1} = 0.$$

(a) It was shown in Section 2.1 that if the initial condition is $d_i^0 = \infty$ for $i = 1, \ldots, N$ and $d_t^0 = 0$, then the iteration (4.1) yields the shortest distances in N steps. Show that if the initial condition is $d_i^0 = 0$, for all $i = 1, \ldots, N, t$, then the iteration (4.1) yields the shortest distances in a finite number of steps.

(b) Assume that the iteration

$$d_i := \min_j \bigl[a_{ij} + d_j\bigr] \tag{4.2}$$

is executed at node i in parallel with the corresponding iteration for d_j at every other node j. However, the times of execution of this iteration at the various nodes are not synchronized. Furthermore, each node i communicates the results of its latest computation of d_i at arbitrary times with potentially large communication delays. Therefore, there is the possibility of a node executing iteration (4.2) several times before receiving a communication from every other neighboring node. Assume that each node never stops executing iteration (4.2) and transmitting the result to the other nodes. Show that the values d_i^T available at time T at the corresponding nodes i are equal to the shortest distances for all T greater than a finite time $\overline{T}$. *Hint*: Let $\overline{d}_i^k$ and $\underline{d}_i^k$ be generated by iteration (4.2) when starting from the first and the second initial conditions in part (a), respectively. Show that for every k there exists a time T_k such that for all $T \geq T_k$ and k, we have $\underline{d}_i^k \leq d_i^T \leq \overline{d}_i^k$. For a detailed analysis of asynchronous iterative algorithms, including algorithms for shortest paths and DP, see [BeT89].

3

Deterministic Continuous-Time Optimal Control

Contents

In this chapter, we provide an introduction to continuous-time deterministic optimal control. We derive the analog of the DP algorithm, which is the Hamilton-Jacobi-Bellman equation. Furthermore, we discuss the Pontryagin Minimum Principle and its variations. We give two different derivations of this result, one of which is based on DP. We also illustrate the result by means of examples.

3.1 CONTINUOUS-TIME OPTIMAL CONTROL

We consider a continuous-time dynamic system

$$\dot{x}(t) = f\big(x(t), u(t)\big), \qquad 0 \leq t \leq T, \tag{1.1}$$
$$x(0) : \text{given},$$

where $x(t) \in \Re^n$ is the state vector at time t, $\dot{x}(t) \in \Re^n$ is the vector of first order time derivatives of the states at time t, $u(t) \in U \subset \Re^m$ is the control vector at time t, U is the control constraint set, and T is the terminal time. The components of f, x, $\dot{x}$, and u will be denoted by f_i, x_i, $\dot{x}_i$, and u_i, respectively. Thus, the system (1.1) represents the n first order differential equations

$$\frac{dx_i(t)}{dt} = f_i\big(x(t), u(t)\big), \qquad i = 1, \ldots, n.$$

We view $\dot{x}(t)$, $x(t)$, and $u(t)$ as column vectors. We assume that the system function f_i is continuously differentiable with respect to x and is continuous with respect to u. The admissible control functions, also called *control trajectories*, are the piecewise continuous functions $\big\{u(t) \mid t \in [0,T]\big\}$ with $u(t) \in U$ for all $t \in [0,T]$.

We should stress at the outset that the subject of this chapter is highly sophisticated, and it is beyond our scope to develop it according to high standards of mathematical rigor. In particular, we assume that, for any admissible control trajectory $\big\{u(t) \mid t \in [0,T]\big\}$, the system of differential equations (1.1) has a unique solution, which is denoted $\big\{x^u(t) \mid t \in [0,T]\big\}$ and is called the corresponding *state trajectory*. In a more rigorous treatment, the issue of existence and uniqueness of this solution would have to be addressed more carefully.

We want to find an admissible control trajectory $\big\{u(t) \mid t \in [0,T]\big\}$, which, together with its corresponding state trajectory $\big\{x^u(t) \mid t \in [0,T]\big\}$, minimizes a cost function of the form

$$h\big(x^u(T)\big) + \int_0^T g\big(x^u(t), u(t)\big)dt,$$

where the functions g and h are continuously differentiable with respect to x and g is continuous with respect to u.

Example 1.1 (Motion Control)

A unit mass moves on a line under the influence of a force u. Let $x_1(t)$ and $x_2(t)$ be the position and velocity of the mass at time t, respectively. From a given $\big(x_1(0), x_2(0)\big)$ we want to bring the mass "near" a given final position-velocity pair $(\overline{x}_1, \overline{x}_2)$ at time T. In particular, we want to

$$\text{minimize } \big|x_1(T) - \overline{x}_1\big|^2 + \big|x_2(T) - \overline{x}_2\big|^2$$

subject to the control constraint

$$|u(t)| \leq 1, \qquad \text{for all } t \in [0, T].$$

The corresponding continuous-time system is

$$\dot{x}_1(t) = x_2(t), \qquad \dot{x}_2(t) = u(t),$$

and the problem fits the general framework given earlier with cost functions given by

$$h\big(x(T)\big) = \big|x_1(T) - \overline{x}_1\big|^2 + \big|x_2(T) - \overline{x}_2\big|^2,$$

$$g\big(x(t), u(t)\big) = 0, \qquad \text{for all } t \in [0, T].$$

There are many variations of the problem; for example, the final position and/or velocity may be fixed. These variations can be handled by various reformulations of the general continuous-time optimal control problem, which will be given later.

Example 1.2 (Resource Allocation)

A producer with production rate $x(t)$ at time t may allocate a portion $u(t)$ of his/her production rate to reinvestment and $1 - u(t)$ to production of a storable good. Thus $x(t)$ evolves according to

$$\dot{x}(t) = \gamma u(t) x(t),$$

where $\gamma > 0$ is a given constant. The producer wants to maximize the total amount of product stored

$$\int_0^T \big(1 - u(t)\big) x(t) dt$$

subject to

$$0 \leq u(t) \leq 1, \qquad \text{for all } t \in [0, T].$$

The initial production rate $x(0)$ is a given positive number.

Example 1.3 (Calculus of Variations Problems)

Calculus of variations problems involve finding (possibly multidimensional) curves $x(t)$ with certain optimality properties. They are among the most celebrated problems of applied mathematics and have been worked on by many of the illustrious mathematicians of the past 400 years (Euler, Lagrange, Bernoulli, Gauss, etc.). We will see that calculus of variations problems can be reformulated as optimal control problems. We illustrate this reformulation by a simple example.

Suppose that we want to find a curve from a given point to a given line that has minimum length. The answer is of course evident, but we want to derive it by using a continuous-time optimal control formulation. Without loss of generality, we let $(0, \alpha)$ be the given point, and we let the given line be the vertical line that passes through $(T, 0)$, as shown in Fig. 3.1.1. Let also $\big(t, x(t)\big)$ be the points of the curve $(0 \leq t \leq T)$. The portion of the curve joining the points $\big(t, x(t)\big)$ and $\big(t + dt, x(t + dt)\big)$ can be approximated, for small dt, by the hypotenuse of a right triangle with sides dt and $\dot{x}(t)dt$. Thus the length of this portion is

$$\sqrt{(dt)^2 + \big(\dot{x}(t)\big)^2 (dt)^2},$$

which is equal to

$$\sqrt{1 + \big(\dot{x}(t)\big)^2}\, dt.$$

The length of the entire curve is the integral over $[0, T]$ of this expression, so the problem is to

$$\begin{aligned} &\text{minimize} \quad \int_0^T \sqrt{1 + \big(\dot{x}(t)\big)^2}\, dt \\ &\text{subject to} \quad x(0) = \alpha. \end{aligned}$$

To reformulate the problem as a continuous-time optimal control problem, we introduce a control u and the system equation

$$\dot{x}(t) = u(t), \qquad x(0) = \alpha.$$

Our problem then becomes

$$\text{minimize} \quad \int_0^T \sqrt{1 + \big(u(t)\big)^2}\, dt.$$

This is a problem that fits our continuous-time optimal control framework.

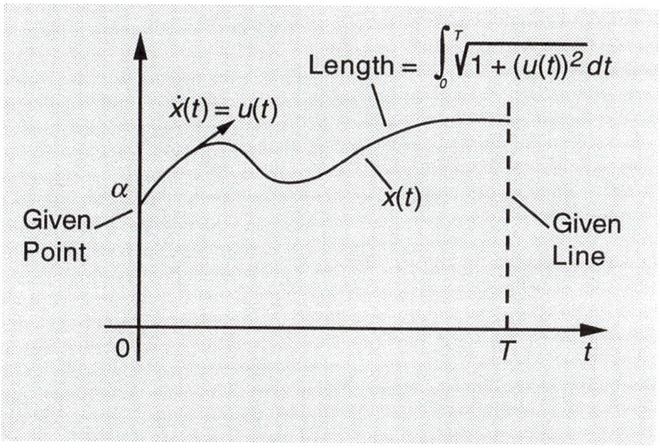

Figure 3.1.1 Problem of finding a curve of minimum length from a given point to a given line and its formulation as a calculus of variations problem.

3.2 THE HAMILTON-JACOBI-BELLMAN EQUATION

We will now derive informally a partial differential equation, which is satisfied by the optimal cost-to-go function. This equation is the continuous-time analog of the DP algorithm, and will be motivated by applying DP to a discrete-time approximation of the continuous-time optimal control problem.

Let us divide the time horizon $[0, T]$ into N pieces using the discretization interval

$$\delta = \frac{T}{N}.$$

We denote

$$x_k = x(k\delta), \qquad k = 0, 1, \ldots, N,$$
$$u_k = u(k\delta), \qquad k = 0, 1, \ldots, N.$$

We approximate the continuous-time system by

$$x_{k+1} = x_k + f(x_k, u_k) \cdot \delta$$

and the cost function by

$$h(x_N) + \sum_{k=0}^{N-1} g(x_k, u_k) \cdot \delta.$$

We now apply DP to the discrete-time approximation. Let

$J^*(t, x)$: Optimal cost-to-go at time t and state x for the continuous-time problem,

$$\tilde{J}^*(t,x): \text{Optimal cost-to-go at time } t \text{ and state } x \text{ for the discrete-time approximation.}$$

The DP equations are

$$\tilde{J}^*(N\delta, x) = h(x),$$

$$\tilde{J}^*(k\delta, x) = \min_{u\in U}\big[g(x,u)\cdot\delta + \tilde{J}^*\big((k+1)\cdot\delta, x + f(x,u)\cdot\delta\big)\big], \quad k = 0,\ldots,N-1.$$

Assuming that $\tilde{J}^*$ has the required differentiability properties, we expand it into a first order Taylor series, obtaining

$$\tilde{J}^*\big((k+1)\cdot\delta, x + f(x,u)\cdot\delta\big) = \tilde{J}^*(k\delta,x) + \nabla_t\tilde{J}^*(k\delta,x)\cdot\delta + \nabla_x\tilde{J}^*(k\delta,x)'f(x,u)\cdot\delta + o(\delta),$$

where $o(\delta)$ represents second order terms satisfying $\lim_{\delta\to 0} o(\delta)/\delta = 0$, ∇_t denotes partial derivative with respect to t, and ∇_x denotes the n-dimensional (column) vector of partial derivatives with respect to x. Substituting in the DP equation, we obtain

$$\tilde{J}^*(k\delta,x) = \min_{u\in U}\big[g(x,u)\cdot\delta + \tilde{J}^*(k\delta,x) + \nabla_t\tilde{J}^*(k\delta,x)\cdot\delta + \nabla_x\tilde{J}^*(k\delta,x)'f(x,u)\cdot\delta + o(\delta)\big].$$

Canceling $\tilde{J}^*(k\delta,x)$ from both sides, dividing by δ, and taking the limit as $\delta\to 0$, while assuming that the discrete-time cost-to-go function yields in the limit its continuous-time counterpart,

$$\lim_{k\to\infty,\ \delta\to 0,\ k\delta=t} \tilde{J}^*(k\delta,x) = J^*(t,x), \qquad \text{for all } t,x,$$

we obtain the following equation for the cost-to-go function $J^*(t,x)$:

$$0 = \min_{u\in U}\big[g(x,u) + \nabla_t J^*(t,x) + \nabla_x J^*(t,x)'f(x,u)\big], \qquad \text{for all } t,x,$$

with the boundary condition $J^*(T,x) = h(x)$.

This is the *Hamilton-Jacobi-Bellman (HJB) equation*. It is a *partial differential equation*, which is satisfied for all time-state pairs (t,x) by the cost-to-go function $J^*(t,x)$. Of course we do not know a priori that $J^*(t,x)$ is differentiable, which was assumed in the preceding informal derivation. However, if we can solve the HJB equation analytically or computationally, then we can obtain an optimal control policy by minimizing its right-hand-side, as shown in the following proposition. This situation is similar to DP; if we can execute the DP algorithm, which may not be possible due to excessive computational requirements, we can find an optimal policy by minimization of the right-hand side.

Proposition 2.1: (Sufficiency Theorem) Suppose $V(t,x)$ is a solution to the HJB equation; that is, V is continuously differentiable in x and t, and is such that

$$0 = \min_{u \in U} \big[g(x,u) + \nabla_t V(t,x) + \nabla_x V(t,x)' f(x,u) \big], \qquad \text{for all } t, x, \tag{2.1}$$

$$V(T,x) = h(x), \qquad \text{for all } x. \tag{2.2}$$

Suppose also that $\mu^*(t,x)$ attains the minimum in Eq. (2.1) for all t and x. Let $\big\{x^*(t) \mid t \in [0,T]\big\}$ be the state trajectory obtained from the given initial condition $x(0)$ when the control trajectory $u^*(t) = \mu^*\big(t, x^*(t)\big)$, $t \in [0,T]$ is used [that is, $x^*(0) = x(0)$ and for all $t \in [0,T]$, $\dot{x}^*(t) = f\big(x^*(t), \mu^*(t, x^*(t))\big)$; we assume that this differential equation has a unique solution starting at any pair (t,x) and that the control trajectory $\big\{\mu^*(t, x^*(t)) \mid t \in [0,T]\big\}$ is piecewise continuous as a function of t]. Then V is the unique solution of the HJB equation and is equal to the optimal cost-to-go function, i.e.,

$$V(t,x) = J^*(t,x), \qquad \text{for all } t, x.$$

Furthermore, the control trajectory $\big\{u^*(t) \mid t \in [0,T]\big\}$ is optimal.

Proof: Let $\{\hat{u}(t) \mid t \in [0,T]\}$ be any admissible control trajectory and let $\{\hat{x}(t) \mid t \in [0,T]\}$ be the corresponding state trajectory. From Eq. (2.1) we have for all $t \in [0,T]$

$$0 \le g\big(\hat{x}(t), \hat{u}(t)\big) + \nabla_t V\big(t, \hat{x}(t)\big) + \nabla_x V\big(t, \hat{x}(t)\big)' f\big(\hat{x}(t), \hat{u}(t)\big).$$

Using the system equation $\dot{\hat{x}}(t) = f\big(\hat{x}(t), \hat{u}(t)\big)$, the right-hand side of the above inequality is equal to the expression

$$g\big(\hat{x}(t), \hat{u}(t)\big) + \frac{d}{dt}\big(V(t, \hat{x}(t))\big),$$

where $d/dt(\cdot)$ denotes total derivative with respect to t. Integrating this expression over $t \in [0,T]$, and using the preceding inequality, we obtain

$$0 \le \int_0^T g\big(\hat{x}(t), \hat{u}(t)\big) dt + V\big(T, \hat{x}(T)\big) - V\big(0, \hat{x}(0)\big).$$

Thus by using the terminal condition $V(T,x) = h(x)$ of Eq. (2.2) and the initial condition $\hat{x}(0) = x(0)$, we have

$$V\big(0, x(0)\big) \le h\big(\hat{x}(T)\big) + \int_0^T g\big(\hat{x}(t), \hat{u}(t)\big) dt.$$

If we use $u^*(t)$ and $x^*(t)$ in place of $\hat{u}(t)$ and $\hat{x}(t)$, respectively, the preceding inequalities becomes equalities, and we obtain

$$V\big(0, x(0)\big) = h\big(x^*(T)\big) + \int_0^T g\big(x^*(t), u^*(t)\big)dt.$$

Therefore the cost corresponding to $\{u^*(t) \mid t \in [0,T]\}$ is $V\big(0, x(0)\big)$ and is no larger than the cost corresponding to any other admissible control trajectory $\{\hat{u}(t) \mid t \in [0,T]\}$. It follows that $\{u^*(t) \mid t \in [0,T]\}$ is optimal and that

$$V\big(0, x(0)\big) = J^*\big(0, x(0)\big).$$

We now note that the preceding argument can be repeated with any initial time $t \in [0,T]$ and any initial state x. We thus obtain

$$V(t,x) = J^*(t,x), \qquad \text{for all } t, x.$$

Q.E.D.

Example 2.1

To illustrate the HJB equation, let us consider a simple example involving the scalar system

$$\dot{x}(t) = u(t),$$

with the constraint $|u(t)| \leq 1$ for all $t \in [0,T]$. The cost is

$$\frac{1}{2}\big(x(T)\big)^2.$$

The HJB equation here is

$$0 = \min_{|u|\leq 1}\big[\nabla_t V(t,x) + \nabla_x V(t,x)u\big], \qquad \text{for all } t, x, \tag{2.3}$$

with the terminal condition

$$V(T,x) = \frac{1}{2}x^2. \tag{2.4}$$

There is an evident candidate for optimality, namely moving the state towards 0 as quickly as possible, and keeping it at 0 once it is at 0. The corresponding control policy is

$$\mu^*(t,x) = -\text{sgn}(x) = \begin{cases} 1 & \text{if } x < 0 \\ 0 & \text{if } x = 0 \\ -1 & \text{if } x > 0. \end{cases} \tag{2.5}$$

For a given initial time t and initial state x, the cost associated with this policy can be calculated to be

$$J^*(t,x) = \frac{1}{2}\Big(\max\big\{0, |x| - (T-t)\big\}\Big)^2. \tag{2.6}$$

This function, which is illustrated in Fig. 3.2.1, satisfies the terminal condition (2.4), since $J^*(T,x) = (1/2)x^2$. Let us verify that this function also satisfies the HJB Eq. (2.3), and that $u = -\text{sgn}(x)$ attains the minimum in the right-hand side of the equation for all t and x. Proposition 2.1 will then guarantee that the state and control trajectories corresponding to the policy $\mu^*(t,x)$ are optimal.

Indeed, we have

$$\nabla_t J^*(t,x) = \max\{0, |x| - (T-t)\},$$

$$\nabla_x J^*(t,x) = \text{sgn}(x) \cdot \max\{0, |x| - (T-t)\}.$$

Substituting these expressions, the HJB Eq. (2.3) becomes

$$0 = \min_{|u|\leq 1} \left[1 + \text{sgn}(x) \cdot u\right] \max\{0, |x| - (T-t)\}, \tag{2.7}$$

which can be seen to hold as an identity for all (t,x). Furthermore, the minimum is attained for $u = -\text{sgn}(x)$. We therefore conclude based on Prop. 2.1 that $J^*(t,x)$ as given by Eq. (2.6) is indeed the optimal cost-to-go function, and that the policy defined by Eq. (2.5) is optimal. Note, however, that the optimal policy is not unique. Based on Prop. 2.1, any policy for which the minimum is attained in Eq. (2.7) is optimal. In particular, when $|x(t)| \leq T-t$, applying any control from the range $[-1,1]$ is optimal.

The preceding derivation generalizes to the case of the cost

$$h\big(x(T)\big),$$

where h is a nonnegative differentiable convex function with $h(0) = 0$. The corresponding optimal cost-to-go function is

$$J^*(t,x) = \begin{cases} h\big(x - (T-t)\big) & \text{if } x > T-t \\ h\big(x + (T-t)\big) & \text{if } x < -(T-t) \\ 0 & \text{if } |x| \leq T-t \end{cases}$$

and can be similarly verified to be a solution of the HJB equation.

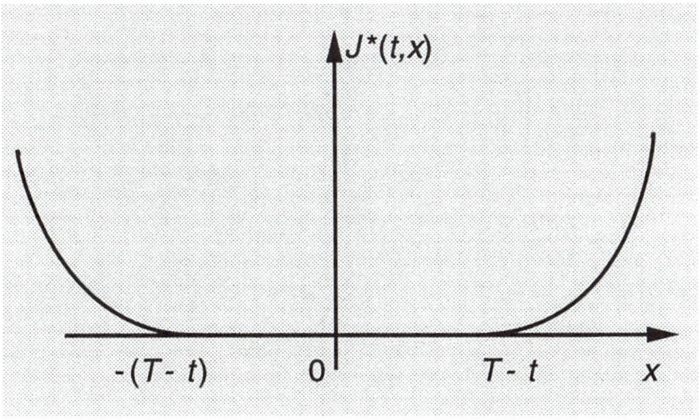

Figure 3.2.1 Optimal cost-to-go function $J^*(t,x)$ for Example 2.1.

Example 2.2 (Linear-Quadratic Problems)

Consider the n-dimensional linear system

$$\dot{x}(t) = Ax(t) + Bu(t),$$

where A and B are given matrices, and the quadratic cost

$$x(T)'Q_T x(T) + \int_0^T \big(x(t)'Qx(t) + u(t)'Ru(t)\big)\,dt,$$

where the matrices Q_T and Q are symmetric positive semidefinite, and the matrix R is symmetric positive definite (Appendix A defines positive definite and semidefinite matrices). The HJB equation is

$$0 = \min_{u\in\Re^m} \big[x'Qx + u'Ru + \nabla_t V(t,x) + \nabla_x V(t,x)'(Ax + Bu)\big], \tag{2.8}$$

$$V(T,x) = x'Q_T x.$$

Let us try a solution of the form

$$V(t,x) = x'K(t)x, \qquad K(t) : n \times n \text{ symmetric},$$

and see if we can solve the HJB equation. We have $\nabla_x V(t,x) = 2K(t)x$ and $\nabla_t V(t,x) = x'\dot{K}(t)x$, where $\dot{K}(t)$ is the matrix with elements the first order derivatives of the elements of $K(t)$ with respect to time. By substituting these expressions in Eq. (2.8), we obtain

$$0 = \min_u \big[x'Qx + u'Ru + x'\dot{K}(t)x + 2x'K(t)Ax + 2x'K(t)Bu\big]. \tag{2.9}$$

The minimum is attained at a u for which the gradient with respect to u is zero, that is,

$$2B'K(t)x + 2Ru = 0$$

or

$$u = -R^{-1}B'K(t)x. \tag{2.10}$$

Substituting the minimizing value of u in Eq. (2.9), we obtain

$$0 = x'\big(\dot{K}(t) + K(t)A + A'K(t) - K(t)BR^{-1}B'K(t) + Q\big)x, \quad \text{for all } (t,x).$$

Therefore, in order for $V(t,x) = x'K(t)x$ to solve the HJB equation, $K(t)$ must satisfy the following matrix differential equation (known as the *continuous-time Riccati equation*)

$$\dot{K}(t) = -K(t)A - A'K(t) + K(t)BR^{-1}B'K(t) - Q \tag{2.11}$$

with the terminal condition

$$K(T) = Q_T. \tag{2.12}$$

Reversing the argument, we see that if $K(t)$ is a solution of the Riccati equation (2.11) with the boundary condition (2.12), then $V(t,x) = x'K(t)x$ is a solution of the HJB equation. Thus, by using Prop. 2.1, we conclude that the optimal cost-to-go function is

$$J^*(t,x) = x'K(t)x.$$

Furthermore, in view of the expression derived for the control that minimizes in the right-hand side of the HJB equation [cf. Eq. (2.10)], an optimal policy is

$$\mu^*(t,x) = -R^{-1}B'K(t)x.$$

3.3 THE PONTRYAGIN MINIMUM PRINCIPLE

In this section we discuss the continuous-time and the discrete-time versions of the Minimum Principle, starting with a DP-based informal argument.

3.3.1 An Informal Derivation Using the HJB Equation

Recall the HJB equation

$$0 = \min_{u \in U} \big[g(x,u) + \nabla_t J^*(t,x) + \nabla_x J^*(t,x)' f(x,u) \big], \quad \text{for all } t, x, \tag{3.1}$$

$$J^*(T,x) = h(x), \qquad \text{for all } x. \tag{3.2}$$

We argued that the optimal cost-to-go function $J^*(t,x)$ satisfies this equation under some conditions. Furthermore, the sufficiency theorem of the preceding section suggests that if for a given initial state $x(0)$, the control trajectory $\{u^*(t) \mid t \in [0,T]\}$ is optimal with corresponding state trajectory $\{x^*(t) \mid t \in [0,T]\}$, then for all $t \in [0,T]$,

$$u^*(t) = \arg\min_{u \in U} \Big[g\big(x^*(t),u\big) + \nabla_x J^*\big(t,x^*(t)\big)' f\big(x^*(t),u\big) \Big]. \tag{3.3}$$

Note that to obtain the optimal control trajectory via this equation, we do not need to know $\nabla_x J^*$ at *all* values of x and t; it is sufficient to know $\nabla_x J^*$ at only *one* value of x for each t, that is, to know only $\nabla_x J^*\big(t,x^*(t)\big)$.

The Minimum Principle is basically Eq. (3.3) above. Its application is facilitated by streamlining the computation of the derivative $\nabla_x J^*\big(t,x^*(t)\big)$. It turns out that we can often calculate $\nabla_x J^*\big(t,x^*(t)\big)$ along the optimal state trajectory far more easily than we can solve the HJB equation. In particular, $\nabla_x J^*\big(t,x^*(t)\big)$ satisfies a certain differential equation, called the *adjoint equation*. We will derive this equation informally by differentiating the HJB equation (3.1). We first need the following lemma, which indicates how to differentiate functions involving minima.

Lemma 3.1: Let $F(t,x,u)$ be a continuously differentiable function of $t \in \Re$, $x \in \Re^n$, and $u \in \Re^m$, and let U be a convex subset of $\Re^m$. Assume that $\mu^*(t,x)$ is a continuously differentiable function such that

$$\mu^*(t,x) = \arg\min_{u \in U} F(t,x,u), \qquad \text{for all } t, x.$$

Then

$$\nabla_t \left\{ \min_{u \in U} F(t,x,u) \right\} = \nabla_t F\big(t,x,\mu^*(t,x)\big), \qquad \text{for all } t,x,$$

$$\nabla_x \left\{ \min_{u \in U} F(t,x,u) \right\} = \nabla_x F\big(t,x,\mu^*(t,x)\big), \qquad \text{for all } t,x.$$

[Note: On the left-hand side, $\nabla_t\{\cdot\}$ and $\nabla_x\{\cdot\}$ denote the gradients of the function $G(t,x) = \min_{u\in U} F(t,x,u)$ with respect to t and x, respectively. On the right-hand side, ∇_t and ∇_x denote the vectors of partial derivatives of F with respect to t and x, respectively, evaluated at $\big(x, \mu^*(t,x)\big)$.]

Proof: For notational simplicity, denote $y = (t,x)$, $F(y,u) = F(t,x,u)$, and $\mu^*(y) = \mu^*(t,x)$. Since $\min_{u\in U} F(y,u) = F\big(y,\mu^*(y)\big)$,

$$\nabla \left\{ \min_{u \in U} F(y,u) \right\} = \nabla_y F\big(y,\mu^*(y)\big) + \nabla \mu^*(y) \nabla_u F\big(y,\mu^*(y)\big).$$

We will prove the result by showing that the second term in the right-hand side above is zero. This is true when $U = \Re^m$, because then $\mu^*(y)$ is an unconstrained minimum of $F(y,u)$ and $\nabla_u F\big(y,\mu^*(y)\big) = 0$. More generally, for every fixed y, we have

$$\big(u - \mu^*(y)\big)' \nabla_u F\big(y,\mu^*(y)\big) \geq 0, \qquad \text{for all } u \in U,$$

[see Eq. (B.2) in Appendix B]. Now by Taylor's Theorem, we have that when y changes to $y + \Delta y$, the minimizing $\mu^*(y)$ changes from $\mu^*(y)$ to some vector $\mu^*(y+\Delta y) = \mu^*(y) + \nabla\mu^*(y)'\Delta y + o\big(\|\Delta y\|\big)$ of U, so

$$\big(\nabla\mu^*(y)'\Delta y + o(\|\Delta y\|)\big)' \nabla_u F\big(y,\mu^*(y)\big) \geq 0, \qquad \text{for all } \Delta y,$$

implying that

$$\nabla \mu^*(y) \nabla_u F\big(y,\mu^*(y)\big) = 0.$$

Q.E.D.

Consider the HJB equation (3.1), and for any (t,x), suppose that $\mu^*(t,x)$ is a control attaining the minimum in the right-hand side. We make the restrictive assumptions that U is a convex set, and that $\mu^*(t,x)$ is continuously differentiable in (t,x), so that we can use Lemma 3.1. (We note, however, that alternative derivations of the Minimum Principle do not require these assumptions; see Section 3.3.2.)

We differentiate the HJB equation with respect to x and with respect to t, and we rely on Lemma 3.1 to disregard the terms involving the derivatives of $\mu^*(t,x)$ with respect to t and x. We obtain for all (t,x),

$$0 = \nabla_x g\big(x, \mu^*(t,x)\big) + \nabla^2_{xt} J^*(t,x) + \nabla^2_{xx} J^*(t,x) f\big(x, \mu^*(t,x)\big) + \nabla_x f\big(x, \mu^*(t,x)\big) \nabla_x J^*(t,x), \tag{3.4}$$

$$0 = \nabla^2_{tt} J^*(t,x) + \nabla^2_{xt} J^*(t,x)' f\big(x, \mu^*(t,x)\big), \tag{3.5}$$

where $\nabla_x f\big(x, \mu^*(t,x)\big)$ is the matrix

$$\nabla_x f = \begin{pmatrix} \frac{\partial f_1}{\partial x_1} & \cdots & \frac{\partial f_n}{\partial x_1} \\ \vdots & \vdots & \vdots \\ \frac{\partial f_1}{\partial x_n} & \cdots & \frac{\partial f_n}{\partial x_n} \end{pmatrix}$$

with the partial derivatives evaluated at the argument $\big(x, \mu^*(t,x)\big)$.

The above equations hold for all (t,x). Let us specialize them along an optimal state and control trajectory $\big\{\big(x^*(t), u^*(t)\big) \mid t \in [0,T]\big\}$, where $u^*(t) = \mu^*\big(t, x^*(t)\big)$ for all $t \in [0,T]$. We have for all t,

$$\dot{x}^*(t) = f\big(x^*(t), u^*(t)\big),$$

so that the term

$$\nabla^2_{xt} J^*\big(t, x^*(t)\big) + \nabla^2_{xx} J^*\big(t, x^*(t)\big) f\big(x^*(t), u^*(t)\big)$$

in Eq. (3.4) is equal to the following total derivative with respect to t

$$\frac{d}{dt}\Big(\nabla_x J^*\big(t, x^*(t)\big)\Big).$$

Similarly, the term

$$\nabla^2_{tt} J^*\big(t, x^*(t)\big) + \nabla^2_{xt} J^*\big(t, x^*(t)\big)' f\big(x^*(t), u^*(t)\big)$$

in Eq. (3.5) is equal to the total derivative

$$\frac{d}{dt}\Big(\nabla_t J^*\big(t, x^*(t)\big)\Big).$$

Thus, by denoting

$$p(t) = \nabla_x J^*\big(t, x^*(t)\big), \tag{3.6}$$

$$p_0(t) = \nabla_t J^*\big(t, x^*(t)\big), \tag{3.7}$$

Eq. (3.4) becomes

$$\dot{p}(t) = -\nabla_x f\big(x^*(t), u^*(t)\big) p(t) - \nabla_x g\big(x^*(t), u^*(t)\big) \tag{3.8}$$

and Eq. (3.5) becomes

$$\dot{p}_0(t) = 0$$

or equivalently,

$$p_0(t) = \text{constant}, \qquad \text{for all } t \in [0, T]. \tag{3.9}$$

Equation (3.8) is a system of n first order differential equations known as the *adjoint equation*. From the boundary condition

$$J^*(T, x) = h(x), \qquad \text{for all } x, \tag{3.10}$$

we have, by differentiation with respect to x, the relation $\nabla_x J^*(T, x) = \nabla h(x)$, and by using the definition $\nabla_x J^*\big(t, x^*(t)\big) = p(t)$, we obtain

$$p(T) = \nabla h\big(x^*(T)\big). \tag{3.11}$$

Thus, we have a terminal boundary condition for the adjoint equation (3.8).

To summarize, along optimal state and control trajectories $x^*(t)$, $u^*(t)$, $t \in [0, T]$, the adjoint equation (3.8) holds together with the boundary condition (3.11), while Eq. (3.3) and the definition of $p(t)$ imply that $u^*(t)$ satisfies

$$u^*(t) = \arg\min_{u \in U} \big[g\big(x^*(t), u\big) + p(t)' f\big(x^*(t), u\big)\big], \qquad \text{for all } t \in [0, T]. \tag{3.12}$$

Hamiltonian Formulation

Motivated by the condition (3.12), we introduce the Hamiltonian function mapping triplets $(x, u, p) \in \Re^n \times \Re^m \times \Re^n$ to real numbers and given by

$$H(x, u, p) = g(x, u) + p' f(x, u).$$

Note that the adjoint equation can be compactly written in terms of the Hamiltonian as

$$\dot{p}(t) = -\nabla_x H\big(x^*(t), u^*(t), p(t)\big).$$

We state the Minimum Principle in terms of the Hamiltonian function.

Proposition 3.1: (Minimum Principle) Let $\{u^*(t) \mid t \in [0, T]\}$ be an optimal control trajectory and let $\{x^*(t) \mid t \in [0, T]\}$ be the corresponding state trajectory, i.e.,

$$\dot{x}^*(t) = f\big(x^*(t), u^*(t)\big) \tag{3.13}$$

$$x^*(0) = x(0) : \text{given}. \tag{3.14}$$

Let also $p(t)$ be the solution of the adjoint equation

$$\dot{p}(t) = -\nabla_x H\big(x^*(t), u^*(t), p(t)\big), \tag{3.16}$$

with the boundary condition

$$p(T) = \nabla h(x^*(T)), \tag{3.17}$$

where $h(\cdot)$ is the terminal cost function. Then, for all $t \in [0, T]$,

$$u^*(t) = \arg\min_{u \in U} H\big(x^*(t), u, p(t)\big). \tag{3.15}$$

Furthermore, there is a constant C such that

$$H\big(x^*(t), u^*(t), p(t)\big) = C, \qquad \text{for all } t \in [0, T].$$

All the assertions of the Minimum Principle have been (informally) derived earlier except for the last assertion. To see why the Hamiltonian function is constant for $t \in [0, T]$ along the optimal state and control trajectories, note that by Eqs. (3.1), (3.6), and (3.7), we have for all $t \in [0, T]$

$$H\big(x^*(t), u^*(t), p(t)\big) = -\nabla_t J^*\big(t, x^*(t)\big) = -p_0(t),$$

and $p_0(t)$ is constant by Eq. (3.9). We should note here that the Hamiltonian function need not be constant along the optimal trajectory if the system and cost are not time-independent, contrary to our assumption thus far (see Section 3.4.4).

It is important to note that the Minimum Principle provides *necessary* optimality conditions, so all optimal control trajectories satisfy these conditions, but if a control trajectory satisfies these conditions, it is not necessarily optimal. Further analysis is needed to guarantee optimality. One method that often works is to prove that an optimal control trajectory exists, and to verify that there is only one control trajectory satisfying the conditions of the Minimum Principle (or that all control trajectories satisfying these conditions have equal cost). Another possibility to conclude optimality arises when the system function f is linear in (x, u), the constraint set U is convex, and the cost functions h and g are convex. Then it can be shown that the conditions of the Minimum Principle are both necessary and sufficient for optimality.

The Minimum Principle can often be used as the basis of a numerical solution. One possibility is the *two-point boundary problem method*. In this method, we use the minimum condition

$$u^*(t) = \arg\min_{u \in U} H\big(x^*(t), u, p(t)\big),$$

to express $u^*(t)$ in terms of $x^*(t)$ and $p(t)$. We then substitute the result into the system and the adjoint equations, to obtain a set of $2n$ first order differential equations in the components of $x^*(t)$ and $p(t)$. These equations can be solved using the split boundary conditions

$$x^*(0) = x(0), \qquad p(T) = \nabla h\big(x^*(T)\big).$$

The number of boundary conditions (which is $2n$) is equal to the number of differential equations, so that we generally expect to be able to solve these differential equations numerically.

Using the Minimum Principle to obtain an analytical solution is possible in many interesting problems, but typically requires considerable creativity. We give some simple examples.

Example 3.1 (Calculus of Variations Continued)

Consider the problem of finding the curve of minimum length from a point $(0, \alpha)$ to the line $\{(T, y) \mid y \in \Re\}$. In Section 3.1 , we formulated this problem as the problem of finding an optimal control trajectory $\{u(t) \mid t \in [0, T]\}$ that minimizes

$$\int_0^T \sqrt{1 + \big(u(t)\big)^2}\, dt$$

subject to

$$\dot{x}(t) = u(t), \qquad x(0) = \alpha.$$

Let us apply the preceding necessary conditions. The Hamiltonian is

$$H(x, u, p) = \sqrt{1 + u^2} + pu,$$

and the adjoint equation is

$$\dot{p}(t) = 0, \qquad p(T) = 0.$$

It follows that

$$p(t) = 0, \qquad \text{for all } t \in [0, T],$$

so minimization of the Hamiltonian gives

$$u^*(t) = \arg\min_{u \in \Re} \sqrt{1 + u^2} = 0, \qquad \text{for all } t \in [0, T].$$

Therefore we have $\dot{x}^*(t) = 0$ for all t, which implies that $x^*(t)$ is constant. Using the initial condition $x^*(0) = \alpha$, it follows that

$$x^*(t) = \alpha, \qquad \text{for all } t \in [0, T].$$

We thus obtain the (a priori obvious) optimal solution, which is the horizontal line passing through $(0, \alpha)$.

Example 3.2 (Resource Allocation Continued)

Consider the optimal production problem (Example 1.2 in Section 3.1). We want to maximize

$$\int_0^T \big(1 - u(t)\big)x(t)dt$$

subject to

$$0 \le u(t) \le 1, \qquad \text{for all } t \in [0, T],$$

$$\dot{x}(t) = \gamma u(t)x(t), \qquad x(0) > 0 : \text{given}.$$

The Hamiltonian is

$$H(x, u, p) = (1 - u)x + p\gamma ux.$$

The adjoint equation is

$$\dot{p}(t) = -\gamma u^*(t)p(t) - 1 + u^*(t),$$
$$p(T) = 0.$$

Maximization of the Hamiltonian over $u \in [0, 1]$ yields

$$u^*(t) = \begin{cases} 0 & \text{if } p(t) < \frac{1}{\gamma}, \\ 1 & \text{if } p(t) \ge \frac{1}{\gamma}. \end{cases}$$

Since $p(T) = 0$, for t close to T we will have $p(t) < 1/\gamma$ and $u^*(t) = 0$. Therefore, for t near T the adjoint equation has the form $\dot{p}(t) = -1$ and $p(t)$ has the form shown in Fig. 3.3.1.

Thus, near $t = T$, $p(t)$ decreases with slope -1. For $t = T - 1/\gamma$, $p(t)$ is equal to $1/\gamma$, so $u^*(t)$ changes to $u^*(t) = 1$. It follows that for $t < T - 1/\gamma$, the adjoint equation is

$$\dot{p}(t) = -\gamma p(t)$$

or

$$p(t) = e^{-\gamma t} + \text{constant}.$$

Piecing together $p(t)$ for t greater and less than $T - 1/\gamma$, we obtain the form shown in Fig. 3.3.2 for $p(t)$ and $u^*(t)$. Note that if $T < 1/\gamma$, the optimal control is $u^*(t) = 0$ for all $t \in [0, T]$; that is, for a short enough horizon, it does not pay to reinvest at any time.

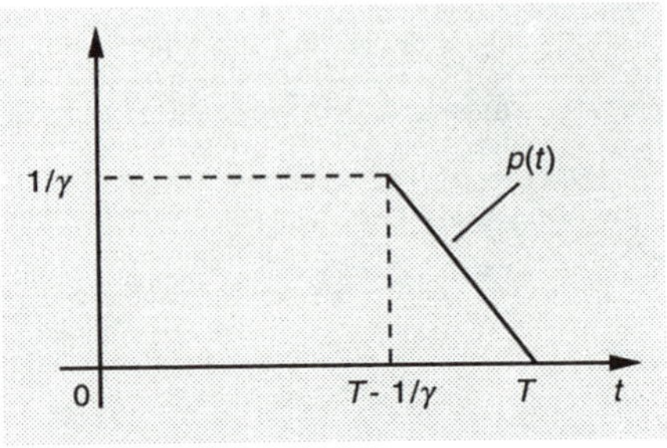

Figure 3.3.1 Form of the adjoint variable $p(t)$ for t near T in the resource allocation example.

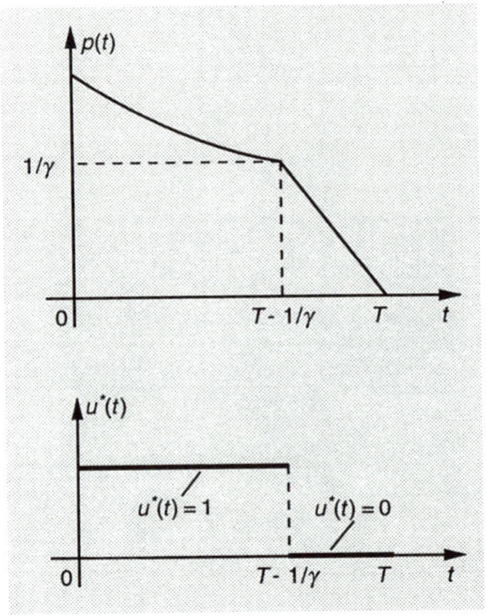

Figure 3.3.2 Form of the adjoint variable $p(t)$ and the optimal control in the resource allocation example.

Example 3.3 (A Linear-Quadratic Problem)

Consider the one-dimensional linear system

$$\dot{x}(t) = ax(t) + bu(t),$$

where a and b are given scalars. We want to find an optimal control over a given interval $[0, T]$ that minimizes the quadratic cost

$$\frac{1}{2}q \cdot \big(x(T)\big)^2 + \frac{1}{2}\int_0^T \big(u(t)\big)^2 dt,$$

where q is a given positive scalar. There are no constraints on the control, so we have a special case of the linear-quadratic problem of Example 2.2. We will solve this problem via the Minimum Principle.

The Hamiltonian here is

$$H(x, u, p) = \frac{1}{2}u^2 + p(ax + bu),$$

and the adjoint equation is

$$\dot{p}(t) = -ap(t),$$

with the terminal condition

$$p(T) = qx^*(T).$$

The optimal control is obtained by minimizing the Hamiltonian with respect to u, yielding

$$u^*(t) = \arg\min_u \left[\frac{1}{2}u^2 + p(t)\big(ax^*(t) + bu\big)\right] = -bp(t). \tag{3.18}$$

We will extract the optimal solution from these conditions using two different approaches.

In the first approach, we solve the two-point boundary value problem discussed following Prop. 3.1. In particular, by eliminating the control from the system equation using Eq. (3.18), we obtain

$$\dot{x}^*(t) = ax^*(t) - b^2 p(t).$$

Also, from the adjoint equation, we see that

$$p(t) = e^{-at}\xi, \qquad \text{for all } t \in [0, T],$$

where $\xi = p(0)$ is an unknown parameter. The last two equations yield

$$\dot{x}^*(t) = ax^*(t) - b^2 e^{-at}\xi. \tag{3.19}$$

This differential equation, together with the given initial condition $x^*(0) = x(0)$ and the terminal condition

$$x^*(T) = \frac{e^{-aT}\xi}{q},$$

(which is the terminal condition for the adjoint equation) can be solved for the unknown variable ξ. In particular, it can be verified that the solution of the differential equation (3.19) is given by

$$x^*(t) = x(0)e^{at} + \frac{b^2\xi}{a}\big(e^{-at} - 1\big),$$

and ξ can be obtained from the last two relations. Given ξ, we obtain $p(t) = e^{-at}\xi$, and from $p(t)$, we can then determine the optimal control trajectory as $u^*(t) = -bp(t)$, $t \in [0, T]$ [cf. Eq. (3.18)].

In the second approach, we basically derive the Riccati equation encountered in Example 2.2. In particular, we hypothesize a linear relation between $x^*(t)$ and $p(t)$, that is,

$$K(t)x^*(t) = p(t), \qquad \text{for all } t \in [0, T],$$

and we show that $K(t)$ can be obtained by solving the Riccati equation. Indeed, from Eq. (3.18) we have

$$u^*(t) = -bK(t)x^*(t),$$

which by substitution in the system equation, yields

$$\dot{x}^*(t) = \big(a - b^2K(t)\big)x^*(t).$$

By differentiating the equation $K(t)x^*(t) = p(t)$ and by also using the adjoint equation, we obtain

$$\dot{K}(t)x^*(t) + K(t)\dot{x}^*(t) = \dot{p}(t) = -ap(t) = -aK(t)x^*(t).$$

By combining the last two relations, we have

$$\dot{K}(t)x^*(t) + K(t)\big(a - b^2K(t)\big)x^*(t) = -aK(t)x^*(t),$$

from which we see that $K(t)$ should satisfy

$$\dot{K}(t) = -2aK(t) + b^2\big(K(t)\big)^2.$$

This is the Riccati equation of Example 2.2, specialized to the problem of the present example. This equation can be solved using the terminal condition

$$K(T) = q,$$

which is implied by the terminal condition $p(T) = qx^*(T)$ for the adjoint equation. Once $K(t)$ is known, the optimal control is obtained in the closed-loop form $u^*(t) = -bK(t)x^*(t)$. By reversing the preceding arguments, this control can then be shown to satisfy all the conditions of the Minimum Principle.

3.3.2 A Derivation Based on Variational Ideas

In this subsection we outline an alternative and more rigorous proof of the Minimum Principle. This proof is primarily directed towards the advanced reader, and is based on a comparison of an optimal trajectory with neighboring trajectories, which are obtained by making small variations in the optimal trajectory.

For convenience, we restrict attention to the case where the cost is

$$h\big(x(T)\big).$$

The more general cost

$$h\big(x(T)\big) + \int_0^T g\big(x(t), u(t)\big)\,dt \tag{3.20}$$

can be reformulated as a terminal cost by introducing a new state variable y and the additional differential equation

$$\dot{y}(t) = g\big(x(t), u(t)\big). \tag{3.21}$$

The cost then becomes

$$h\big(x(T)\big) + y(T), \tag{3.22}$$

and the Minimum Principle corresponding to this terminal cost yields the Minimum Principle for the general cost (3.20).

We introduce some assumptions:

Convexity Assumption: For every state x the set

$$D = \{f(x,u) \mid u \in U\}$$

is convex.

The convexity assumption is satisfied if U is a convex set and f is linear in u [and g is linear in u in the case where there is an integral cost of the form (3.20), which is reformulated as a terminal cost by using the additional state variable y of Eq. (3.21)]. Thus the convexity assumption is quite restrictive. However, the Minimum Principle typically holds without the convexity assumption, because even when the set $D = \{f(x,u) \mid u \in U\}$ is nonconvex, any vector in the convex hull of D can be generated by quick alternation between vectors from D (for an example, see Exercise 3.10). This involves the complicated mathematical concept of *randomized* or *relaxed* controls and will not be discussed further.

Regularity Assumption: Let $u(t)$ and $u^*(t)$, $t \in [0,T]$, be any two admissible control trajectories and let $\{x^*(t) \mid t \in [0,T]\}$ be the state trajectory corresponding to $u^*(t)$. For any $\epsilon \in [0,1]$, the solution $\{x_\epsilon(t) \mid t \in [0,T]\}$ of the system

$$\dot{x}_\epsilon(t) = (1-\epsilon)f\big(x_\epsilon(t), u^*(t)\big) + \epsilon f\big(x_\epsilon(t), u(t)\big), \tag{3.23}$$

with $x_\epsilon(0) = x^*(0)$, satisfies

$$x_\epsilon(t) = x^*(t) + \epsilon\xi(t) + o(\epsilon), \tag{3.24}$$

where $\{\xi(t) \mid t \in [0,T]\}$ is the solution of the linear differential system

$$\dot{\xi}(t) = \nabla_x f\big(x^*(t), u^*(t)\big)\xi(t) + f\big(x^*(t), u(t)\big) - f\big(x^*(t), u^*(t)\big), \quad (3.25)$$

with initial condition $\xi(0) = 0$.

The regularity assumption "typically" holds because from Eq. (3.23) we have

$$\begin{aligned}\dot{x}_\epsilon(t) - \dot{x}^*(t) = {} & f\big(x_\epsilon(t), u^*(t)\big) - f\big(x^*(t), u^*(t)\big)\\ & + \epsilon\Big(f\big(x_\epsilon(t), u(t)\big) - f\big(x_\epsilon(t), u^*(t)\big)\Big),\end{aligned}$$

so from a first order Taylor series expansion we obtain

$$\begin{aligned}\delta\dot{x}(t) = {} & \nabla f\big(x^*(t), u^*(t)\big)'\delta x(t) + o\big(\|\delta x(t)\|\big)\\ & + \epsilon\Big(f\big(x_\epsilon(t), u(t)\big) - f\big(x_\epsilon(t), u^*(t)\big)\Big),\end{aligned}$$

where

$$\delta x(t) = x_\epsilon(t) - x^*(t).$$

Dividing by ϵ and taking the limit as $\epsilon \to 0$, we see that the function

$$\xi(t) = \lim_{\epsilon \to 0} \delta x(t)/\epsilon, \qquad t \in [0, T],$$

should "typically" solve the linear system of differential equations (3.25), while satisfying Eq. (3.24).

Suppose now that $\{u^*(t) \mid t \in [0, T]\}$ is an optimal control trajectory, and let $\{x^*(t) \mid t \in [0, T]\}$ be the corresponding state trajectory. Then for any other admissible control trajectory $\{u(t) \mid t \in [0, T]\}$ and any $\epsilon \in [0, 1]$, the convexity assumption guarantees that for each t, there exists a control $\overline{u}(t) \in U$ such that

$$f\big(x_\epsilon(t), \overline{u}(t)\big) = (1 - \epsilon) f\big(x_\epsilon(t), u^*(t)\big) + \epsilon f\big(x_\epsilon(t), u(t)\big).$$

Thus, the state trajectory $\{x_\epsilon(t) \mid t \in [0, T]\}$ of Eq. (3.23) corresponds to the admissible control trajectory $\{\overline{u}(t) \mid t \in [0, T]\}$. Hence, using the optimality of $\{x^*(t) \mid t \in [0, T]\}$ and the regularity assumption, we have

$$\begin{aligned}h\big(x^*(T)\big) & \le h\big(x_\epsilon^*(T)\big)\\ & = h\big(x^*(T) + \epsilon\xi(T) + o(\epsilon)\big)\\ & = h\big(x^*(T)\big) + \epsilon\nabla h\big(x^*(T)\big)'\xi(T) + o(\epsilon),\end{aligned}$$

which implies that

$$\nabla h\big(x^*(T)\big)'\xi(T) \ge 0. \quad (3.26)$$

Using a standard result in the theory of linear differential equations (see e.g. [CoL65]), the solution of the linear differential system (3.25) can be written in closed form as

$$\xi(t) = \Phi(t,\tau)\xi(\tau) + \int_\tau^t \Phi(t,\tau)\big(f(x^*(\tau),u(\tau)) - f(x^*(\tau),u^*(\tau))\big)d\tau, \quad (3.27)$$

where the square matrix Φ satisfies for all t and τ,

$$\frac{\partial \Phi(t,\tau)}{\partial \tau} = -\Phi(t,\tau)\nabla_x f\big(x^*(\tau),u^*(\tau)\big)', \quad (3.28)$$

$$\Phi(t,t) = I.$$

Since $\xi(0) = 0$, we have from Eq. (3.27),

$$\xi(T) = \int_0^T \Phi(T,t)\big(f(x^*(t),u(t)) - f(x^*(t),u^*(t))\big)dt. \quad (3.29)$$

Define

$$p(T) = \nabla h\big(x^*(T)\big), \qquad p(t) = \Phi(T,t)'p(T), \quad t \in [0,T]. \quad (3.30)$$

By differentiating with respect to t, we obtain

$$\dot{p}(t) = \frac{\partial \Phi(T,t)'}{\partial t} p(T).$$

Combining this equation with Eqs. (3.28) and (3.30), we see that $p(t)$ is generated by the differential equation

$$\dot{p}(t) = -\nabla_x f\big(x^*(t),u^*(t)\big)p(t), \quad (3.31)$$

with the terminal condition

$$p(T) = \nabla h\big(x^*(T)\big). \quad (3.32)$$

This is the adjoint equation corresponding to $\big\{\big(x^*(t),u^*(t)\big) \mid t \in [0,T]\big\}$.

Now, to obtain the Minimum Principle, we note that from Eqs. (3.26), (3.29), and (3.30) we have

$$\begin{aligned} 0 &\le p(T)'\xi(T) \\ &= p(T)' \int_0^T \Phi(T,t)\big(f(x^*(t),u(t)) - f(x^*(t),u^*(t))\big)dt \\ &= \int_0^T p(t)'\big(f(x^*(t),u(t)) - f(x^*(t),u^*(t))\big)dt, \end{aligned} \quad (3.33)$$

from which it can be shown that for all t at which $u^*(\cdot)$ is continuous, we have

$$p(t)'f\big(x^*(t),u^*(t)\big) \le p(t)'f\big(x^*(t),u\big), \qquad \text{for all } u \in U. \tag{3.34}$$

Indeed, if for some $\hat{u} \in U$ and $t_0 \in [0,T]$, we have

$$p(t_0)'f\big(x^*(t_0),u^*(t_0)\big) > p(t_0)'f\big(x^*(t_0),\hat{u}\big),$$

while $\big\{u^*(t) \mid t \in [0,T]\big\}$ is continuous at t_0, we would also have

$$p(t)'f\big(x^*(t),u^*(t)\big) > p(t)'f\big(x^*(t),\hat{u}\big),$$

for all t in some nontrivial interval I containing t_0. By taking

$$u(t) = \begin{cases} \hat{u} & \text{for } t \in I \\ u^*(t) & \text{for } t \notin I, \end{cases}$$

we would then obtain a contradiction of Eq. (3.33). We have thus proved the Minimum Principle (3.34) under the convexity and regularity assumptions.

To prove the Minimum Principle for the more general integral cost function (3.20), one can apply the preceding development to the system of differential equations $\dot{x} = f(x,u)$ augmented by the additional Eq. (3.21) and the equivalent terminal cost (3.22).

3.3.3 Minimum Principle for Discrete-Time Problems

In this subsection we briefly derive a version of the Minimum Principle for discrete-time deterministic optimal control problems. Interestingly, it is essential to make some convexity assumptions in order for the Minimum Principle to hold. For continuous-time problems these convexity assumptions are typically not needed, because, as mentioned earlier, the differential system can generate any $\dot{x}(t)$ in the convex hull of the set of possible vectors $f\big(x(t),u(t)\big)$ by randomization (see for example Exercise 3.10).

Suppose that we want to find a control sequence $(u_0, u_1, \ldots, u_{N-1})$ and a corresponding state sequence $(x_0, x_1, \ldots, x_N)$, which minimize

$$J(u) = g_N(x_N) + \sum_{k=0}^{N-1} g_k(x_k, u_k),$$

subject to the discrete-time system constraints

$$x_{k+1} = f_k(x_k, u_k), \quad k = 0, \ldots, N-1, \qquad x_0 : \text{ given},$$

and the control constraints

$$u_k \in U_k \subset \Re^m, \qquad k = 0, \ldots, N-1.$$

We first develop an expression for the gradient $\nabla J(u_0, \ldots, u_{N-1})$. We have, using the chain rule,

$$\begin{aligned}\nabla_{u_k} J(u_0, \ldots, u_{N-1}) = {} & \nabla_{u_k} f_k \cdot \nabla_{x_{k+1}} f_{k+1} \cdots \nabla_{x_{N-1}} f_{N-1} \cdot \nabla g_N \\ & + \nabla_{u_k} f_k \cdot \nabla_{x_{k+1}} f_{k+1} \cdots \nabla_{x_{N-2}} f_{N-2} \cdot \nabla_{x_{N-1}} g_{N-1} \\ & \cdots \\ & + \nabla_{u_k} f_k \cdot \nabla_{x_{k+1}} g_{k+1} \\ & + \nabla_{u_k} g_k,\end{aligned}$$

where all gradients are evaluated along the control trajectory $(u_0, \ldots, u_{N-1})$ and the corresponding state trajectory. To streamline the computation, we introduce the *discrete-time adjoint equation*

$$p_k = \nabla_{x_k} f_k \cdot p_{k+1} + \nabla_{x_k} g_k, \qquad k = 1, \ldots, N-1,$$

with terminal condition

$$p_N = \nabla g_N.$$

Then it is straightforward to verify that

$$\nabla_{u_k} J(u_0, \ldots, u_{N-1}) = \nabla_{u_k} H_k(x_k, u_k, p_{k+1}), \tag{3.35}$$

where H_k is the Hamiltonian function defined by

$$H_k(x_k, u_k, p_{k+1}) = g_k(x_k, u_k) + p'_{k+1} f_k(x_k, u_k).$$

We will assume that the constraint sets U_k are convex, so that we can apply the optimality condition

$$\sum_{k=0}^{N-1} \nabla_{u_k} J(u_0^*, \ldots, u_{N-1}^*)'(u_k - u_k^*) \geq 0,$$

for all feasible $(u_0, \ldots, u_{N-1})$ (see Appendix B). This condition can be decomposed into the N conditions

$$\nabla_{u_k} J(u_0^*, \ldots, u_{N-1}^*)'(u_k - u_k^*) \geq 0, \qquad \text{for all } u_k \in U_k,\ k = 0, \ldots, N-1. \tag{3.36}$$

We thus obtain:

Proposition 3.2. (Discrete-Time Minimum Principle) Suppose that $(u_0^*, u_1^* \dots, u_{N-1}^*)$ is an optimal control trajectory and that $(x_0^*, x_1^*, \dots, x_N^*)$ is the corresponding state trajectory. Assume also that the constraint sets U_k are convex. Then for all $k = 0, \dots, N-1$, we have

$$\nabla_{u_k} H_k(x_k^*, u_k^*, p_{k+1})'(u_k - u_k^*) \geq 0, \qquad \text{for all } u_k \in U_k, \tag{3.37}$$

where the vectors $p_1, \dots, p_N$ are obtained from the adjoint equation

$$p_k = \nabla_{x_k} f_k \cdot p_{k+1} + \nabla_{x_k} g_k, \qquad k = 1, \dots, N-1, \tag{3.38}$$

with the terminal condition

$$p_N = \nabla g_N(x_N^*). \tag{3.39}$$

The partial derivatives above are evaluated along the optimal state and control trajectories. If, in addition, the Hamiltonian H_k is a convex function of u_k for any fixed x_k and p_{k+1}, we have

$$u_k^* = \arg \min_{u_k \in U_k} H_k(x_k^*, u_k, p_{k+1}), \qquad \text{for all } k = 0, \dots, N-1. \tag{3.40}$$

Proof: Equation (3.37) is a restatement of the necessary condition (3.36) using the expression (3.35) for the gradient of J. If H_k is convex with respect to u_k, Eq. (3.36) is a sufficient condition for the minimum condition (3.39) to hold (see Appendix B). **Q.E.D.**

3.4 EXTENSIONS OF THE MINIMUM PRINCIPLE

We now consider some variations of the continuous-time optimal control problem and derive corresponding variations of the Minimum Principle.

3.4.1 Fixed Terminal State

Suppose that in addition to the initial state $x(0)$, the final state $x(T)$ is given. Then the preceding informal derivations still hold except that the terminal condition $J^*(T, x) = h(x)$ is not true anymore. In effect, here we have

$$J^*(T,x) = \begin{cases} 0 & \text{if } x = x(T) \\ \infty & \text{otherwise.} \end{cases}$$

Thus $J^*(T,x)$ cannot be differentiated with respect to x, and the terminal boundary condition $p(T) = \nabla h\big(x^*(T)\big)$ for the adjoint equation does not hold. However, as compensation, we have the extra condition

$$x(T) : \text{given},$$

thus maintaining the balance between boundary conditions and unknowns.

If only *some* of the terminal states are fixed, that is,

$$x_i(T) : \text{given}, \qquad \text{for all } i \in I,$$

where I is some index set, we have the partial boundary condition

$$p_j(T) = \frac{\partial h\big(x^*(T)\big)}{\partial x_j}, \qquad \text{for all } j \notin I,$$

for the adjoint equation.

Example 4.1

Consider the problem of finding the curve of minimum length connecting two points $(0,\alpha)$ and (T,β). This is a fixed endpoint variation of Example 3.1 in the preceding section. We have

$$\dot{x}(t) = u(t),$$

$$x(0) = \alpha, \qquad x(T) = \beta,$$

and the cost is

$$\int_0^T \sqrt{1 + \big(u(t)\big)^2}\, dt.$$

The adjoint equation is

$$\dot{p}(t) = 0,$$

implying that

$$p(t) = \text{constant}, \qquad \text{for all } t \in [0,T].$$

Minimization of the Hamiltonian,

$$\min_{u \in \Re} \left[\sqrt{1+u^2} + p(t)u\right],$$

yields

$$u^*(t) = \text{constant}, \qquad \text{for all } \ t \in [0,T].$$

Thus the optimal trajectory $\big\{x^*(t) \mid t \in [0,T]\big\}$ is a straight line. Since this trajectory must pass through $(0,\alpha)$ and (T,β), we obtain the (a priori obvious) optimal solution shown in Fig. 3.4.1.

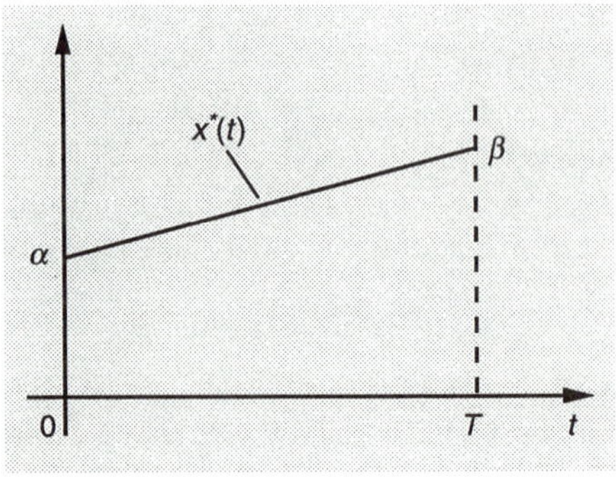

Figure 3.4.1 Optimal solution of the problem of connecting the two points $(0, \alpha)$ and (T, β) with a minimum length curve (cf. Example 4.1).

Example 4.2 (The Brachistochrone Problem)

In 1696 Johann Bernoulli challenged the mathematical world of his time with a problem that played an instrumental role in the development of the calculus of variations: Given two points A and B, find a curve connecting A and B such that a body moving along the curve under the force of gravity reaches B in minimum time (see Fig. 3.4.2). Let A be $(0, 0)$ and B be $(T, -b)$ with $b > 0$. Then it can be seen that the problem is to find $\{x(t) \mid t \in [0, T]\}$ with $x(0) = 0$ and $x(T) = b$, which minimizes

$$\int_0^T \frac{\sqrt{1 + \big(\dot{x}(t)\big)^2}}{\sqrt{2\gamma x(t)}}\, dt,$$

where γ is the acceleration due to gravity. Here $\{\big(t, -x(t)\big) \mid t \in [0, T]\}$, is the desired curve, the term $\sqrt{1 + \big(\dot{x}(t)\big)^2}\,dt$ is the length of the curve from $x(t)$ to $x(t + dt)$, and the term $\sqrt{2\gamma x(t)}$ is the velocity of the body upon reaching the level $x(t)$ [if m and v denote the mass and the velocity of the body, the kinetic energy is $mv^2/2$, which at level $x(t)$ must be equal to the change in potential energy, which is $m\gamma x(t)$; this yields $v = \sqrt{2\gamma x(t)}$].

We introduce the system $\dot{x} = u$, and we obtain a fixed terminal state problem [$x(0) = 0$ and $x(T) = b$]. Letting

$$g(x, u) = \frac{\sqrt{1 + u^2}}{\sqrt{2\gamma x}},$$

the Hamiltonian is

$$H(x, u, p) = g(x, u) + pu.$$

We minimize the Hamiltonian by setting to zero its derivative with respect to u:

$$p(t) = -\nabla_u g\big(x^*(t), u^*(t)\big).$$

We know from the Minimum Principle that the Hamiltonian is constant along an optimal trajectory, i.e.,

$$g\big(x^*(t),u^*(t)\big) - \nabla_u g\big(x^*(t),u^*(t)\big)u^*(t) = \text{constant}, \qquad \text{for all } t \in [0,T].$$

Using the expression for g, this can be written as

$$\frac{\sqrt{1+\big(u^*(t)\big)^2}}{\sqrt{2\gamma x^*(t)}} - \frac{\big(u^*(t)\big)^2}{\sqrt{1+\big(u^*(t)\big)^2}\sqrt{2\gamma x^*(t)}} = \text{constant}, \qquad \text{for all } t \in [0,T],$$

or equivalently

$$\frac{1}{\sqrt{1+\big(u^*(t)\big)^2}\sqrt{2\gamma x^*(t)}} = \text{constant}, \qquad \text{for all } t \in [0,T].$$

Using the relation $\dot{x}^*(t) = u^*(t)$, this yields

$$x^*(t)\big(1+\dot{x}^*(t)^2\big) = C, \qquad \text{for all } t \in [0,T],$$

for some constant C. Thus an optimal trajectory satisfies the differential equation

$$\dot{x}^*(t) = \sqrt{\frac{C - x^*(t)}{x^*(t)}}, \qquad \text{for all } t \in [0,T].$$

The solution of this differential equation was known at Bernoulli's time to be a *cycloid*; see Fig. 3.4.2. The unknown parameters of the cycloid are determined by the boundary conditions $x^*(0) = 0$ and $x^*(T) = b$.

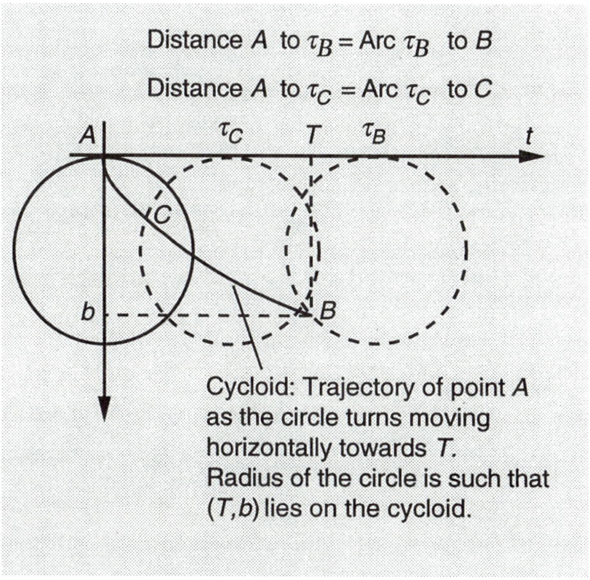

Figure 3.4.2 Formulation and optimal solution of the brachistochrone problem.

3.4.2 Free Initial State

If the initial state $x(0)$ is not fixed but is subject to optimization, we have

$$J^*\big(0, x^*(0)\big) \leq J^*(0, x), \qquad \text{for all } x \in \Re^n,$$

yielding

$$\nabla_x J^*\big(0, x^*(0)\big) = 0$$

and the extra boundary condition for the adjoint equation

$$p(0) = 0.$$

Also if there is a cost $\ell\big(x(0)\big)$ on the initial state, i.e., the cost is

$$\ell\big(x(0)\big) + \int_0^T g\big(x(t), u(t)\big)dt + h\big(x(T)\big),$$

the boundary condition becomes

$$p(0) = -\nabla \ell\big(x^*(0)\big).$$

This follows by setting to zero the gradient with respect to x of $\ell(x) + J(0, x)$, i.e.,

$$\nabla_x \big\{\ell(x) + J(0, x)\big\}\big|_{x=x^*(0)} = 0.$$

3.4.3 Free Terminal Time

Suppose the initial state and/or the terminal state are given, but the terminal time T is subject to optimization.

Let $\big\{\big(x^*(t), u^*(t)\big) \mid t \in [0, T]\big\}$ be an optimal state-control trajectory pair and let T^* be the optimal terminal time. Then if the terminal time were fixed at T^*, $\big\{\big(u^*(t), x^*(t)\big) \mid t \in [0, T]\big\}$ would satisfy the conditions of the Minimum Principle. In particular,

$$u^*(t) = \arg\min_{u \in U} H\big(x^*(t), u, p(t)\big), \qquad \text{for all } \ t \in [0, T^*],$$

where $p(t)$ is the solution of the adjoint equation. What we lose with the terminal time being free, we gain with an extra condition derived as follows.

We argue that if the terminal time were fixed at T^* and the initial state were fixed at the given $x(0)$, but instead the initial time were subject to optimization, it would be optimal to start at $t = 0$. This means that the first order variation of the optimal cost with respect to the initial time must be zero; i.e.,

$$\nabla_t J^*\big(t, x^*(t)\big)\big|_{t=0} = 0.$$

The HJB equation can be written along the optimal trajectory as

$$\nabla_t J^*(t, x^*(t)) = -H\big(x^*(t), u^*(t), p(t)\big), \qquad \text{for all } t \in [0, T^*]$$

[cf. Eqs. (3.1) and (3.6)], so the last two equations yield

$$H\big(x^*(0), u^*(0), p(0)\big) = 0.$$

Since the Hamiltonian was shown earlier to be constant along the optimal trajectory, we obtain for the case of a free terminal time

$$H\big(x^*(t), u^*(t), p(t)\big) = 0, \qquad \text{for all } t \in [0, T^*].$$

Example 4.3 (Minimum-Time Problem)

A unit mass object moves horizontally under the influence of a force $u(t)$, that is,

$$\ddot{y}(t) = u(t),$$

where $y(t)$ is the position of the object at time t. Given the object's initial position $y(0)$ and initial velocity $\dot{y}(0)$, it is required to bring the object to rest (zero velocity) at a given position, say zero, while using at most unit magnitude force,

$$-1 \le u(t) \le 1, \qquad \text{for all } t.$$

We want to accomplish this transfer in minimum time. Thus, we want to

$$\text{minimize } T = \int_0^T 1 dt.$$

Note that the integral cost, $g\big(x(t), u(t)\big) \equiv 1$, is unusual here; it does not depend on the state or the control. However, the theory does not preclude this possibility, and the problem is still meaningful because the terminal time T is free and subject to optimization.

Let the state variables be

$$x_1(t) = y(t), \qquad x_2(t) = \dot{y}(t),$$

so the system equation is

$$\dot{x}_1(t) = x_2(t), \qquad \dot{x}_2(t) = u(t).$$

The initial state $\big(x_1(0), x_2(0)\big)$ is given and the terminal state is also given

$$x_1(T) = 0, \qquad x_2(T) = 0.$$

If $\big\{u^*(t) \mid t \in [0, T]\big\}$ is an optimal control trajectory, $u^*(t)$ must minimize the Hamiltonian for each t, i.e.,

$$u^*(t) = \arg \min_{-1 \le u \le 1} \big[1 + p_1(t) x_2^*(t) + p_2(t) u\big].$$

Therefore

$$u^*(t) = \begin{cases} 1 & \text{if } p_2(t) < 0, \\ -1 & \text{if } p_2(t) \geq 0. \end{cases}$$

The adjoint equation is

$$\dot{p}_1(t) = 0, \qquad \dot{p}_2(t) = -p_1(t),$$

so

$$p_1(t) = c_1, \qquad p_2(t) = c_2 - c_1 t,$$

where c_1 and c_2 are constants. It follows that $\{p_2(t) \mid t \in [0,T]\}$ has one of the four forms shown in Fig. 3.4.3(a); that is, $\{p_2(t) \mid t \in [0,T]\}$ switches at most once in going from negative to positive or reversely. [Note that it is not possible for $p_2(t)$ to be equal to 0 for all t because this implies that $p_1(t)$ is also equal to 0 for all t, so that the Hamiltonian is equal to 1 for all t; the necessary conditions require that the Hamiltonian be 0 along the optimal trajectory.] The corresponding control trajectories are shown in Fig. 3.4.3(b). The conclusion is that, for each t, $u^*(t)$ is either $+1$ or -1, and $\{u^*(t) \mid t \in [0,T]\}$ has at most one switching point in the interval $[0,T]$.

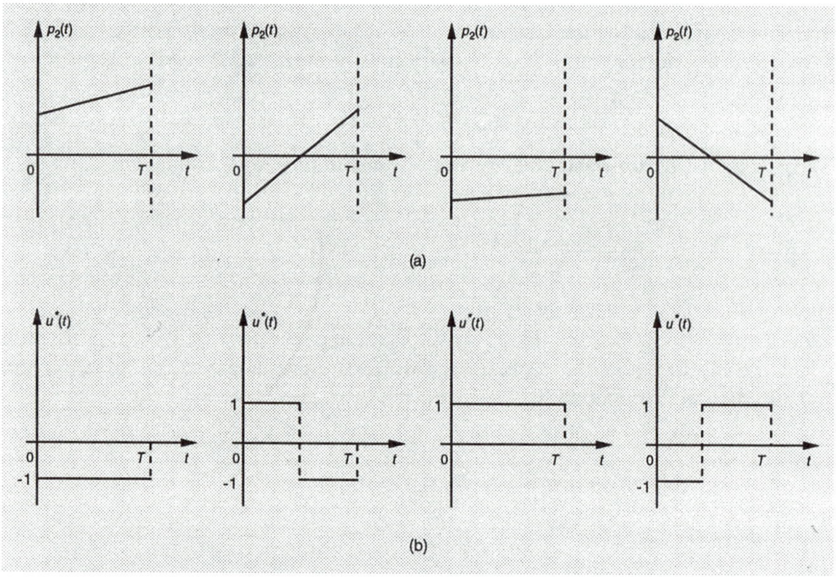

Figure 3.4.3 (a) Possible forms of the adjoint variable $p_2(t)$. (b) Corresponding forms of the optimal control trajectory.

To determine the precise form of the optimal control trajectory, we use the given initial and final states. For $u(t) \equiv \zeta$, where $\zeta = \pm 1$, the system evolves according to

$$x_1(t) = x_1(0) + x_2(0)t + \frac{\zeta}{2}t^2, \qquad x_2(t) = x_2(0) + \zeta t.$$

By eliminating the time t in these two equations, we see that for all t

$$x_1(t) - \frac{1}{2}\zeta\big(x_2(t)\big)^2 = x_1(0) - \frac{1}{2}\zeta\big(x_2(0)\big)^2.$$

Thus for intervals where $u(t) \equiv 1$, the system moves along the curves where $x_1(t) - \frac{1}{2}\big(x_2(t)\big)^2$ is constant, shown in Fig. 3.4.4(a). For intervals where $u(t) \equiv -1$, the system moves along the curves where $x_1(t) + \frac{1}{2}\big(x_2(t)\big)^2$ is constant, shown in Fig. 3.4.4(b).

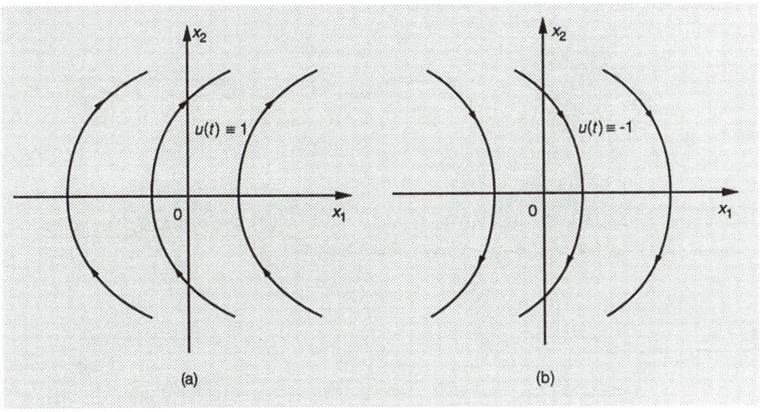

Figure 3.4.4 State trajectories when the control is $u(t) \equiv 1$ [Fig. (a)] and when the control is $u(t) \equiv -1$ [Fig. (b)].

To bring the system from the initial state $\big(x_1(0), x_2(0)\big)$ to the origin with at most one switch in the value of control, we must follow the following rules involving the *switching curve* shown in Fig. 3.4.5.

(a) If the initial state lies *above* the switching curve, use $u^*(t) \equiv -1$ until the state hits the switching curve; then use $u^*(t) \equiv 1$ until reaching the origin.

(b) If the initial state lies *below* the switching curve, use $u^*(t) \equiv 1$ until the state hits the switching curve; then use $u^*(t) \equiv -1$ until reaching the origin.

(c) If the initial state lies on the top (bottom) part of the switching curve, use $u^*(t) \equiv -1$ [$u^*(t) \equiv 1$, respectively] until reaching the origin.

3.4.4 Time-Varying System and Cost

If the system equation and the integral cost depend on the time t, i.e.,

$$\dot{x}(t) = f\big(x(t), u(t), t\big),$$

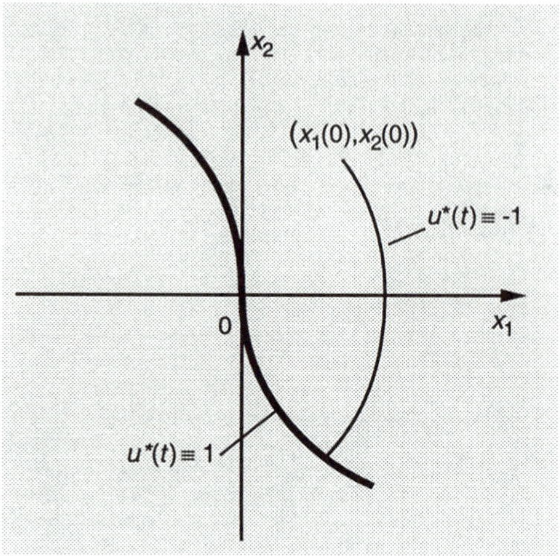

Figure 3.4.5 Switching curve (shown with a thick line) and closed-loop optimal control for the minimum time example.

$$\text{cost} = h\big(x(T)\big) + \int_0^T g\big(x(t), u(t), t\big)dt,$$

we can convert the problem to one involving a time-independent system and cost by introducing an extra state variable $y(t)$ representing time:

$$\dot{y}(t) = 1, \qquad y(0) = 0,$$

$$\dot{x}(t) = f\big(x(t), u(t), y(t)\big), \qquad x(0) : \text{given},$$

$$\text{cost} = h\big(x(T)\big) + \int_0^T g\big(x(t), u(t), y(t)\big)dt.$$

After working out the corresponding optimality conditions, we see that they are the same as when the system and cost are time-independent. The only difference is that the Hamiltonian depends explicitly on time and need not be constant along the optimal trajectory.

3.4.5 Singular Problems

In some cases, the minimum condition

$$u^*(t) = \arg\min_{u \in U} H\big(x^*(t), u, p(t), t\big) \tag{4.1}$$

is insufficient to determine $u^*(t)$ for all t, because the values of $x^*(t)$ and $p(t)$ are such that $H\big(x^*(t), u, p(t), t\big)$ is independent of u over a nontrivial interval of time. Such problems are called *singular*. Their optimal trajectories consist of portions, called *regular arcs*, where $u^*(t)$ can be determined from the minimum condition (4.1), and other portions, called *singular arcs*, which can be determined from the condition that the Hamiltonian is independent of u. The following is an example.

Example 4.4 (Road Construction)

Suppose that we want to construct a road over a one-dimensional terrain whose ground elevation (altitude measured from some reference point) is known and is given by $z(t)$, $t \in [0,T]$. The elevation of the road is denoted by $x(t)$, $t \in [0,T]$, and the difference $x(t) - z(t)$ must be made up by fill-in or excavation. It is desired to minimize

$$\frac{1}{2}\int_0^T \big(x(t) - z(t)\big)^2 dt,$$

subject to the constraint that the gradient of the road $\dot{x}(t)$ lies between $-a$ and a, where a is a specified maximum allowed slope. Thus we have the constraint

$$|u(t)| \leq a, \qquad t \in [0,T],$$

where

$$\dot{x}(t) = u(t), \qquad t \in [0,T].$$

The adjoint equation here is

$$\dot{p}(t) = -x^*(t) + z(t),$$

with the terminal condition

$$p(T) = 0.$$

Minimization of the Hamiltonian

$$H\big(x^*(t), u, p(t), t\big) = \frac{1}{2}\big(x^*(t) - z(t)\big)^2 + p(t)u$$

with respect to u yields

$$u^*(t) = \arg\min_{|u|\leq a} p(t)u,$$

for all t, and shows that optimal trajectories are obtained by concatenation of three types of arcs:

(a) Regular arcs where $p(t) > 0$ and $u^*(t) = -a$ (maximum downhill slope arcs).

(b) Regular arcs where $p(t) < 0$ and $u^*(t) = a$ (maximum uphill slope arcs).

(c) Singular arcs where $p(t) = 0$ and $u^*(t)$ can take any value in $[-a, a]$ that maintains the condition $p(t) = 0$. From the adjoint equation we see that singular arcs are those along which $p(t) = 0$ and $x^*(t) = z(t)$, i.e., the road follows the ground elevation (no fill-in or excavation). Along such arcs we must have $\dot{z}(t) = u^*(t) \in [-a, a]$.

Optimal solutions can be obtained by a graphical method using the above observations. Consider the *sharply uphill intervals* $\overline{I}$ such that $\dot{z}(t) \geq a$ for all $t \in \overline{I}$, and the *sharply downhill intervals* $\underline{I}$ such that $\dot{z}(t) \leq -a$ for all $t \in \underline{I}$. Clearly, within each sharply uphill interval $\overline{I}$ the optimal slope is $u^*(t) = a$, but the optimal slope is also a within a larger maximum uphill slope interval $\overline{V} \supset \overline{I}$, which is such that $p(t) < 0$ within $\overline{V}$ and $p(t_1) = p(t_2) = 0$ at the endpoints t_1 and t_2 of $\overline{V}$. In view of the form of the adjoint equation, we see that the endpoints t_1 and t_2 of $\overline{V}$ should be such that

$$\int_{t_1}^{t_2} \big(z(t) - x^*(t)\big)\,dt = 0;$$

that is, the *total fill-in should be equal to the total excavation within* $\overline{V}$, (see Fig. 3.4.6). Similarly, each sharply downhill interval $\underline{I}$ should be contained within a larger maximum downhill slope interval $\underline{V} \supset \underline{I}$, which is such that $p(t) > 0$ within $\underline{V}$, while the *total fill-in should be equal to the total excavation within* $\underline{V}$, (see Fig. 3.4.6). Thus the regular arcs consist of the intervals $\overline{V}$ and $\underline{V}$ described above. Between the regular arcs there can be one or more singular arcs where $x^*(t) = z(t)$. The optimal solution can be pieced together starting at the endpoint $t = T$ [where we know that $p(T) = 0$], and proceeding backwards.

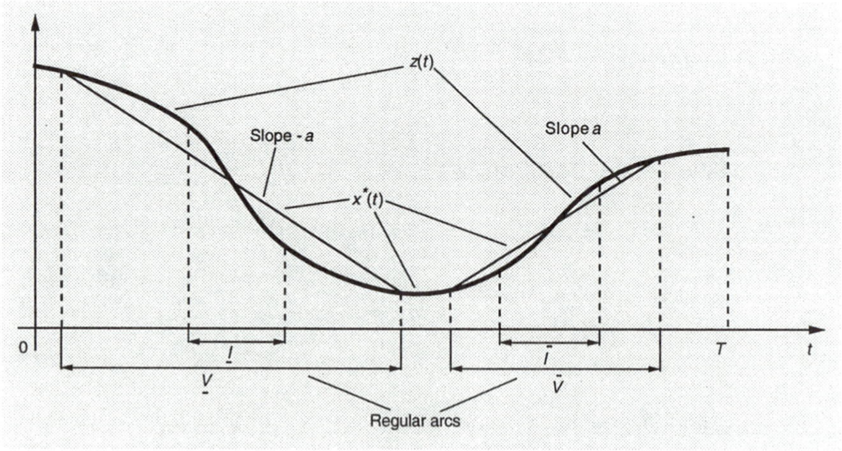

Figure 3.4.6 Graphical method for solving the road construction example. The sharply uphill (downhill) intervals $\overline{I}$ (respectively, $\underline{I}$) are first identified, and are then embedded within maximum uphill (respectively, downhill) slope regular arcs $\overline{V}$ (respectively, $\underline{V}$) within which the total fill-in is equal to the total excavation. The regular arcs are joined by singular arcs where there is no fill-in or excavation. The graphical process is started at the endpoint $t = T$.

3.5 NOTES, SOURCES, AND EXERCISES

The calculus of variations is a classical subject that originated with the works of the great mathematicians of the 17th and 18th centuries. Its rigorous development (by modern mathematical standards) took place in the 1930s and 1940s, with the work of a group of mathematicians that originated mostly from the University of Chicago; Bliss, McShane, and Hestenes are some of the most prominent members of this group. Curiously, this development preceded the development of nonlinear programming by many years.† The modern theory of deterministic optimal control has its roots primarily in the work of Pontryagin and his students in the 1950s. The theoretical and applications literature on the subject is very extensive. We give three representative references: the book by Athans and Falb [AtF66] (a classical extensive text that includes engineering applications), the book by Hestenes [Hes66] (a rigorous mathematical treatment), and the book by Luenberger [Lue69] (which deals with optimal control within a broader infinite dimensional context). The book by Cannon, Cullum, and Polak [CCP70] gives a detailed treatment of discrete-time optimal control that includes proofs of the Minimum Principle under weaker convexity assumptions than the one given here.

EXERCISES

3.1

Solve the problem of Example 2.1 for the case where the cost function is

$$\big(x(T)\big)^2 + \int_0^T \big(u(t)\big)^2 dt.$$

---

† In the 30s and 40s journal space was at a premium, and finite-dimensional optimization research was thought to be a simple special case of the calculus of variations, thus insufficiently challenging or novel for publication. Indeed the modern optimality conditions of finite-dimensional optimization subject to equality and inequality constraints were first developed in the 1939 Master's thesis by Karush but first appeared in a journal quite a few years later under the names of other researchers.

3.2

An egocentric young man has inherited a large sum S and plans to spend it so as to maximize his enjoyment through the rest of his life without working. He estimates that he will live exactly T more years and that his capital $x(t)$ should be reduced to zero at time T, i.e., $x(T) = 0$. Also he models the evolution of his capital by the differential equation

$$\frac{dx(t)}{dt} = \alpha x(t) - u(t),$$

where $x(0) = S$ is his initial capital, $\alpha > 0$ is a given interest rate, and $u(t) \geq 0$ is his rate of expenditure. The total enjoyment he will obtain is given by

$$\int_0^T e^{-\beta t} \sqrt{u(t)}\, dt.$$

Here $\beta \geq 0$ is a given discount factor for future enjoyment. Find the optimal $\{u(t) \mid t \in [0, T]\}$.

3.3

Consider the system of reservoirs shown in Fig. 3.5.1. The system equations are

$$\dot{x}_1(t) = -x_1(t) + u(t),$$
$$\dot{x}_2(t) = x_1(t),$$

and the control constraint is $0 \leq u(t) \leq 1$ for all t. Initially

$$x_1(0) = x_2(0) = 0.$$

We want to maximize $x_2(1)$ subject to the constraint $x_1(1) = 0.5$. Solve the problem.

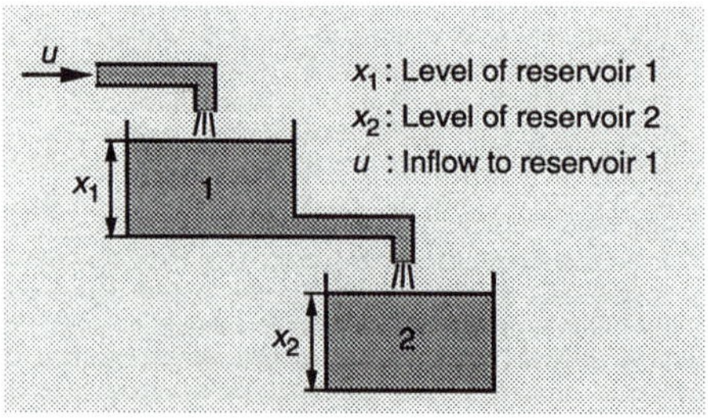

Figure 3.5.1 Reservoir system for Exercise 3.3.

3.4

Work out the minimum-time problem (Example 4.3) for the case where there is friction and the object's position moves according to

$$\ddot{y}(t) = -a\dot{y}(t) + u(t),$$

where $a > 0$ is given. *Hint*: The solution of the system

$$\dot{p}_1(t) = 0,$$
$$\dot{p}_2(t) = -p_1(t) + ap_2(t),$$

is

$$p_1(t) = p_1(0),$$
$$p_2(t) = \frac{1}{a}(1 - e^{at})p_1(0) + e^{at}p_2(0).$$

The trajectories of the system for $u(t) \equiv -1$ and $u(t) \equiv 1$ are sketched in Fig. 3.5.2.

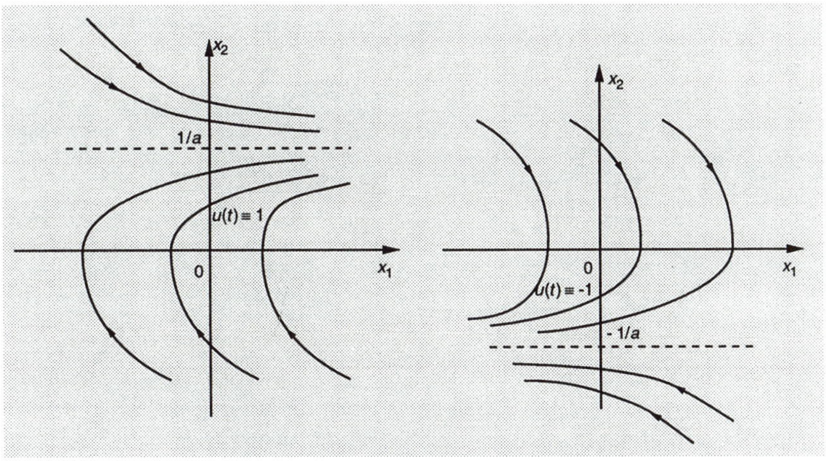

Figure 3.5.2 State trajectories of the system of Exercise 3.4 for $u(t) \equiv -1$ and $u(t) \equiv 1$.

3.5 (Isoperimetric Problem)

Analyze the problem of finding a curve $\{x(t) \mid t \in [0,T]\}$ that maximizes the area under x,

$$\int_0^T x(t)dt,$$

subject to the constraints

$$x(0) = a, \qquad x(T) = b, \qquad \int_0^T \sqrt{1 + \big(\dot{x}(t)\big)^2}\, dt = L,$$

where a, b, and L are given positive scalars. The last constraint is known as an isoperimetric constraint; it requires that the length of the curve be L. *Hint*: Introduce the system $\dot{x}_1 = u$, $\dot{x}_2 = \sqrt{1+u^2}$, and view the problem as a fixed terminal state problem. Show that the optimal $u^*(t)$ depends linearly on t. Under some assumptions on a, b, and L, the optimal curve is a circular arc.

3.6 (L'Hôpital's Problem)

Let a, b, and T be positive scalars, and let $A = (0, a)$ and $B = (T, b)$ be two points in a medium within which the velocity of propagation of light is proportional to the vertical coordinate. Thus the time it takes for light to propagate from A to B along a curve $\big\{x(t) \mid t \in [0, T]\big\}$ is

$$\int_0^T \frac{\sqrt{1 + \big(\dot{x}(t)\big)^2}}{Cx(t)}\, dt,$$

where C is a given positive constant. Find the curve of minimum travel time of light from A to B, and show that it is an arc of a circle of the form

$$x(t)^2 + (t - d)^2 = D,$$

where d and D are some constants.

3.7

A boat moves with constant unit velocity in a stream moving at constant velocity s. The problem is to find the steering angle $u(t)$, $0 \leq t \leq T$, which minimizes the time T required for the boat to move between the point $(0, 0)$ to a given point (a, b). The equations of motion are

$$\dot{x}_1(t) = s + \cos u(t),$$

$$\dot{x}_2(t) = \sin u(t),$$

where $x_1(t)$ and $x_2(t)$ are the positions of the boat parallel and perpendicular to the stream velocity, respectively. Show that the optimal solution is to steer at a constant angle.

3.8

A unit mass object moves on a straight line from a given initial position $x_1(0)$ and velocity $x_2(0)$. Find the force $\{u(t) \mid t \in [0,1]\}$ that brings the object at time 1 to rest $[x_2(1) = 0]$ at position $x_1(1) = 0$, and minimizes

$$\int_0^1 \big(u(t)\big)^2 dt.$$

3.9

Use the Minimum Principle to solve the linear-quadratic problem of Example 2.2. *Hint*: Follow the lines of Example 3.3.

3.10 (On the Need for Convexity Assumptions)

Solve the continuous-time problem involving the system $\dot{x}(t) = u(t)$, the terminal cost $\big(x(T)\big)^2$, and the control constraint $u(t) = -1$ or 1 for all t, and show that the solution satisfies the Minimum Principle. Show that, depending on the initial state x_0, this may not be true for the discrete-time version involving the system $x_{k+1} = x_k + u_k$, the terminal cost x_N^2, and the control constraint $u_k = -1$ or 1 for all k.

3.11

Use the discrete-time Minimum Principle to solve Exercise 1.14 of Chapter 1, assuming that each w_k is fixed at a known deterministic value.

3.12

Use the discrete-time Minimum Principle to solve Exercise 1.15 of Chapter 1, assuming that γ_k and δ_k are fixed at known deterministic values.

4

Problems with Perfect State Information

Contents

In this chapter we consider a number of applications of discrete-time stochastic optimal control with perfect state information. These applications are special cases of the basic problem of Section 1.2 and can be addressed via the DP algorithm. In all these applications the stochastic nature of the disturbances is significant. For this reason, in contrast with the deterministic problems of the preceding two chapters, the use of closed-loop control is essential to achieve optimal performance.

4.1 LINEAR SYSTEMS AND QUADRATIC COST

In this section we consider the special case of a linear system

$$x_{k+1} = A_k x_k + B_k u_k + w_k, \qquad k = 0, 1, \ldots, N-1,$$

and the quadratic cost

$$\mathop{E}_{\substack{w_k \\ k=0,1,\ldots,N-1}} \left\{ x_N' Q_N x_N + \sum_{k=0}^{N-1} (x_k' Q_k x_k + u_k' R_k u_k) \right\}.$$

In these expressions, x_k and u_k are vectors of dimension n and m, respectively, and the matrices A_k, B_k, Q_k, R_k are given and have appropriate dimension. We assume that the matrices Q_k are positive semidefinite symmetric, and the matrices R_k are positive definite symmetric. The controls u_k are unconstrained. The disturbances w_k are independent random vectors with given probability distributions that do not depend on x_k and u_k. Furthermore, each w_k has zero mean and finite second moment.

The problem described above is a popular formulation of a regulation problem whereby we want to keep the state of the system close to the origin. Such problems are common in the theory of automatic control of a motion or a process. The quadratic cost function is often reasonable because it induces a high penalty for large deviations of the state from the origin but a relatively small penalty for small deviations. Also, the quadratic cost is frequently used, even when it is not entirely justified, because it leads to a nice analytical solution. A number of variations and generalizations have similar solutions. For example, the disturbances w_k could have nonzero means and the quadratic cost could have the form

$$E\left\{ (x_N - \overline{x}_N)' Q_N (x_N - \overline{x}_N) + \sum_{k=0}^{N-1} \big((x_k - \overline{x}_k)' Q_k (x_k - \overline{x}_k) + u_k' R_k u_k \big) \right\},$$

which expresses a desire to keep the state of the system close to a given trajectory $(\overline{x}_0, \overline{x}_1, \ldots, \overline{x}_N)$ rather than close to the origin. Another generalized version of the problem arises when A_k, B_k are independent random matrices, rather than being known. This case is considered at the end of this section.

Applying now the DP algorithm, we have

$$J_N(x_N) = x_N' Q_N x_N, \tag{1.1}$$

$$J_k(x_k) = \min_{u_k} E\{x_k' Q_k x_k + u_k' R_k u_k + J_{k+1}(A_k x_k + B_k u_k + w_k)\}. \tag{1.2}$$

It turns out that the cost-to-go functions J_k are quadratic and as a result the optimal control law is a linear function of the state. These facts can be verified by straightforward induction. We write Eq. (1.2) for $k = N-1$,

$$\begin{aligned} J_{N-1}(x_{N-1}) = \min_{u_{N-1}} E\{ & x_{N-1}' Q_{N-1} x_{N-1} + u_{N-1}' R_{N-1} u_{N-1} \\ & + (A_{N-1} x_{N-1} + B_{N-1} u_{N-1} + w_{N-1})' Q_N \\ & \cdot (A_{N-1} x_{N-1} + B_{N-1} u_{N-1} + w_{N-1})\}, \end{aligned}$$

and we expand the last quadratic form in the right-hand side. We then use the fact $E\{w_{N-1}\} = 0$ to eliminate the term $E\{w_{N-1}' Q_N (A_{N-1} x_{N-1} + B_{N-1} u_{N-1})\}$, and we obtain

$$\begin{aligned} J_{N-1}(x_{N-1}) = x_{N-1}' Q_{N-1} x_{N-1} + \min_{u_{N-1}} \big[& u_{N-1}' R_{N-1} u_{N-1} \\ & + u_{N-1}' B_{N-1}' Q_N B_{N-1} u_{N-1} + 2 x_{N-1}' A_{N-1}' Q_N B_{N-1} u_{N-1} \big] \\ & + x_{N-1}' A_{N-1}' Q_N A_{N-1} x_{N-1} + E\{w_{N-1}' Q_N w_{N-1}\}. \end{aligned}$$

By differentiating with respect to u_{N-1} and by setting the derivative equal to zero, we obtain

$$(R_{N-1} + B_{N-1}' Q_N B_{N-1}) u_{N-1} = -B_{N-1}' Q_N A_{N-1} x_{N-1}.$$

The matrix multiplying u_{N-1} on the left is positive definite (and hence invertible), since R_{N-1} is positive definite and $B_{N-1}' Q_N B_{N-1}$ is positive semidefinite. As a result, the minimizing control vector is given by

$$u_{N-1}^* = -(R_{N-1} + B_{N-1}' Q_N B_{N-1})^{-1} B_{N-1}' Q_N A_{N-1} x_{N-1}.$$

By substitution into the expression for J_{N-1}, we have

$$J_{N-1}(x_{N-1}) = x_{N-1}' K_{N-1} x_{N-1} + E\{w_{N-1}' Q_N w_{N-1}\},$$

where by straightforward calculation, the matrix K_{N-1} is verified to be

$$\begin{aligned} K_{N-1} = {} & A_{N-1}' (Q_N - Q_N B_{N-1} (B_{N-1}' Q_N B_{N-1} + R_{N-1})^{-1} B_{N-1}' Q_N) A_{N-1} \\ & + Q_{N-1}. \end{aligned}$$

The matrix K_{N-1} is clearly symmetric. It is also positive semidefinite. To see this, note that from the preceding calculation we have for $x \in \Re^n$

$$x'K_{N-1}x = \min_u \big[x'Q_{N-1}x + u'R_{N-1}u + (A_{N-1}x + B_{N-1}u)'Q_N(A_{N-1}x + B_{N-1}u)\big].$$

Since Q_{N-1}, R_{N-1}, and Q_N are positive semidefinite, the expression within brackets is nonnegative. Minimization over u preserves nonnegativity, so it follows that $x'K_{N-1}x \geq 0$ for all $x \in \Re^n$. Hence K_{N-1} is positive semidefinite.

Since J_{N-1} is a positive semidefinite quadratic function (plus an inconsequential constant term), we may proceed similarly and obtain from the DP equation (1.2) the optimal control law for stage $N-2$. As earlier, we show that J_{N-2} is a positive semidefinite quadratic function, and by proceeding sequentially, we obtain the optimal control law for every k. It has the form

$$\mu_k^*(x_k) = L_k x_k, \tag{1.3}$$

where the gain matrices L_k are given by the equation

$$L_k = -(B_k'K_{k+1}B_k + R_k)^{-1}B_k'K_{k+1}A_k, \tag{1.4}$$

and where the symmetric positive semidefinite matrices K_k are given recursively by the algorithm

$$K_N = Q_N, \tag{1.5}$$

$$K_k = A_k'\big(K_{k+1} - K_{k+1}B_k(B_k'K_{k+1}B_k + R_k)^{-1}B_k'K_{k+1}\big)A_k + Q_k. \tag{1.6}$$

Just like DP, this algorithm starts at the terminal time N and proceeds backwards. The optimal cost is given by

$$J_0(x_0) = x_0'K_0x_0 + \sum_{k=0}^{N-1} E\{w_k'K_{k+1}w_k\}.$$

The control law (1.3) is simple and attractive for implementation in engineering applications: the current state x_k is being fed back as input through the linear feedback gain matrix L_k as shown in Fig. 4.1.1. This accounts in part for the popularity of the linear-quadratic formulation. As we will see in Chapter 5, the linearity of the control law is still maintained even for problems where the state x_k is not completely observable (imperfect state information).

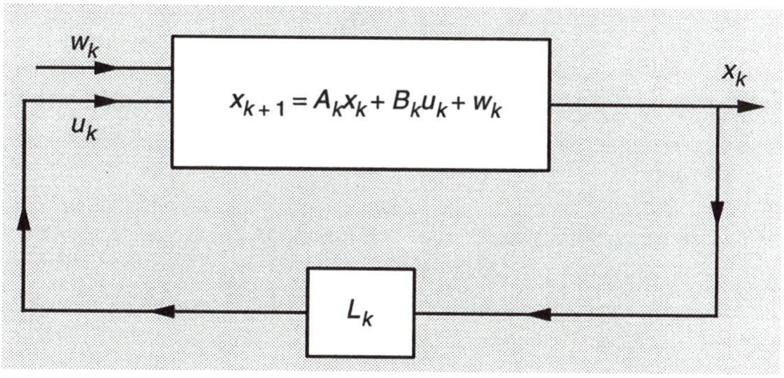

Figure 4.1.1 Linear feedback structure of the optimal controller for the linear-quadratic problem.

The Riccati Equation and Its Asymptotic Behavior

Equation (1.6) is called the *discrete-time Riccati equation.* It plays an important role in control theory. Its properties have been studied extensively and exhaustively. One interesting property of the Riccati equation is that whenever the matrices A_k, B_k, Q_k, R_k are constant and equal to A, B, Q, R, respectively, then the solution K_k converges as $k \to -\infty$ (under mild assumptions) to a steady-state solution K satisfying the *algebraic Riccati equation*

$$K = A'\big(K - KB(B'KB + R)^{-1}B'K\big)A + Q. \tag{1.7}$$

This property, to be proved shortly, indicates that for the system

$$x_{k+1} = Ax_k + Bu_k + w_k, \qquad k = 0, 1, \ldots, N-1, \tag{1.8}$$

and a large number of stages N, one can reasonably approximate the control law (1.3) by the control law $\{\mu^*, \mu^*, \ldots, \mu^*\}$, where

$$\mu^*(x) = Lx, \tag{1.9}$$

$$L = -(B'KB + R)^{-1}B'KA, \tag{1.10}$$

and K solves the algebraic Riccati equation (1.7). This control law is *stationary*; that is, it does not change over time.

We now turn to proving convergence of the sequence of matrices $\{K_k\}$ generated by the Riccati equation (1.6). We first introduce the notions of controllability and observability, which are very important in control theory.

Definition 1.1: A pair (A, B), where A is an $n \times n$ matrix and B is an $n \times m$ matrix, is said to be *controllable* if the $n \times nm$ matrix

$$[B, AB, A^2B, \ldots, A^{n-1}B]$$

has full rank (i.e., has linearly independent rows). A pair (A, C), where A is an $n \times n$ matrix and C an $m \times n$ matrix, is said to be *observable* if the pair (A', C') is controllable, where A' and C' denote the transposes of A and C, respectively.

One may show that if the pair (A, B) is controllable, then for any initial state x_0 there exists a sequence of control vectors $u_0, u_1, \ldots, u_{n-1}$ that force the state x_n of the system

$$x_{k+1} = Ax_k + Bu_k \tag{1.11}$$

to be equal to zero at time n. Indeed, by successively applying the above equation for $k = n-1, n-2, \ldots, 0$, we obtain

$$x_n = A^n x_0 + Bu_{n-1} + ABu_{n-2} + \cdots + A^{n-1}Bu_0$$

or equivalently

$$x_n - A^n x_0 = (B, AB, \ldots, A^{n-1}B) \begin{pmatrix} u_{n-1} \\ u_{n-2} \\ \vdots \\ u_0 \end{pmatrix}. \tag{1.12}$$

If (A, B) is controllable, the matrix $(B, AB, \ldots, A^{n-1}B)$ has full rank and as a result the right-hand side of Eq. (1.12) can be made equal to any vector in $\Re^n$ by appropriate selection of $(u_0, u_1, \ldots, u_{n-1})$. In particular, one can choose $(u_0, u_1, \ldots, u_{n-1})$ so that the right-hand side of Eq. (1.12) is equal to $-A^n x_0$, which implies $x_n = 0$. This property explains the name "controllable pair" and in fact is often used to define controllability.

The notion of observability has an analogous interpretation in the context of estimation problems; that is, given measurements $z_0, z_1, \ldots, z_{n-1}$ of the form $z_k = Cx_k$, it is possible to infer the initial state x_0 of the system $x_{k+1} = Ax_k$, in view of the relation

$$\begin{pmatrix} z_{n-1} \\ \vdots \\ z_1 \\ z_0 \end{pmatrix} = \begin{pmatrix} CA^{n-1} \\ \vdots \\ CA \\ C \end{pmatrix} x_0.$$

Alternatively, it can be seen that observability is equivalent to the property that, in the absence of control, if $x_k \to 0$ then $Cx_k \to 0$.

The notion of stability is of paramount importance in control theory. In the context of our problem it is important that the stationary control law (1.9) results in a stable closed-loop system; that is, in the absence of input disturbance, the state of the system

$$x_{k+1} = (A + BL)x_k, \qquad k = 0, 1, \ldots,$$

tends to zero as $k \to \infty$. Since $x_k = (A + BL)^k x_0$, it follows that the closed-loop system is stable if and only if $(A + BL)^k \to 0$, or equivalently (see Appendix A), if and only if the eigenvalues of the matrix $(A + BL)$ are strictly within the unit circle.

The following proposition shows that for a stationary controllable system and constant matrices Q and R, the solution of the Riccati equation (1.6) converges to a positive definite symmetric matrix K for an arbitrary positive semidefinite symmetric initial matrix. In addition, the proposition shows that the corresponding closed-loop system is stable. The proposition also requires an observability assumption, namely, that Q can be written as $C'C$, where the pair (A, C) is observable. Note that if r is the rank of Q, there exists an $r \times n$ matrix C of rank r such that $Q = C'C$ (see Appendix A). The implication of the observability assumption is that in the absence of control, if the state cost per stage $x_k' Q x_k$ tends to zero or equivalently $Cx_k \to 0$, then also $x_k \to 0$.

To simplify notation, we reverse the time indexing of the Riccati equation in the following proposition. Thus P_k in the following Eq. (1.13) corresponds to K_{N-k} in Eq. (1.6). A graphical proof of the proposition for the case of a scalar system is given in Fig. 4.1.2.

Proposition 4.1: Let A be an $n \times n$ matrix, B be an $n \times m$ matrix, Q be an $n \times n$ positive semidefinite symmetric matrix, and R be an $m \times m$ positive definite symmetric matrix. Consider the discrete-time Riccati equation

$$P_{k+1} = A'\big(P_k - P_k B(B'P_k B + R)^{-1} B'P_k\big)A + Q, \qquad k = 0, 1, \ldots, \tag{1.13}$$

where the initial matrix P_0 is an arbitrary positive semidefinite symmetric matrix. Assume that the pair (A, B) is controllable. Assume also that Q may be written as $C'C$, where the pair (A, C) is observable. Then:

(a) There exists a positive definite symmetric matrix P such that for every positive semidefinite symmetric initial matrix P_0 we have

$$\lim_{k \to \infty} P_k = P.$$

Furthermore, P is the unique solution of the algebraic matrix equation

$$P = A'\big(P - PB(B'PB + R)^{-1}B'P\big)A + Q \tag{1.14}$$

within the class of positive semidefinite symmetric matrices.

(b) The corresponding closed-loop system is stable; that is, the eigenvalues of the matrix

$$D = A + BL, \tag{1.15}$$

where

$$L = -(B'PB + R)^{-1}B'PA, \tag{1.16}$$

are strictly within the unit circle.

Proof: The proof proceeds in several steps. First we show convergence of the sequence generated by Eq. (1.13) when the initial matrix P_0 is equal to zero. Next we show that the corresponding matrix D of Eq. (1.15) satisfies $D^k \to 0$. Then we show the convergence of the sequence generated by Eq. (1.13) when P_0 is any positive semidefinite symmetric matrix, and finally we show uniqueness of the solution of Eq. (1.14).

Initial Matrix $P_0 = 0$. Consider the optimal control problem of finding $u_0, u_1, \ldots, u_{k-1}$ that minimize

$$\sum_{i=0}^{k-1}(x_i'Qx_i + u_i'Ru_i) \tag{1.17}$$

subject to

$$x_{i+1} = Ax_i + Bu_i, \qquad i = 0, 1, \ldots, k-1, \tag{1.18}$$

where x_0 is given. The optimal value of this problem, according to the theory of this section, is

$$x_0'P_k(0)x_0,$$

where $P_k(0)$ is given by the Riccati equation (1.13) with $P_0 = 0$. For any control sequence $(u_0, u_1, \ldots, u_k)$ we have

$$\sum_{i=0}^{k-1}(x_i'Qx_i + u_i'Ru_i) \leq \sum_{i=0}^{k}(x_i'Qx_i + u_i'Ru_i)$$

and hence

$$\begin{aligned} x_0'P_k(0)x_0 &= \min_{\substack{u_i \\ i=0,\ldots,k-1}} \sum_{i=0}^{k-1}(x_i'Qx_i + u_i'Ru_i) \\ &\leq \min_{\substack{u_i \\ i=0,\ldots,k}} \sum_{i=0}^{k}(x_i'Qx_i + u_i'Ru_i) = x_0'P_{k+1}(0)x_0, \end{aligned}$$

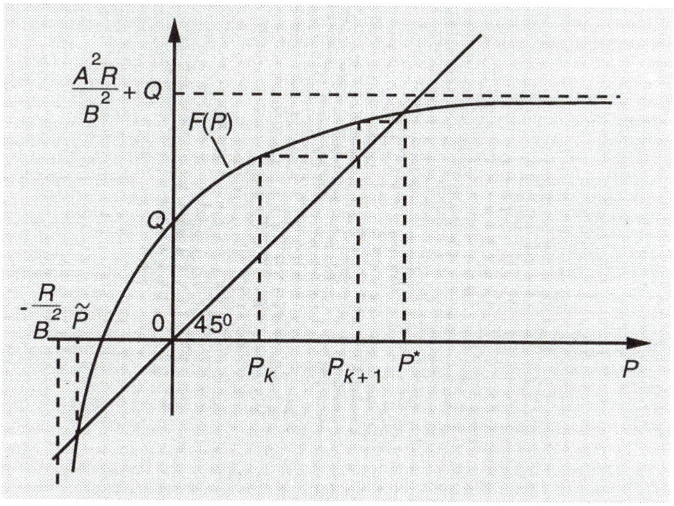

Figure 4.1.2 Graphical proof of Prop. 4.1 for the case of a scalar stationary system (one-dimensional state and control), assuming that $A \neq 0$, $B \neq 0$, $Q > 0$, and $R > 0$. The Riccati equation (1.13) is given by

$$P_{k+1} = A^2\left(P_k - \frac{B^2P_k^2}{B^2P_k + R}\right) + Q,$$

which can be equivalently written as

$$P_{k+1} = F(P_k),$$

where the function F is given by

$$F(P) = \frac{A^2RP}{B^2P + R} + Q.$$

Because F is concave and monotonically increasing in the interval $(-R/B^2, \infty)$, as shown in the figure, the equation $P = F(P)$ has one positive solution P^* and one negative solution $\tilde{P}$. The Riccati iteration $P_{k+1} = F(P_k)$ converges to P^* starting anywhere in the interval $(\tilde{P}, \infty)$ as shown in the figure.

where both minimizations are subject to the system equation constraint $x_{i+1} = Ax_i + Bu_i$. Furthermore, for a fixed x_0 and for every k, $x_0'P_k(0)x_0$ is bounded from above by the cost corresponding to a control sequence that forces x_0 to the origin in n steps and applies zero control after that. Such a sequence exists by the controllability assumption. Thus the sequence $\{x_0'P_k(0)x_0\}$ is nondecreasing with respect to k and bounded from above, and therefore converges to some real number for every $x_0 \in \Re^n$. It follows that the sequence $\{P_k(0)\}$ converges to some matrix P in the sense that each of the sequences of the elements of $P_k(0)$ converges to the corresponding elements of P. To see this, take $x_0 = (1, 0, \dots, 0)$. Then $x_0' P_k(0) x_0$ is equal to the first diagonal element of $P_k(0)$, so it follows that the sequence of first diagonal elements of $P_k(0)$ converges; the limit of this sequence is the first diagonal element of P. Similarly, by taking $x_0 = (0, \dots, 0, 1, 0, \dots, 0)$ with the 1 in the ith coordinate, for $i = 2, \dots, n$, it follows that all the diagonal elements of $P_k(0)$ converge to the corresponding diagonal elements of P. Next take $x_0 = (1, 1, 0, \dots, 0)$ to show that the second elements of the first row converge. Continuing similarly, we obtain

$$\lim_{k\to\infty} P_k(0) = P,$$

where $P_k(0)$ are generated by Eq. (1.13) with $P_0 = 0$. Furthermore, since $P_k(0)$ is positive semidefinite and symmetric, so is the limit matrix P. Now by taking the limit in Eq. (1.13) it follows that P satisfies

$$P = A'\big(P - PB(B'PB + R)^{-1}B'P\big)A + Q. \tag{1.19}$$

In addition, by direct calculation we can verify the following useful equality

$$P = D'PD + Q + L'RL, \tag{1.20}$$

where D and L are given by Eqs. (1.15) and (1.16). An alternative way to derive this equality is to observe that from the DP algorithm corresponding to a finite horizon N we have for all states x_{N-k}

$$x_{N-k}' P_{k+1}(0) x_{N-k} = x_{N-k}' Q x_{N-k} + \mu_{N-k}^*(x_{N-k})' R \mu_{N-k}^*(x_{N-k}) + x_{N-k+1}' P_k(0) x_{N-k+1}.$$

By using the optimal controller expression $\mu_{N-k}^*(x_{N-k}) = L_{N-k} x_{N-k}$ and the closed-loop system equation $x_{N-k+1} = (A + BL_{N-k}) x_{N-k}$, we thus obtain

$$P_{k+1}(0) = Q + L_{N-k}' R L_{N-k} + (A + BL_{N-k})' P_k(0) (A + BL_{N-k}). \tag{1.21}$$

Equation (1.20) then follows by taking the limit as $k \to \infty$ in Eq. (1.21). Q.E.D.

Stability of the Closed-Loop System. Consider the system

$$x_{k+1} = (A + BL)x_k = Dx_k \tag{1.22}$$

for an arbitrary initial state x_0. We will show that $x_k \to 0$ as $k \to \infty$. We have for all k, by using Eq. (1.20),

$$x_{k+1}' P x_{k+1} - x_k' P x_k = x_k' (D'PD - P) x_k = -x_k' (Q + L'RL) x_k.$$

Hence

$$x_{k+1}' P x_{k+1} = x_0' P x_0 - \sum_{i=0}^{k} x_i' (Q + L'RL) x_i. \tag{1.23}$$

The left-hand side of this equation is bounded below by zero, so it follows that

$$\lim_{k\to\infty} x_k'(Q + L'RL)x_k = 0.$$

Since R is positive definite and Q may be written as $C'C$, we obtain

$$\lim_{k\to\infty} Cx_k = 0, \qquad \lim_{k\to\infty} Lx_k = \lim_{k\to\infty} \mu^*(x_k) = 0. \tag{1.24}$$

The preceding relations imply that as the control asymptotically becomes negligible, we have $\lim_{k\to\infty} Cx_k = 0$, and in view of the observability assumption, this implies that $x_k \to 0$. To express this argument more precisely, let us use the relation $x_{k+1} = (A + BL)x_k$ [cf. Eq. (1.22)], to write

$$\begin{pmatrix} C\left(x_{k+n-1} - \sum_{i=1}^{n-1} A^{i-1}BLx_{k+n-i-1}\right) \\ C\left(x_{k+n-2} - \sum_{i=1}^{n-2} A^{i-1}BLx_{k+n-i-2}\right) \\ \vdots \\ C(x_{k+1} - BLx_k) \\ Cx_k \end{pmatrix} = \begin{pmatrix} CA^{n-1} \\ CA^{n-2} \\ \vdots \\ CA \\ C \end{pmatrix} x_k. \tag{1.25}$$

Since $Lx_k \to 0$ by Eq. (1.24), the left-hand side tends to zero and hence the right-hand side tends to zero also. By the observability assumption, however, the matrix multiplying x_k on the right side of (1.25) has full rank. It follows that $x_k \to 0$.

Positive Definiteness of P. Assume the contrary, i.e., there exists some $x_0 \neq 0$ such that $x_0'Px_0 = 0$. Since P is positive semidefinite, from Eq. (1.23) we obtain

$$x_k'(Q + L'RL)x_k = 0, \qquad k = 0, 1, \ldots$$

Since $x_k \to 0$, we obtain $x_k'Qx_k = x_k'C'Cx_k = 0$ and $x_k'L'RLx_k = 0$, or

$$Cx_k = 0, \qquad Lx_k = 0, \qquad k = 0, 1, \ldots$$

Thus all the controls $\mu^*(x_k) = Lx_k$ of the closed-loop system are zero while we have $Cx_k = 0$ for all k. Based on the observability assumption, we will show that this implies $x_0 = 0$, thereby reaching a contradiction. Indeed, consider Eq. (1.25) for $k = 0$. By the preceding equalities, the left-hand side is zero and hence

$$0 = \begin{pmatrix} CA^{n-1} \\ \vdots \\ CA \\ C \end{pmatrix} x_0.$$

Since the matrix multiplying x_0 above has full rank by the observability assumption, we obtain $x_0 = 0$, which contradicts the hypothesis $x_0 \neq 0$ and proves that P is positive definite.

Arbitrary Initial Matrix P_0. Next we show that the sequence of matrices $\{P_k(P_0)\}$, defined by Eq. (1.13) when the starting matrix is an arbitrary positive semidefinite symmetric matrix P_0, converges to $P = \lim_{k\to\infty} P_k(0)$. Indeed, the optimal cost of the problem of minimizing

$$x_k' P_0 x_k + \sum_{i=0}^{k-1} (x_i' Q x_i + u_i' R u_i) \tag{1.26}$$

subject to the system equation $x_{i+1} = Ax_i + Bu_i$ is equal to $x_0' P_k(P_0) x_0$. Hence we have for every $x_0 \in \Re^n$

$$x_0' P_k(0) x_0 \le x_0' P_k(P_0) x_0.$$

Consider now the cost (1.26) corresponding to the controller $\mu(x_k) = u_k = Lx_k$, where L is defined by Eq. (1.20). This cost is

$$x_0' \left(D^{k'} P_0 D^k + \sum_{i=0}^{k-1} D^{i'} (Q + L'RL) D^i \right) x_0$$

and is greater or equal to $x_0' P_k(P_0) x_0$, which is the optimal value of the cost (1.26). Hence we have for all k and $x \in \Re^n$

$$x' P_k(0) x \le x' P_k(P_0) x \le x' \left(D^{k'} P_0 D^k + \sum_{i=0}^{k-1} D^{i'} (Q + L'RL) D^i \right) x. \tag{1.27}$$

Now we have proved that

$$\lim_{k\to\infty} P_k(0) = P. \tag{1.28}$$

We also have, using the fact $\lim_{k\to\infty} D^{k'} P_0 D^k = 0$, and the relation $Q + L'RL = P - D'PD$ [cf. Eq. (1.20)],

$$\begin{aligned} \lim_{k\to\infty} & \left\{ D^{k'} P_0 D_k + \sum_{i=0}^{k-1} D^{i'} (Q + L'RL) D^i \right\} \\ &= \lim_{k\to\infty} \left\{ \sum_{i=0}^{k-1} D^{i'} (Q + L'RL) D^i \right\} \\ &= \lim_{k\to\infty} \left\{ \sum_{i=0}^{k-1} D^{i'} (P - D'PD) D^i \right\} \\ &= P. \end{aligned} \tag{1.29}$$

Combining the preceding three equations, we obtain

$$\lim_{k\to\infty} P_k(P_0) = P,$$

for an arbitrary positive semidefinite symmetric initial matrix P_0.

Uniqueness of Solution. If $\tilde{P}$ is another positive semidefinite symmetric solution of the algebraic Riccati equation (1.14), we have $P_k(\tilde{P}) = \tilde{P}$ for all $k = 0, 1, \ldots$ From the convergence result just proved, we then obtain

$$\lim_{k \to \infty} P_k(\tilde{P}) = P,$$

implying that $\tilde{P} = P$. **Q.E.D.**

The assumptions of the preceding proposition can be relaxed somewhat. Suppose that, instead of controllability of the pair (A, B), we assume that the system is *stabilizable* in the sense that there exists an $m \times n$ feedback gain matrix G such that the closed-loop system $x_{k+1} = (A + BG)x_k$ is stable. Then the proof of convergence of $P_k(0)$ to some positive semidefinite P given previously carries through. [We use the stationary control law $\mu(x) = Gx$ for which the closed-loop system is stable to ensure that $x_0' P_k(0) x_0$ is bounded.] Suppose that, instead of observability of the pair (A, C), the system is assumed *detectable* in the sense that A is such that if $u_k \to 0$ and $Cx_k \to 0$ then it follows that $x_k \to 0$. (This essentially means that instability of the system can be detected by looking at the measurement sequence $\{z_k\}$ with $z_k = Cx_k$.) Then Eq. (1.24) implies that $x_k \to 0$ and that the system $x_{k+1} = (A + BL)x_k$ is stable. The other parts of the proof of the proposition follow similarly, with the exception of positive definiteness of P, which cannot be guaranteed anymore. (As an example, take $A = 0$, $B = 0$, $C = 0$, $R > 0$. Then both the stabilizability and the detectability assumptions are satisfied, but $P = 0$.)

To summarize, if the controllability and observability assumptions of the proposition are replaced by the preceding stabilizability and detectability assumptions, the conclusions of the proposition hold with the exception of positive definiteness of the limit matrix P, which can now only be guaranteed to be positive semidefinite.

Random System Matrices

We consider now the case where $\{A_0, B_0\}, \ldots, \{A_{N-1}, B_{N-1}\}$ are not known but rather are independent random matrices that are also independent of $w_0, w_1, \ldots, w_{N-1}$. Their probability distributions are given, and they are assumed to have finite second moments. This problem falls again within the framework of the basic problem by considering as disturbance at each time k the triplet (A_k, B_k, w_k). The DP algorithm is written as

$$J_N(x_N) = x_N' Q_N x_N,$$

$$J_k(x_k) = \min_{u_k} \mathop{E}_{w_k, A_k, B_k} \{ x_k' Q_k x_k + u_k' R_k u_k + J_{k+1}(A_k x_k + B_k u_k + w_k) \}.$$

Calculations very similar to those for the case where A_k, B_k are not random show that the optimal control law has the form

$$\mu_k^*(x_k) = L_k x_k, \tag{1.30}$$

where the gain matrices L_k are given by

$$L_k = -\big(R_k + E\{B_k' K_{k+1} B_k\}\big)^{-1} E\{B_k' K_{k+1} A_k\}, \tag{1.31}$$

and where the matrices K_k are given by the recursive equation

$$K_N = Q_N, \tag{1.32}$$

$$\begin{aligned} K_k = {} & E\{A_k' K_{k+1} A_k\} \\ & - E\{A_k' K_{k+1} B_k\}\big(R_k + E\{B_k' K_{k+1} B_k\}\big)^{-1} E\{B_k' K_{k+1} A_k\} + Q_k. \end{aligned} \tag{1.33}$$

In the case of a stationary system and constant matrices Q_k and R_k it is not necessarily true that the above equation converges to a steady-state solution. This is demonstrated in Fig. 4.1.3 for a scalar system, where it is shown that if the expression

$$T = E\{A^2\}E\{B^2\} - \big(E\{A\}\big)^2\big(E\{B\}\big)^2$$

exceeds a certain threshold, the matrices K_k diverge to ∞ starting from any nonnegative initial condition. A possible interpretation is that if there is a lot of uncertainty about the system, as quantified by T, optimization over a long horizon is meaningless. This phenomenon has been called the *uncertainty threshold principle*; see [AGK77], [KuA77].

On Certainty Equivalence

We close this section by making an observation about the nature of the quadratic cost. Consider the minimization over u of

$$\mathop{E}_{w}\{(ax + bu + w)^2\},$$

where a and b are given scalars, x is known, and w is a random variable. The optimum is attained for

$$u^* = -\left(\frac{a}{b}\right)x - \left(\frac{1}{b}\right)E\{w\}.$$

Thus u^* depends on the probability distribution of w only through the mean $E\{w\}$. In particular, the result of the optimization is the same as for the corresponding deterministic problem where w is replaced by $E\{w\}$.

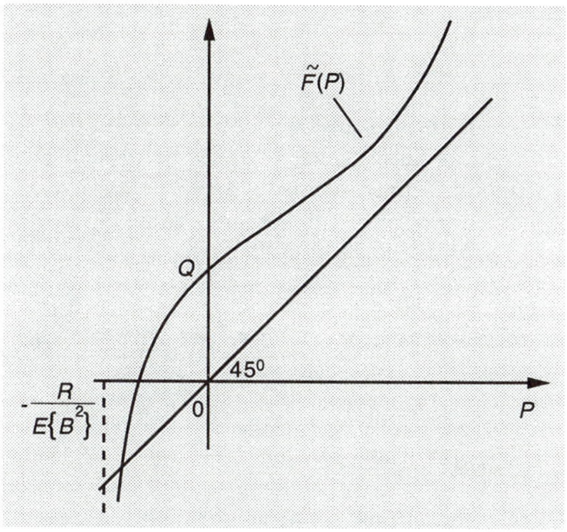

Figure 4.1.3 Graphical illustration of the asymptotic behavior of the generalized Riccati equation (1.33) in the case of a scalar stationary system (one-dimensional state and control). Using P_k in place of K_{N-k}, this equation is written as

$$P_{k+1} = \tilde{F}(P_k),$$

where the function $\tilde{F}$ is given by

$$\tilde{F}(P) = \frac{E\{A^2\}RP}{E\{B^2\}P + R} + Q + \frac{TP^2}{E\{B^2\}P + R},$$

$$T = E\{A^2\}E\{B^2\} - \big(E\{A\}\big)^2\big(E\{B\}\big)^2.$$

If $T = 0$, as in the case where A and B are not random, the Riccati equation becomes identical with the one of Fig. 4.1.2 and converges to a steady-state. Convergence also occurs when T has a small positive value. However, as illustrated in the figure, for T large enough, the graph of the function $\tilde{F}$ and the 45-degree line that passes through the origin do not intersect at a positive value of P, and the Riccati equation diverges to infinity.

This property is called the *certainty equivalence principle* and appears in various forms in many (but not all) stochastic control problems involving linear systems and quadratic cost. For the first problem of this section, where A_k, B_k are known, certainty equivalence holds because the optimal control law (1.3) is the same as the one that would be obtained from the corresponding deterministic problem where w_k is not random but rather is known and is equal to zero (its expected value). However, for the problem where A_k, B_k are random, the certainty equivalence principle does not hold, since if one replaces A_k, B_k with their expected values in Eq. (1.33), the resulting control law need not be optimal.

4.2 INVENTORY CONTROL

We consider now the inventory control problem discussed in Sections 1.1 and 1.2. We assume that excess demand at each period is backlogged and is filled when additional inventory becomes available. This is represented by negative inventory in the system equation

$$x_{k+1} = x_k + u_k - w_k, \qquad k = 0, 1, \ldots, N-1.$$

We assume that the demands w_k take values within some bounded interval and are independent. We will analyze the problem for the case of a holding/shortage cost of the form

$$r(x) = p\max(0, -x) + h\max(0, x),$$

where p and h are given nonnegative scalars. Thus the total expected cost to be minimized is

$$\mathop{E}_{\substack{w_k \\ k=0,\ldots,N-1}} \left\{ \sum_{k=0}^{N-1} \big(cu_k + p\max(0, w_k - x_k - u_k) + h\max(0, x_k + u_k - w_k)\big) \right\}$$

We assume that the purchase cost per unit stock c is positive and that $p > c$. The last assumption is necessary for the problem to be well posed; if c, the purchase cost per unit, were greater than p, the depletion cost per unit, it would never be optimal to buy new stock at the last period and possibly in earlier periods. Much of the subsequent analysis generalizes to the case where r is a convex function that grows to infinity with asymptotic slopes p and h as its argument tends to ∞ and $-\infty$, respectively.

By applying the DP algorithm, we have

$$J_N(x_N) = 0 \tag{2.1}$$

$$J_k(x_k) = \min_{u_k \geq 0} \big[cu_k + H(x_k + u_k) + E\big\{J_{k+1}(x_k + u_k - w_k)\big\}\big], \tag{2.2}$$

where the function H is defined by

$$H(y) = pE\big\{\max(0, w_k - y)\big\} + hE\big\{\max(0, y - w_k)\big\}.$$

Actually, H depends on k whenever the probability distribution of w_k depends on k. To simplify notation, we do not show this dependence and assume that all demands are identically distributed, but the following analysis carries through even when the demand statistics are time-varying.

By introducing the variable $y_k = x_k + u_k$, we can write the right-hand side of the DP equation (2.2) as

$$\min_{y_k \geq x_k} \big[cy_k + H(y_k) + E\big\{J_{k+1}(y_k - w_k)\big\}\big] - cx_k.$$

The function H can be seen to be convex, since the functions $\max(0, w_k - y)$ and $\max(0, y - w_k)$ are convex in y for each fixed w_k, and taking expectation over w_k preserves convexity. We will prove shortly that J_{k+1} is convex, but for the moment let us proceed by assuming this fact. Then the function in brackets above is convex. Suppose that this function has an unconstrained minimum with respect to y_k, denoted by S_k. Then it is seen that (in view of the constraint $y_k \geq x_k$) a minimizing y_k equals S_k if $x_k < S_k$, and equals x_k otherwise. Using the reverse transformation $u_k = y_k - x_k$, we see that the minimum in the DP equation (2.2) is attained at $u_k = S_k - x_k$ if $x_k < S_k$, and at $u_k = 0$ otherwise. An optimal policy is determined by the sequence of scalars $\{S_0, S_1, \ldots, S_{N-1}\}$ and has the form

$$\mu_k^*(x_k) = \begin{cases} S_k - x_k & \text{if } x_k < S_k, \\ 0 & \text{if } x_k \geq S_k. \end{cases} \tag{2.3}$$

For each k, the scalar S_k minimizes the function

$$G_k(y) = cy + H(y) + E\big\{J_{k+1}(y - w)\big\}. \tag{2.4}$$

Thus, the optimality of the policy (2.3) will be proved if we can show that the cost-to-go functions J_k [and hence also the functions G_k of Eq. (2.4)] are convex, and furthermore $\lim_{|y| \to \infty} G_k(y) = \infty$, so that the minimizing scalars S_k exist. We proceed to show these properties inductively.

We have that J_N is the zero function, so it is convex. Since $c < p$ and the derivative of $H(y)$ tends to $-p$ as $y \to -\infty$, we see that $G_{N-1}(y)$ [which is $cy + H(y)$] has a derivative that becomes negative as $y \to -\infty$ and becomes positive as $y \to \infty$ (see Fig. 4.2.1). Therefore

$$\lim_{|y| \to \infty} G_{N-1}(y) = \infty.$$

Thus, as shown above, an optimal policy at time $N - 1$ is given by

$$\mu_{N-1}^*(x_{N-1}) = \begin{cases} S_{N-1} - x_{N-1} & \text{if } x_{N-1} < S_{N-1}, \\ 0 & \text{if } x_{N-1} \geq S_{N-1}. \end{cases}$$

Furthermore, from the DP equation (2.1) we have

$$J_{N-1}(x_{N-1}) = \begin{cases} c(S_{N-1} - x_{N-1}) + H(S_{N-1}) & \text{if } x_{N-1} < S_{N-1}, \\ H(x_{N-1}) & \text{if } x_{N-1} \geq S_{N-1}, \end{cases}$$

which is a convex function because H is convex and S_{N-1} minimizes $cy + H(y)$ (see Fig. 4.2.1). Thus, given the convexity of J_N, we were able to prove the convexity of J_{N-1}. Furthermore,

$$\lim_{|y| \to \infty} J_{N-1}(y) = \infty.$$

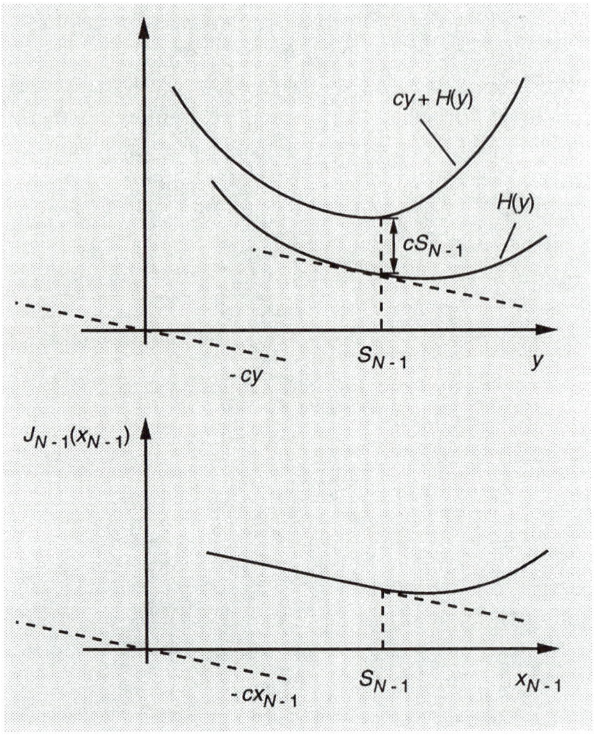

Figure 4.2.1 Structure of the cost-to-go functions when the fixed cost is zero.

This argument can be repeated to show that for all $k = N-2, \ldots, 0$, if J_{k+1} is convex, $\lim_{|y|\to\infty} J_{k+1}(y) = \infty$, and $\lim_{|y|\to\infty} G_k(y) = \infty$, then we have

$$J_k(x_k) = \begin{cases} c(S_k - x_k) + H(S_k) + E\{J_{k+1}(S_k - w_k)\} & \text{if } x_k < S_k, \\ H(x_k) + E\{J_{k+1}(x_k - w_k)\} & \text{if } x_k \geq S_k, \end{cases}$$

where S_k minimizes $cy + H(y) + E\{J_{k+1}(y-w)\}$. Furthermore, J_k is convex, $\lim_{|y|\to\infty} J_k(y) = \infty$, and $\lim_{|y|\to\infty} G_{k-1}(y) = \infty$. Thus, the optimality proof of the policy (2.3) is completed.

Positive Fixed Cost and (s, S) Policies

We now turn to the more complicated case where there is a positive fixed cost K associated with a positive inventory order. Thus the cost for ordering inventory $u \geq 0$ is

$$C(u) = \begin{cases} K + cu & \text{if } u > 0, \\ 0 & \text{if } u = 0. \end{cases}$$

The DP algorithm takes the form

$$J_N(x_N) = 0,$$

$$J_k(x_k) = \min_{u_k \geq 0} \left[C(u_k) + H(x_k + u_k) + E\{J_{k+1}(x_k + u_k - w_k)\} \right],$$

with H defined as earlier by

$$H(y) = pE\{\max(0, w - y)\} + hE\{\max(0, y - w)\}.$$

Consider again the functions

$$G_k(y) = cy + H(y) + E\{J_{k+1}(y - w)\}. \tag{2.5}$$

Then J_k is written as

$$J_k(x_k) = \min \left[G_k(x_k),\ \min_{u_k > 0} \left[K + cu_k + G_k(x_k + u_k) \right] \right],$$

or equivalently, through the change of variable $y_k = x_k + u_k$,

$$J_k(x_k) = \min \left[cx_k + G(x_k),\ \min_{y_k > x_k} \left[K + cy_k + G_k(y_k) \right] \right] - cx_k.$$

If, as in the case where $K = 0$, we could prove that the functions G_k are convex, then it would not be difficult to verify [see part (d) of the following Lemma 2.1] that the policy

$$\mu_k^*(x_k) = \begin{cases} S_k - x_k & \text{if } x_k < s_k, \\ 0 & \text{if } x_k \geq s_k, \end{cases} \tag{2.6}$$

is optimal, where S_k is a value of y that minimizes $G_k(y)$ and s_k is the smallest value of y for which $G_k(y) = K + G_k(S_k)$. A policy of the form (2.6) is known as a *multiperiod* (s, S) *policy*.

Unfortunately, when $K > 0$ it is not necessarily true that the functions G_k are convex. This opens the possibility of G_k having the form shown in Fig. 4.2.2, where the optimal policy is to order $(S - x)$ in interval I, zero in intervals II and IV, and $(\tilde{S} - x)$ in interval III. However, we will show that even though the functions G_k may not be convex, they have the property

$$K + G_k(z + y) \geq G_k(y) + z\left(\frac{G_k(y) - G_k(y - b)}{b}\right), \qquad \text{for all } z \geq 0, b > 0, y. \tag{2.7}$$

This property is called *K-convexity* and was first used by Scarf [Sca60] to show the optimality of multiperiod (s, S) policies. Now if K-convexity holds, the situation shown in Fig. 4.2.2 is impossible; if y_0 is the local maximum in the interval III, then we must have, for sufficiently small $b > 0$,

$$\frac{G_k(y_0) - G_k(y_0 - b)}{b} \geq 0,$$

and from Eq. (2.7) it follows that

$$K + G_k(\tilde{S}) \geq G_k(y_0),$$

which contradicts the construction shown in Fig. 4.2.2. More generally, it will be shown by using part (d) of the following Lemma 2.1 that if the K-convexity Eq. (2.7) holds, then an optimal policy takes the (s, S) form (2.6).

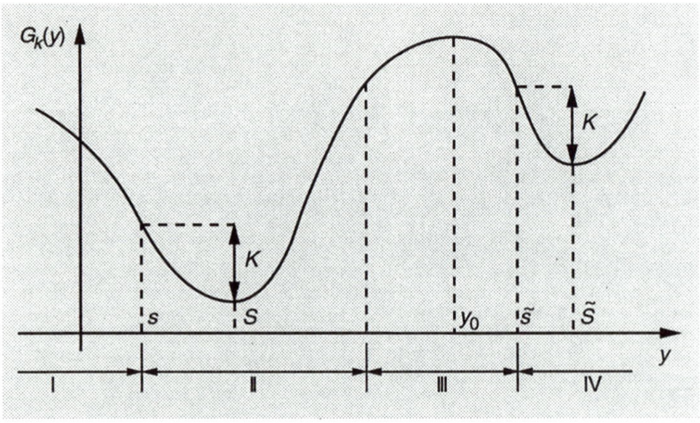

Figure 4.2.2 Potential form of the function G_k when the fixed cost is nonzero. If G_k had indeed the form shown in the figure, the optimal policy would be to order $(S - x)$ in interval I, zero in intervals II and IV, and $(\tilde{S} - x)$ in interval III. The use of K-convexity allows us to show that the form of G_k shown in the figure is impossible.

Definition 2.1: We say that a real-valued function g is *K-convex*, where $K \geq 0$, if

$$K + g(z + y) \geq g(y) + z\left(\frac{g(y) - g(y - b)}{b}\right), \qquad \text{for all } z \geq 0, b > 0, y.$$

Some properties of K-convex functions are provided in the following lemma. Part (d) of the lemma essentially proves the optimality of the (s, S) policy (2.6) when the functions G_k are K-convex.

Lemma 2.1:

(a) A real-valued convex function g is also 0-convex and hence also K-convex for all $K \geq 0$.

(b) If $g_1(y)$ and $g_2(y)$ are K-convex and L-convex ($K \geq 0$, $L \geq 0$), respectively, then $\alpha g_1(y) + \beta g_2(y)$ is $(\alpha K + \beta L)$-convex for all $\alpha > 0$ and $\beta > 0$.

(c) If $g(y)$ is K-convex and w is a random variable, then $E_w\{g(y - w)\}$ is also K-convex, provided $E_w\{|g(y-w)|\} < \infty$ for all y.

(d) If g is a continuous K-convex function and $g(y) \to \infty$ as $|y| \to \infty$, then there exist scalars s and S with $s \leq S$ such that

(i) $g(S) \leq g(y)$, for all scalars y;

(ii) $g(S) + K = g(s) < g(y)$, for all $y < s$;

(iii) $g(y)$ is a decreasing function on $(-\infty, s)$;

(iv) $g(y) \leq g(z) + K$ for all y, z with $s \leq y \leq z$.

Proof: Part (a) follows from elementary properties of convex functions, and parts (b) and (c) follow from the definition of a K-convex function. We will thus concentrate on proving part (d).

Since g is continuous and $g(y) \to \infty$ as $|y| \to \infty$, there exists a minimizing point of g. Let S be such a point. Also let s be the smallest scalar z for which $z \leq S$ and $g(S) + K = g(z)$. For all y with $y < s$, we have from the definition of K-convexity

$$K + g(S) \geq g(s) + \frac{S - s}{s - y}\big(g(s) - g(y)\big).$$

Since $K + g(S) - g(s) = 0$, we obtain $g(s) - g(y) \leq 0$. Since $y < s$ and s is the smallest scalar for which $g(S) + K = g(s)$, we must have $g(s) < g(y)$ and (ii) is proved. To prove (iii), note that for $y_1 < y_2 < s$, we have

$$K + g(S) \geq g(y_2) + \frac{S - y_2}{y_2 - y_1}\big(g(y_2) - g(y_1)\big).$$

Also from (ii),

$$g(y_2) > g(S) + K,$$

and by adding these two inequalities we have

$$0 > \frac{S - y_2}{y_2 - y_1}\big(g(y_2) - g(y_1)\big),$$

from which we obtain $g(y_1) > g(y_2)$, thus proving (iii). To prove (iv), we note that it holds for $y = z$ as well as for either $y = S$ or $y = s$. There remain two other possibilities: $S < y < z$ and $s < y < S$. If $S < y < z$, then by K-convexity

$$K + g(z) \geq g(y) + \frac{z - y}{y - S}\big(g(y) - g(S)\big) \geq g(y),$$

and (iv) is proved. If $s < y < S$, then by K-convexity

$$g(s) = K + g(S) \geq g(y) + \frac{S - y}{y - s}\big(g(y) - g(s)\big),$$

from which

$$\left(1 + \frac{S - y}{y - s}\right) g(s) \geq \left(1 + \frac{S - y}{y - s}\right) g(y),$$

and $g(s) \geq g(y)$. Noting that

$$g(z) + K \geq g(S) + K = g(s),$$

it follows that $g(z) + K \geq g(y)$. Thus (iv) is proved for this case as well. **Q.E.D.**

Consider now the function G_{N-1} of Eq. (2.5):

$$G_{N-1}(y) = cy + H(y).$$

Clearly, G_{N-1} is convex and hence by part (a) of the previous lemma, it is also K-convex. We have

$$J_{N-1}(x) = \min\left[cx + G_{N-1}(x),\ \min_{y>x}\big[K + cy + G_{N-1}(y)\big]\right] - cx.$$

and it can be seen that

$$J_{N-1}(x) = \begin{cases} K + G_{N-1}(S_{N-1}) - cx & \text{for } x < s_{N-1}, \\ G_{N-1}(x) - cx & \text{for } x \geq s_{N-1}, \end{cases} \tag{2.8}$$

where S_{N-1} minimizes $G_{N-1}(y)$ and s_{N-1} is the smallest value of y for which $G_{N-1}(y) = K + G_{N-1}(S_{N-1})$. Note that since $K > 0$, we have $s_{N-1} \neq S_{N-1}$ and furthermore the derivative of G_{N-1} at s_{N-1} is negative. As a result the left derivative of J_{N-1} at s_{N-1} is greater than the right derivative, as shown in Fig. 4.2.3, and J_{N-1} is not convex. However, we will show that J_{N-1} is K-convex based on the fact that G_{N-1} is K-convex. To this end we must verify that for all $z \geq 0$, $b > 0$, and y, we have

$$K + J_{N-1}(y + z) \geq J_{N-1}(y) + z\left(\frac{J_{N-1}(y) - J_{N-1}(y - b)}{b}\right). \tag{2.9}$$

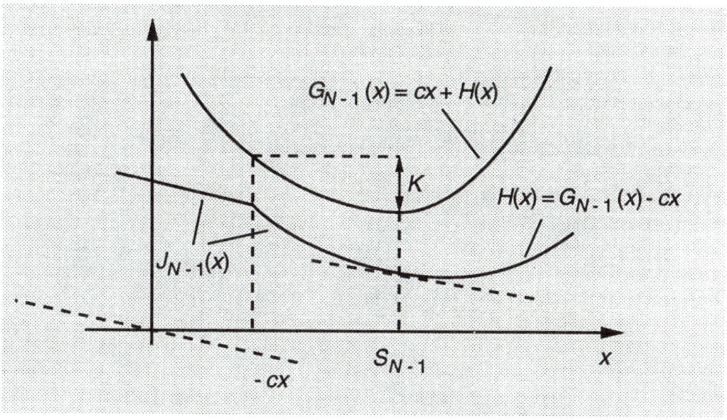

Figure 4.2.3 Structure of the cost-to-go function when fixed cost is nonzero.

We distinguish three cases:

Case 1: $y \geq s_{N-1}$. If $y - b \geq s_{N-1}$, then in this region of values of z, b, and y, the function J_{N-1}, by Eq. (2.8), is the sum of a K-convex function and a linear function. Hence by part (b) of Lemma 2.1, it is K-convex and Eq. (2.9) holds. If $y - b < s_{N-1}$, then in view of Eq. (2.8) we can write Eq. (2.9) as

$$K + G_{N-1}(y+z) - c(y+z) \geq G_{N-1}(y) - cy + z\left(\frac{G_{N-1}(y) - cy - G_{N-1}(s_{N-1}) + c(y-b)}{b}\right)$$

or equivalently

$$K + G_{N-1}(y+z) \geq G_{N-1}(y) + z\left(\frac{G_{N-1}(y) - G_{N-1}(s_{N-1})}{b}\right). \quad (2.10)$$

Now if y is such that $G_{N-1}(y) \geq G_{N-1}(s_{N-1})$, then by K-convexity of G_{N-1} we have

$$\begin{aligned} K + G_{N-1}(y+z) &\geq G_{N-1}(y) + z\left(\frac{G_{N-1}(y) - G_{N-1}(s_{N-1})}{y - s_{N-1}}\right) \\ &\geq G_{N-1}(y) + z\left(\frac{G_{N-1}(y) - G_{N-1}(s_{N-1})}{b}\right). \end{aligned}$$

Thus Eq. (2.10) and hence also Eq. (2.9) hold. If y is such that $G_{N-1}(y) < G_{N-1}(s_{N-1})$, then we have

$$\begin{aligned} K + G_{N-1}(y+z) &\geq K + G_{N-1}(S_{N-1}) \\ &= G_{N-1}(s_{N-1}) \\ &> G_{N-1}(y) \\ &\geq G_{N-1}(y) + z\left(\frac{G_{N-1}(y) - G_{N-1}(s_{N-1})}{b}\right). \end{aligned}$$

So for this case, Eq. (2.10) holds, and hence also the desired K-convexity Eq. (2.9) holds.

Case 2: $y \leq y + z \leq s_{N-1}$. In this region, by Eq. (2.8), the function J_{N-1} is linear and hence the K-convexity Eq. (2.9) holds.

Case 3: $y < s_{N-1} < y+z$. For this case, in view of Eq. (2.8), we can write the K-convexity Eq. (2.9) as

$$K + G_{N-1}(y+z) - c(y+z) \geq G_{N-1}(s_{N-1}) - cy + z\left(\frac{G_{N-1}(s_{N-1}) - cy - G_{N-1}(s_{N-1}) + c(y-b)}{b}\right),$$

or equivalently

$$K + G_{N-1}(y+z) \geq G_{N-1}(s_{N-1}),$$

which holds by the definition of s_{N-1}.

We have thus proved that K-convexity and continuity of G_{N-1}, together with the fact that $G_{N-1}(y) \to \infty$ as $|y| \to \infty$, imply K-convexity of J_{N-1}. In addition, J_{N-1} can be seen to be continuous. Now using Lemma 2.1, it follows from Eq. (2.5) that G_{N-2} is a K-convex function. Furthermore, by using the boundedness of w_{N-2}, it follows that G_{N-2} is continuous and, in addition, $G_{N-2}(y) \to \infty$ as $|y| \to \infty$. Repeating the preceding argument, we obtain that J_{N-2} is K-convex, and proceeding similarly, we prove K-convexity and continuity of the functions G_k for all k, as well as that $G_k(y) \to \infty$ as $|y| \to \infty$. At the same time [by using part (d) of Lemma 2.1] we prove optimality of the multiperiod (s, S) policy of Eq. (2.6).

The optimality of policies of the (s, S) type can be proved for several other inventory problems (see Exercises 4.3 to 4.10).

4.3 DYNAMIC PORTFOLIO ANALYSIS

Portfolio theory deals with the question of how to invest a certain amount of wealth among a collection of assets, perhaps over a long time interval. One approach, to be discussed in this section, is to assume that an investor makes a decision in each of several successive time periods with the objective of maximizing final wealth. We will start with an analysis of a single-period model and then extend the results to the multiperiod case.

Let x_0 denote the initial wealth of the investor and assume that there are n risky assets, with corresponding random rates of return $e_1, \ldots, e_n$ among which the investor can allocate his wealth. The investor can also invest in a riskless asset offering a sure rate of return s. If we denote by

$u_1, \ldots, u_n$ the corresponding amounts invested in the n risky assets and by $(x_0 - u_1 - \cdots - u_n)$ the amount invested in the riskless asset, the wealth at the end of the first period is given by

$$x_1 = s(x_0 - u_1 - \cdots - u_n) + \sum_{i=1}^{n} e_i u_i,$$

or equivalently

$$x_1 = s x_0 + \sum_{i=1}^{n} (e_i - s) u_i. \tag{3.1}$$

The objective is to maximize over $u_1, \ldots, u_n$,

$$E\{U(x_1)\},$$

where U is a known function, referred to as the *utility function* for the investor. We assume that U is concave and twice continuously differentiable, and that the given expected value is well defined and finite for all x_0 and u_i. We will not impose constraints on $u_1, \ldots, u_n$. This is necessary in order to obtain the results in convenient form. A few additional assumptions will be made later.

Let us denote by $u_i^* = \mu^{i^*}(x_0)$, $i = 1, \ldots, n$, the optimal amounts to be invested in the n risky assets when the initial wealth is x_0. We will show that when the utility function satisfies

$$-\frac{U'(x)}{U''(x)} = a + bx, \qquad \text{for all } x, \tag{3.3}$$

where U' and U'' denote the first and second derivatives of U, respectively, and a and b are some scalars, then the optimal portfolio is given by the linear policy

$$\mu^{i*}(x_0) = \alpha^i (a + b s x_0), \qquad i = 1, \ldots, n, \tag{3.4}$$

where α^i are some constant scalars. Furthermore, if $J(x_0)$ is the optimal value of the problem

$$J(x_0) = \max_{u_i} E\{U(x_1)\}, \tag{3.5}$$

then we have

$$-\frac{J'(x_0)}{J''(x_0)} = \frac{a}{s} + b x_0, \qquad \text{for all } x_0. \tag{3.6}$$

It can be verified that the following utility functions $U(x)$ satisfy condition (3.3):

$$\begin{aligned}
&\text{exponential}: \quad e^{-x/a}, \ \text{for } b = 0,\\
&\text{logarithmic}: \ln(x + a), \ \text{for } b = 1,\\
&\text{power}: \big(1/(b-1)\big)(a + bx)^{1-(1/b)}, \ \text{for } b \neq 0, \ b \neq 1.
\end{aligned}$$

Only concave utility functions from this class are admissible for our problem. Furthermore, if a utility function that is not defined over the whole real line is used, the problem should be formulated in a way that ensures that all possible values of the resulting final wealth are within the domain of definition of the utility function.

To show the desired relations, let us hypothesize that an optimal portfolio exists and is of the form

$$\mu^{i*}(x_0) = \alpha^i(x_0)(a + bsx_0),$$

where $\alpha^i(x_0)$, $i = 1, \ldots, n$, are some differentiable functions. We will prove that $d\alpha^i(x_0)/dx_0 = 0$ for all x_0, implying that the functions α^i must be constant, so the optimal portfolio has the linear form (3.4).

We have for every x_0 and $i = 1, \ldots, n$, by the optimality of $\mu^{i^*}(x_0)$,

$$\begin{aligned}\frac{dE\{U(x_1)\}}{du_i} &= E\left\{U'\left(sx_0 + \sum_{j=1}^{n}(e_j - s)\alpha^j(x_0)(a + bsx_0)\right)(e_i - s)\right\}\\ &= 0.\end{aligned} \tag{3.7}$$

Differentiating each of the n equations above with respect to x_0 yields

$$\begin{aligned}&E\left\{\begin{pmatrix}(e_1 - s)^2 \cdots (e_1 - s)(e_n - s)\\ \vdots\\ (e_n - s)(e_1 - s) \cdots (e_n - s)^2\end{pmatrix} U''(x_1)(a + bsx_0)\right\}\begin{pmatrix}\frac{d\alpha^1(x_0)}{dx_0}\\ \vdots\\ \frac{d\alpha^n(x_0)}{dx_0}\end{pmatrix}\\ &\quad = -\begin{pmatrix}E\big\{U''(x_1)(e_1 - s)s\big(1 + \sum_{i=1}^{n}(e_i - s)\alpha^i(x_0)b\big)\big\}\\ \vdots\\ E\big\{U''(x_1)(e_n - s)s\big(1 + \sum_{i=1}^{n}(e_i - s)\alpha^i(x_0)b\big)\big\}\end{pmatrix}.\end{aligned} \tag{3.8}$$

Using relation (3.4), we have

$$\begin{aligned}U''(x_1) &= -\frac{U'(x_1)}{a + b\big(sx_0 + \sum_{i=1}^{n}(e_i - s)\alpha^i(x_0)(a + bsx_0)\big)}\\ &= -\frac{U'(x_1)}{(a + bsx_0)\big(1 + \sum_{i=1}^{n}(e_i - s)\alpha^i(x_0)b\big)}.\end{aligned} \tag{3.9}$$

Substituting in Eq. (3.8) and using Eq. (3.7), we have that the right-hand side of Eq. (3.8) is the zero vector. The matrix on the left in Eq. (3.8), except for degenerate cases, can be shown to be nonsingular. Assuming that it is indeed nonsingular, we obtain

$$\frac{d\alpha^i(x_0)}{dx_0} = 0, \qquad i = 1, \ldots, n,$$

and $\alpha^i(x_0) = \alpha^i$, where α^i are some constants, thus proving the optimality of the linear policy (3.4).

We now prove Eq. (3.6). We have

$$J(x_0) = E\{U(x_1)\}$$
$$= E\left\{U\left(s\left(1+\sum_{i=1}^{n}(e_i - s)\alpha^i b\right)x_0 + \sum_{i=1}^{n}(e_i - s)\alpha^i a\right)\right\}$$

and hence

$$J'(x_0) = E\left\{U'(x_1)s\left(1+\sum_{i=1}^{n}(e_i - s)\alpha^i b\right)\right\},$$

$$J''(x_0) = E\left\{U''(x_1)s^2\left(1+\sum_{i=1}^{n}(e_i - s)\alpha^i b\right)^2\right\}. \tag{3.10}$$

The last relation after some calculation using Eq. (3.9) yields

$$J''(x_0) = -\frac{E\{U'(x_1)s\big(1+\sum_{i=1}^{n}(e_i - s)\alpha^i b\big)\}s}{a + bsx_0}. \tag{3.11}$$

By combining Eqs. (3.10) and (3.11), we obtain the desired result:

$$-\frac{J'(x_0)}{J''(x_0)} = \frac{a}{s} + bx_0. \tag{3.12}$$

The Multiperiod Problem

We now extend the preceding one-period analysis to the multiperiod case. We will assume that the current wealth can be reinvested at the beginning of each of N consecutive time periods. We denote

x_k: the wealth of the investor at the start of the kth period,

u_i^k: the amount invested at the start of the kth period in the ith risky asset,

e_i^k: the rate of return of the ith risky asset during the kth period,

s_k: the rate of return of the riskless asset during the kth period.

We have the system equation

$$x_{k+1} = s_k x_k + \sum_{i=1}^{n}(e_i^k - s_k)u_i^k, \qquad k = 0, 1, \ldots, N-1. \tag{3.13}$$

We assume that the vectors $e^k = (e_1^k, \ldots, e_n^k)$, $k = 0, \ldots, N-1$, are independent with given probability distributions that yield finite expected values throughout the following analysis. The objective is to maximize $E\{U(x_N)\}$, the expected utility of the terminal wealth x_N, where we assume that U satisfies

$$-\frac{U'(x)}{U''(x)} = a + bx, \qquad \text{for all } x.$$

Applying the DP algorithm to this problem, we have

$$J_N(x_N) = U(x_N), \tag{3.14}$$

$$J_k(x_k) = \max_{u_1^k, \ldots, u_n^k} E\left\{ J_{k+1}\left(s_k x_k + \sum_{i=1}^{n} (e_i^k - s_k) u_i^k \right) \right\}. \tag{3.15}$$

From the solution of the one-period problem, we have that the optimal policy at period $N-1$ has the form

$$\mu_{N-1}^*(x_{N-1}) = \alpha_{N-1}(a + b s_{N-1} x_{N-1}),$$

where α_{N-1} is an appropriate n-dimensional vector. Furthermore, we have

$$-\frac{J'_{N-1}(x)}{J''_{N-1}(x)} = \frac{a}{s_{N-1}} + bx. \tag{3.16}$$

Hence, applying the one-period result in the DP Eq. (3.15) for the next to the last period, we obtain the optimal policy

$$\mu_{N-2}^*(x_{N-2}) = \alpha_{N-2}\left(\frac{a}{s_{N-1}} + b s_{N-2} x_{N-2} \right),$$

where α_{N-2} is again an appropriate n-dimensional vector.

Proceeding similarly, we have for the kth period

$$\mu_k^*(x_k) = \alpha_k \left(\frac{a}{s_{N-1} \cdots s_{k+1}} + b s_k x_k \right), \tag{3.17}$$

where α_k is an n-dimensional vector that depends on the probability distributions of the rates of return e_i^k of the risky assets and are determined by maximization in the DP Eq. (3.15). The corresponding cost-to-go functions satisfy

$$-\frac{J'_k(x)}{J''_k(x)} = \frac{a}{s_{N-1} \cdots s_k} + bx, \qquad k = 0, 1, \ldots, N-1. \tag{3.18}$$

Thus it is seen that the investor, when faced with the opportunity to sequentially reinvest his wealth, uses a policy similar to that of the single-period case. Carrying the analysis one step further, it is seen that if the utility function U is such that $a = 0$, that is, U has one of the forms

$$\ln x, \qquad \text{for } b = 1,$$
$$\left(\frac{1}{b-1}\right)(bx)^{1-(1/b)}, \qquad \text{for } b \neq 0,\ b \neq 1,$$

then it follows from Eq. (3.17) that the investor acts at each stage k as if he were faced with a *single-period* investment characterized by the rates of return s_k and e_i^k, and the objective function $E\{U(x_{k+1})\}$. This policy whereby the investor can ignore the fact that he will have the opportunity to reinvest his wealth is called a *myopic policy*.

Note that a myopic policy is also optimal when $s_k = 1$ for all k, which means that wealth is discounted at the rate of return of the riskless asset. Furthermore, it can be proved that when $a = 0$ a myopic policy is optimal even in the more general case where the rates of return s_k are independent random variables [Mos68], and for the case where forecasts on the probability distributions of the rates of return e_i^k of the risky assets become available during the investment process (see Exercise 4.14).

It turns out that even in the more general case where $a \neq 0$ only a small amount of foresight is required on the part of the decision maker. It can be seen [compare Eqs. (3.15) to (3.18)] that the optimal policy (3.17) at period k is the one that the investor would use in a single-period problem to maximize over u_i^k, $i = 1, \ldots, n$,

$$E\{U(s_{N-1} \cdots s_{k+1} x_{k+1})\}$$

subject to

$$x_{k+1} = s_k x_k + \sum_{i=1}^{n} (e_i^k - s_k) u_i^k.$$

In other words, the investor at period k should maximize the expected utility of wealth that results if amounts u_i^k are invested in the risky assets in period k and the resulting wealth x_{k+1} is subsequently invested *exclusively* in the riskless asset during the remaining periods $k+1, \ldots, N-1$. This is known as a *partially myopic policy*. Such a policy can also be shown to be optimal when forecasts on the probability distributions of the rates of return of the risky assets become available during the investment process (see Exercise 4.14).

Another interesting aspect of the case where $a \neq 0$ is that if $s_k > 1$ for all k, then as the horizon becomes increasingly long ($N \to \infty$), the policy in the initial stages approaches a myopic policy [compare Eqs. (3.17) and (3.18)]. Thus, for $s_k > 1$, a partially myopic policy becomes asymptotically myopic as the horizon tends to infinity.

4.4 OPTIMAL STOPPING PROBLEMS

Optimal stopping problems of the type that we will consider in this and subsequent sections are characterized by the availability, at each state, of a control that stops the evolution of the system. Thus at each stage the decision maker observes the current state of the system and decides whether to continue the process (perhaps at a certain cost) or stop the process and incur a certain loss. If the decision is to continue, a control must be selected from a given set of available choices. If there is only one choice other than stopping, then each policy is characterized at each period by the *stopping set*, that is, the set of states where the policy stops the system.

Asset Selling

As a first example, consider a person having an asset (say a piece of land) for which he is offered an amount of money from period to period. We assume that the offers, denoted $w_0, w_1, \ldots, w_{N-1}$, are random and independent, and take values within some bounded interval. We consider a horizon of N stages and assume that if the person accepts the offer, he can invest the money at a fixed rate of interest $r > 0$, and if he rejects the offer, he waits until the next period to consider the next offer. Offers rejected are not renewed, and we assume that the last offer w_{N-1} must be accepted if every prior offer has been rejected. The objective is to find a policy for accepting and rejecting offers that maximizes the revenue of the person at the Nth period.

The DP algorithm for this problem can be derived by elementary reasoning. As a modeling exercise, however, we will try to embed the problem in the framework of the basic problem by defining the state space, control space, disturbance space, system equation, and cost function. We consider as disturbance at time k the random offer w_k and as corresponding disturbance space the real line. The control space consists of two elements u^1 and u^2, which correspond to the decisions "sell" and "do not sell," respectively. We define the state space to be the real line, augmented with an additional state (call it T), which is a *termination state*. The system moves into the termination state as soon as the asset is sold. By writing that the system is at a state $x_k \neq T$ at time k, we mean that the asset has not been sold as yet and the current offer under consideration is equal to x_k. By writing that the system is at state $x_k = T$ at time k, we mean that the asset has already been sold.

With these conventions we may write a system equation of the form

$$x_{k+1} = f_k(x_k, u_k, w_k), \qquad k = 0, 1, \ldots, N-1,$$

where the state space is the real line augmented by the termination state T, and the function f_k is defined via the relation

$$x_{k+1} = \begin{cases} T & \text{if } x_k = T, \text{ or if } x_k \neq T \text{ and } u_k = u^1 \text{ (sell)}, \\ w_k & \text{otherwise.} \end{cases}$$

The corresponding reward function may be written as

$$\mathop{E}_{\substack{w_k \\ k=0,1,\ldots,N-1}} \left\{ g_N(x_N) + \sum_{k=0}^{N-1} g_k(x_k, u_k, w_k) \right\}$$

where

$$g_N(x_N) = \begin{cases} x_N & \text{if } x_N \neq T, \\ 0 & \text{otherwise,} \end{cases}$$

$$g_k(x_k, u_k, w_k) = \begin{cases} (1+r)^{N-k} x_k & \text{if } x_k \neq T \text{ and } u_k = u^1 \text{ (sell)}, \\ 0 & \text{otherwise.} \end{cases}$$

Based on this formulation we can write the corresponding DP algorithm:

$$J_N(x_N) = \begin{cases} x_N & \text{if } x_N \neq T, \\ 0 & \text{if } x_N = T, \end{cases} \tag{4.1}$$

$$J_k(x_k) = \begin{cases} \max\big[(1+r)^{N-k} x_k,\ E\{J_{k+1}(w_k)\}\big] & \text{if } x_k \neq T, \\ 0 & \text{if } x_k = T. \end{cases} \tag{4.2}$$

In the above equation, $(1+r)^{N-k} x_k$ is the revenue resulting from decision u^1 (sell) when the offer is x_k, and $E\{J_{k+1}(w_k)\}$ represents the expected revenue corresponding to the decision u^2 (do not sell). Thus, the optimal policy is to accept an offer if it is greater than $E\{J_{k+1}(w_k)\}/(1+r)^{N-k}$, which represents expected revenue discounted to the present time:

$$\text{accept the offer } x_k \qquad \text{if } x_k > \alpha_k,$$

$$\text{reject the offer } x_k \qquad \text{if } x_k < \alpha_k,$$

where

$$\alpha_k = \frac{E\{J_{k+1}(w_k)\}}{(1+r)^{N-k}}.$$

When $x_k = \alpha_k$, both acceptance and rejection are optimal. Thus the optimal policy is determined by the scalar sequence $\{\alpha_k\}$ (see Fig. 4.4.1).

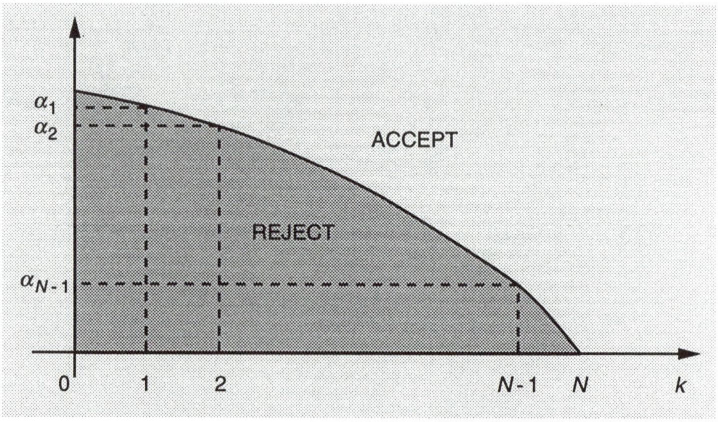

Figure 4.4.1 Optimal policy for accepting offers in the asset selling problem.

Properties of the Optimal Policy

We will now derive some properties of the optimal policy with some further analysis. Let us assume that the offers w_k are identically distributed, and to simplify notation, let us drop the time index k and denote by $E_w\{\cdot\}$ the expected value of the corresponding expression over w_k, for all k. We will then show that

$$\alpha_k \geq \alpha_{k+1}, \qquad \text{for all } k,$$

which expresses the intuitive fact that if an offer is good enough to be acceptable at time k, it should also be acceptable at time $k+1$ when there will be one less chance for improvement. We will also obtain an equation for the limit of α_k as $k \to -\infty$.

Let us introduce the functions

$$V_k(x_k) = \frac{J_k(x_k)}{(1+r)^{N-k}}, \qquad x_k \neq T.$$

It can be seen from Eqs. (4.1) and (4.2) that

$$V_N(x_N) = x_N, \tag{4.3}$$

$$V_k(x_k) = \max\left[x_k,\ (1+r)^{-1} \mathop{E}_{w}\{V_{k+1}(w)\}\right], \qquad k = 0, 1, \ldots, N-1, \tag{4.4}$$

and that

$$\alpha_k = \frac{E_w\{V_{k+1}(w)\}}{1+r}.$$

To prove that $\alpha_k \geq \alpha_{k+1}$, note that from Eqs. (4.3) and (4.4), we have

$$V_{N-1}(x) \geq V_N(x), \qquad \text{for all } x \geq 0.$$

Applying Eq. (4.4) for $k = N-2$ and $k = N-1$, and using the preceding inequality, we obtain for all $x \geq 0$

$$\begin{aligned} V_{N-2}(x) &= \max\left[x,\ (1+r)^{-1} \underset{w}{E}\{V_{N-1}(w)\}\right] \\ &\geq \max\left[x,\ (1+r)^{-1} \underset{w}{E}\{V_N(w)\}\right] \\ &= V_{N-1}(x). \end{aligned}$$

Continuing in the same manner, we see that

$$V_k(x) \geq V_{k+1}(x), \qquad \text{for all } x \geq 0 \text{ and } k.$$

Since $\alpha_k = E_w\{V_{k+1}(w)\}/(1+r)$, we obtain $\alpha_k \geq \alpha_{k+1}$, as desired.

Let us now see what happens when the horizon N is very large. From the algorithm (4.3) and (4.4) we have

$$V_k(x_k) = \max(x_k, \alpha_k). \tag{4.5}$$

Hence we obtain

$$\begin{aligned} \alpha_k &= \frac{1}{1+r} E_w\{V_{k+1}(w)\} \\ &= \frac{1}{1+r} \int_0^{\alpha_{k+1}} \alpha_{k+1} dP(w) + \frac{1}{1+r} \int_{\alpha_{k+1}}^{\infty} w dP(w), \end{aligned}$$

where the function P is defined for all scalars λ by

$$P(\lambda) = \text{Prob}\{w < \lambda\}.$$

The difference equation for α_k may also be written as

$$\alpha_k = \frac{P(\alpha_{k+1})}{1+r} \alpha_{k+1} + \frac{1}{1+r} \int_{\alpha_{k+1}}^{\infty} w dP(w), \qquad \text{for all } k, \tag{4.6}$$

with $\alpha_N = 0$.

Now since we have

$$0 \leq \frac{P(\alpha)}{1+r} \leq \frac{1}{1+r} < 1, \qquad \text{for all } \ \alpha \geq 0,$$

$$0 \leq \frac{1}{1+r} \int_{\alpha_{k+1}}^{\infty} w dP(w) \leq \frac{E\{w\}}{1+r}, \qquad \text{for all } k,$$

it can be seen, using the property $\alpha_k \geq \alpha_{k+1}$, that the sequence $\{\alpha_k\}$ generated (backward) by the difference equation (4.6) converges (as $k \to -\infty$) to a constant $\overline{\alpha}$ satisfying

$$(1+r)\overline{\alpha} = P(\overline{\alpha})\overline{\alpha} + \int_{\overline{\alpha}}^{\infty} w dP(w).$$

This equation is obtained from Eq. (4.6) by taking limits as $k \to -\infty$ and by using the fact that P is continuous from the left.

Thus, when the horizon N tends to become longer, the optimal policy for every fixed $k \geq 1$ can be approximated by the stationary policy

$$\text{accept the offer } x_k \qquad \text{if } x_k > \overline{\alpha},$$
$$\text{reject the offer } x_k \qquad \text{if } x_k < \overline{\alpha}.$$

The optimality of this policy for the corresponding infinite horizon problem will be shown in Section 7.3.

Purchasing with a Deadline

Let us consider another stopping problem that has similar nature. Assume that a certain quantity of raw material is required by a certain time. If the price of this material fluctuates, there arises the problem of deciding whether to purchase at the current price or wait a further period, during which the price may go up or down. We thus want to minimize the expected price of purchase. We assume that successive prices w_k are independent and identically distributed with distribution $P(w_k)$, and that the purchase must be made within N time periods.

This problem and the earlier asset selling problem have obvious similarities. Let us denote by

$$x_{k+1} = w_k$$

the price prevailing at the beginning of period $k+1$. We have similar to the earlier problem the DP algorithm

$$J_N(x_N) = x_N,$$

$$J_k(x_k) = \min\big[x_k,\ E\{J_{k+1}(w_k)\}\big].$$

Note that $J_k(x_k)$ is the optimal cost-to-go when the current price is x_k *and* the material has not been purchased yet. To be strictly formal, we should introduce a termination state T, to which the system moves following a purchasing decision and at which the system subsequently stays at no cost. A nonzero cost is incurred only when moving from x_k to T; this cost is equal to x_k. Thus the cost-to-go from the termination state T is 0, and for this reason it was neglected in the preceding DP equation.

The optimal policy is given by

$$\text{purchase if } x_k < \alpha_k,$$

$$\text{do not purchase if } x_k > \alpha_k,$$

where

$$\alpha_k = E\{J_{k+1}(w_k)\}.$$

Similar to the asset selling problem, the thresholds $\alpha_1, \alpha_2, \ldots, \alpha_{N-1}$ can be obtained from the discrete-time equation

$$\alpha_k = \alpha_{k+1}\big(1 - P(\alpha_{k+1})\big) + \int_0^{\alpha_{k+1}} w dP(w),$$

with the terminal condition

$$\alpha_{N-1} = \int_0^{\infty} w dP(w) = E\{w\}.$$

The Case of Correlated Prices

Consider now a variation of the purchasing problem where we do not assume that the successive prices $w_0, \ldots, w_{N-1}$ are independent. Instead, we assume that they are correlated and can be represented as the state of a linear system driven by independent disturbances (cf. Section 1.4). In particular, we have

$$w_k = x_{k+1}, \qquad k = 0, 1, \ldots, N-1,$$

with

$$x_{k+1} = \lambda x_k + \xi_k, \qquad x_0 = 0,$$

where λ is a scalar with $0 \leq \lambda < 1$ and $\xi_0, \xi_1, \ldots, \xi_{N-1}$ are independent identically distributed random variables taking positive values with given probability distribution. As discussed in Section 1.4, the DP algorithm under these circumstances has the form

$$J_N(x_N) = x_N,$$

$$J_k(x_k) = \min\big[x_k,\, E\{J_{k+1}(\lambda x_k + \xi_k)\}\big],$$

where the cost associated with the purchasing decision is x_k and the cost associated with the waiting decision is $E\{J_{k+1}(\lambda x_k + \xi_k)\}$.

We will show that in this case the optimal policy is of the same type as the one for independent prices. Indeed, we have

$$J_{N-1}(x_{N-1}) = \min\big[x_{N-1},\, \lambda x_{N-1} + \bar{\xi}\,\big],$$

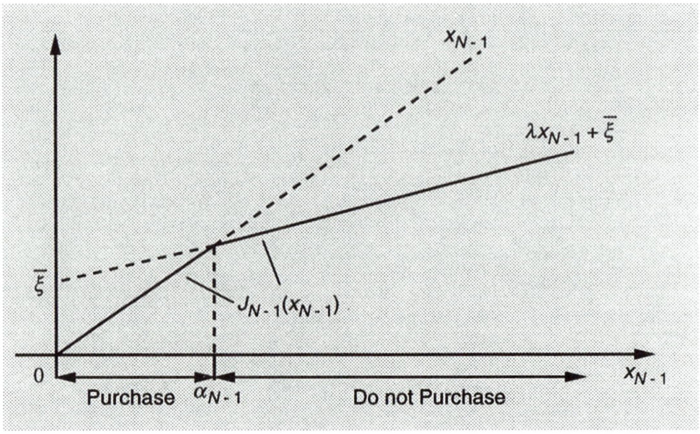

Figure 4.4.2 Structure of the cost-to-go function $J_{N-1}(x_{N-1})$ when prices are correlated.

where $\overline{\xi} = E\{\xi_{N-1}\}$. As shown in Fig. 4.4.2, an optimal policy at time $N-1$ is given by

$$\text{purchase if } x_{N-1} < \alpha_{N-1},$$

$$\text{do not purchase if } x_{N-1} > \alpha_{N-1},$$

where α_{N-1} is defined from the equation $\alpha_{N-1} = \lambda\alpha_{N-1} + \overline{\xi}$; that is,

$$\alpha_{N-1} = \frac{\overline{\xi}}{1-\lambda}.$$

Note that

$$J_{N-1}(x) \le J_N(x), \qquad \text{for all } x,$$

and that J_{N-1} is concave and increasing in x. Using this fact in the DP algorithm, one may show (the monotonicity property of DP; Exercise 1.23 in Chapter 1) that

$$J_k(x) \le J_{k+1}(x), \qquad \text{for all } x \text{ and } k,$$

and that J_k is concave and increasing in x for all k. Furthermore, since $\overline{\xi} = E\{\xi_k\} > 0$ for all k, one can show that

$$E\big\{J_{k+1}(\xi_k)\big\} > 0, \qquad \text{for all } k.$$

These facts imply (as illustrated in Fig. 4.4.3) that the optimal policy for every period k is of the form

$$\text{purchase if } x_k < \alpha_k,$$

$$\text{do not purchase if } x_k > \alpha_k,$$

where the scalar α_k is obtained as the unique positive solution of the equation

$$x = E\{J_{k+1}(\lambda x + \xi_k)\}.$$

Note that the relation $J_k(x) \leq J_{k+1}(x)$ for all x and k implies that

$$\alpha_{k-1} \leq \alpha_k \leq \alpha_{k+1}, \qquad \text{for all } k,$$

and hence (as one would expect) the threshold price to purchase increases as the deadline gets closer.

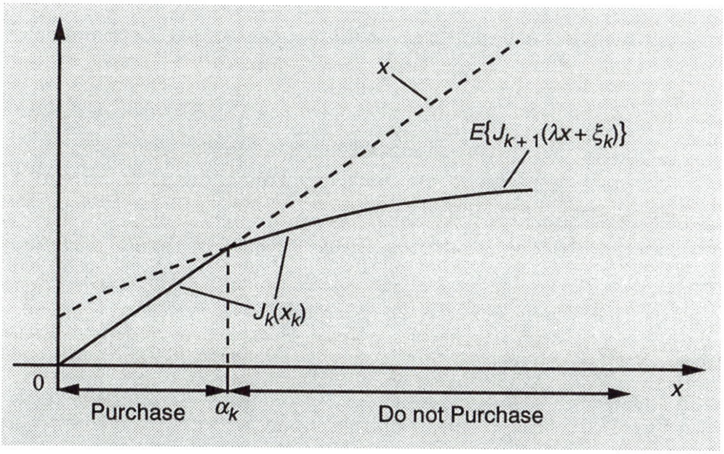

Figure 4.4.3 Determining the optimal policy when prices are correlated.

General Stopping Problems and the One-Step-Look-Ahead Rule

We now formulate a general type of N-stage problem where stopping is mandatory at or before stage N. Consider the stationary version of the basic problem of Chapter 1 (state, control, and disturbance spaces, disturbance distribution, control constraint set, and cost per stage are the same for all times). Assume that at each state x_k and at time k there is available, in addition to the controls $u_k \in U(x_k)$, a stopping action that forces the system to enter a termination state at a cost $t(x_k)$ and subsequently remain there at no cost. The terminal cost, assuming stopping has not occurred by the last stage, is $t(x_N)$. Thus, in effect, we assume that the termination cost will always be incurred either at the last stage or earlier.

The DP algorithm is given by

$$J_N(x_N) = t(x_N) \tag{4.7}$$

$$J_k(x_k) = \min \left[t(x_k),\ \min_{u_k \in U(x_k)} E\big\{g(x_k, u_k, w_k) + J_{k+1}\big(f(x_k, u_k, w_k)\big)\big\} \right], \tag{4.8}$$

and it is optimal to stop at time k for states x in the set

$$T_k = \left\{ x \;\middle|\; t(x) \le \min_{u \in U(x)} E\big\{g(x, u, w) + J_{k+1}\big(f(x, u, w)\big)\big\} \right\}.$$

We have from Eqs. (4.7) and (4.8)

$$J_{N-1}(x) \le J_N(x), \qquad \text{for all } x,$$

and using this fact in the DP equation (4.8), we obtain inductively

$$J_k(x) \le J_{k+1}(x), \qquad \text{for all } x \text{ and } k.$$

[We are making use here of the stationarity of the problem and the monotonicity property of DP (Exercise 1.23 in Chapter 1).] Using this fact and the definition of T_k we see that

$$T_0 \subset \ldots \subset T_k \subset T_{k+1} \subset \ldots \subset T_{N-1}. \tag{4.9}$$

We will now consider a condition guaranteeing that all the stopping sets T_k are equal. Suppose that the set T_{N-1} is *absorbing* in the sense that if a state belongs to T_{N-1} and termination is not selected, the next state will also be in T_{N-1}; that is,

$$f(x, u, w) \in T_{N-1}, \qquad \text{for all } x \in T_{N-1}, \quad u \in U(x), \quad w. \tag{4.10}$$

We will show that equality holds in Eq. (4.9) and for all k we have

$$T_k = T_{N-1} = \left\{ x \in S \;\middle|\; t(x) \le \min_{u \in U(x)} E\big\{g(x, u, w) + t\big(f(x, u, w)\big)\big\} \right\}.$$

To see this, note that by the definition of T_{N-1}, we have

$$J_{N-1}(x) = t(x), \qquad \text{for all } x \in T_{N-1},$$

and using Eq. (4.10) we obtain for $x \in T_{N-1}$

$$\begin{aligned} \min_{u \in U(x)} & E\big\{g(x, u, w) + J_{N-1}\big(f(x, u, w)\big)\big\} \\ &= \min_{u \in U(x)} E\big\{g(x, u, w) + t\big(f(x, u, w)\big)\big\} \\ &\ge t(x). \end{aligned}$$

Therefore, stopping is optimal for all $x_{N-2} \in T_{N-1}$ or equivalently $T_{N-1} \subset T_{N-2}$. This together with Eq. (4.9) implies $T_{N-2} = T_{N-1}$. Proceeding similarly, we obtain $T_k = T_{N-1}$ for all k.

In conclusion, *if condition (4.10) holds (the one-step stopping set T_{N-1} is absorbing), then the stopping sets T_k are all equal to the set of states for which it is better to stop rather than continue for one more stage and then stop.* A policy of this type is known as a *one-step-look-ahead policy*. Such a policy turns out to be optimal in several types of applications. We provide next some examples. Additional examples are given in the exercises, and in Vol. II, Chapter 3, where the stopping problem will be reexamined in an infinite horizon context.

Example 4.1 (Asset Selling with Past Offers Retained)

Consider the asset selling problem discussed earlier in this section with the difference that rejected offers can be accepted at a later time. Then if the asset is not sold at time k the state evolves according to

$$x_{k+1} = \max(x_k, w_k)$$

instead of $x_{k+1} = w_k$. The DP equations (4.3) and (4.4) then become

$$V_N(x_N) = x_N,$$

$$V_k(x_k) = \max\left[x_k, (1+r)^{-1}E\left\{V_{k+1}\big(\max(x_k, w_k)\big)\right\}\right].$$

The one-step stopping set is

$$T_{N-1} = \left\{x \mid x \geq (1+r)^{-1}E\left\{\max(x, w)\right\}\right\}.$$

It is seen [compare with Eqs. (4.5) and (4.6)] that an alternative characterization is

$$T_{N-1} = \{x \mid x \geq \overline{\alpha}\}, \tag{4.11}$$

where $\overline{\alpha}$ is obtained from the equation

$$(1+r)\overline{\alpha} = P(\overline{\alpha})\overline{\alpha} + \int_{\overline{\alpha}}^{\infty} w dp(w).$$

Since past offers can be accepted at a later date, the effective offer available cannot decrease with time, and it follows that the one-step stopping set (4.11) is absorbing in the sense of Eq. (4.10). Therefore, the one-step-look-ahead stopping rule that accepts the first offer that equals or exceeds $\overline{\alpha}$ is optimal. Note that this policy is independent of the horizon length N.

Example 4.2 (The Rational Burglar [Whi82])

A burglar may at any night k choose to retire with his accumulated earnings x_k or enter a house and bring home an amount w_k. However, in the latter case he gets caught with probability p, and then he is forced to terminate his activities and forfeit his earnings thus far. The amounts w_k are independent, identically distributed with mean $\overline{w}$. The problem is to find a policy that maximizes the burglar's expected earnings over N nights.

We can formulate this problem as a stopping problem with two actions (retire or continue) and a state space consisting of the real line, the retirement state, and a special state corresponding to the burglar getting caught. The DP algorithm is given by

$$J_N(x_N) = x_N$$

$$J_k(x_k) = \max\left[x_k,\ (1-p)E\left\{J_{k+1}(x_k + w_k)\right\}\right].$$

The one-step stopping set is

$$T_{N-1} = \{x \mid x \geq (1-p)(x+\overline{w})\} = \left\{x \;\middle|\; x \geq \frac{(1-p)\overline{w}}{p}\right\},$$

(more accurately this set together with the special state corresponding to the burglar's arrest). Since this set is absorbing in the sense of Eq. (4.10), we see that the one-step-look-ahead policy by which the burglar retires when his earnings reach or exceed $(1-p)\overline{w}/p$ is optimal. The optimality of this policy for the corresponding infinite horizon problem will be demonstrated in Vol. II, Chapter 3.

4.5 SCHEDULING AND THE INTERCHANGE ARGUMENT

Suppose one has a collection of tasks to perform but the ordering of the tasks is subject to optimal choice. As examples, consider the ordering of operations in a construction project so as to minimize construction time or the scheduling of jobs in a workshop so as to minimize machine idle time. In such problems a useful technique is to start with some schedule and then to interchange two adjacent tasks and see what happens. We first provide some examples, and we then formalize mathematically the technique.

Example 5.1 (A Quiz Contest)

Consider a quiz contest where a person is given a list of N questions and can answer these questions in any order he chooses. Question i will be answered correctly with probability p_i, and the person will then receive a reward R_i. At the first incorrect answer, the quiz terminates and the person is allowed to keep his previous rewards. The problem is to choose the ordering of questions so as to maximize expected rewards.

Let i and j be the kth and $(k+1)$st questions in an optimally ordered list

$$L = (i_0, \ldots, i_{k-1}, i, j, i_{k+2}, \ldots, i_{N-1}).$$

Consider the list

$$L' = (i_0, \ldots, i_{k-1}, j, i, i_{k+2}, \ldots, i_{N-1})$$

obtained from L by interchanging the order of questions i and j. We compare the expected rewards of L and L'. We have

$$\begin{aligned} E\{\text{reward of } L\} = {} & E\big\{\text{reward of } \{i_0, \ldots, i_{k-1}\}\big\} \\ & + p_{i_0} \cdots p_{i_{k-1}} (p_i R_i + p_i p_j R_j) \\ & + p_{i_0} \cdots p_{i_{k-1}} p_i p_j E\big\{\text{reward of } \{i_{k+2}, \ldots, i_{N-1}\}\big\} \end{aligned}$$

$$\begin{aligned} E\{\text{reward of } L'\} = E\{&\text{reward of } \{i_0, \ldots, i_{k-1}\}\} \\ &+ p_{i_0} \cdots p_{i_{k-1}} (p_j R_j + p_j p_i R_i) \\ &+ p_{i_0} \cdots p_{i_{k-1}} p_j p_i E\{\text{reward of } \{i_{k+2}, \ldots, i_{N-1}\}\}. \end{aligned}$$

Since L is optimally ordered, we have

$$E\{\text{reward of } L\} \geq E\{\text{reward of } L'\},$$

so it follows from these equations that

$$p_i R_i + p_i p_j R_j \geq p_j R_j + p_j p_i R_i$$

or equivalently

$$\frac{p_i R_i}{1 - p_i} \geq \frac{p_j R_j}{1 - p_j}.$$

Therefore, to maximize expected rewards, questions should be answered in decreasing order of $p_i R_i / (1 - p_i)$.

Example 5.2 (Job Scheduling on a Single Processor)

Suppose we have N jobs to process in sequential order with the ith job requiring a random time T_i for its execution. The times $T_1, \ldots, T_N$ are independent. If job i is completed at time t, the reward is $\alpha^t R_i$, where α is a discount factor with $0 < \alpha < 1$. The problem is to find a schedule that maximizes the total expected reward.

It can be seen that the state for this problem is just the collection of jobs yet to be processed. Indeed, because the execution times T_i are independent, and also because future costs are multiplicatively affected through discounting by the completion times of preceding jobs, the optimization of the scheduling of future jobs is unaffected by the completion times of preceding jobs. As a result these times need not be included in the state; this would not be so if either the times T_i were correlated or if the reward for completing job i at time t were not $\alpha^t R_i$ but instead had a general dependence on t. Now, given that the state is the collection of jobs yet to be processed, it is clear that an optimal policy can be mapped into an optimal job schedule $(i_0, \ldots, i_{N-1})$.

Suppose that $L = (i_0, \ldots, i_{k-1}, i, j, i_{k+2}, \ldots, i_{N-1})$ is an optimal job schedule, and consider the schedule $L' = (i_0, \ldots, i_{k-1}, j, i, i_{k+2}, \ldots, i_{N-1})$ obtained by interchanging i and j. Let t_k be the time of completion of job i_{k-1}. We compare the rewards of the schedules L and L', similar to the preceding example. Since the reward for completing the remaining jobs $i_{k+2}, \ldots, i_{N-1}$ is independent of the order in which jobs i and j are executed, we obtain

$$E\{\alpha^{t_k + T_i} R_i + \alpha^{t_k + T_i + T_j} R_j\} \geq E\{\alpha^{t_k + T_j} R_j + \alpha^{t_k + T_j + T_i} R_i\}.$$

Since t_k, T_i, and T_j are independent, this relation can be written as

$$\begin{aligned} E\{\alpha^{t_k}\}&\big(E\{\alpha^{T_i}\} R_i + E\{\alpha^{T_i}\} E\{\alpha^{T_j}\} R_j\big) \\ &\geq E\{\alpha^{t_k}\}\big(E\{\alpha^{T_j}\} R_j + E\{\alpha^{T_j}\} E\{\alpha^{T_i}\} R_i\big), \end{aligned}$$

from which we finally obtain

$$\frac{E\{\alpha^{T_i}\}R_i}{1 - E\{\alpha^{T_i}\}} \geq \frac{E\{\alpha^{T_j}\}R_j}{1 - E\{\alpha^{T_j}\}}.$$

It follows that scheduling jobs in order of decreasing $E\{\alpha^{T_i}\}R_i/\big(1 - E\{\alpha^{T_i}\}\big)$ maximizes expected rewards. The structure of the optimal policy is identical with the one we derived for the preceding quiz contest example (identify $E\{\alpha^{T_i}\}$ with the probability p_i of answering correctly question i).

Example 5.3 (Job Scheduling on Two Processors in Series)

Consider the scheduling of N jobs on two processors A and B, such that B accepts the output of A as input. Job i requires known times a_i and b_i for processing in A and B, respectively. The problem is to find a schedule that minimizes the total processing time.

To formulate the problem into the form of the basic problem, we increment discrete time at the moments when processing of a job is completed at machine A and the next job is started. We take as state at time k the collection of jobs X_k that remain to be processed at A, together with the backlog of work τ_k at machine B, that is, the amount of time that the jobs currently at B need to clear B. Thus if (X_k, τ_k) is the state at stage k and job i is completed at machine A, the state changes to (X_{k+1}, τ_{k+1}) given by

$$X_{k+1} = X_k - \{i\}, \qquad \tau_{k+1} = b_i + \max(0, \tau_k - a_i).$$

The corresponding DP algorithm is

$$J_k(X_k, \tau_k) = \min_{i \in X_k} \big[a_i + J_{k+1}\big(X_k - \{i\},\, b_i + \max(0, \tau_k - a_i)\big)\big]$$

with the terminal condition

$$J_N(\emptyset, \tau_N) = \tau_N,$$

where $\emptyset$ is the empty set.

Since the problem is deterministic, there exists an optimal open-loop schedule

$$\{i_0, \ldots, i_{k-1}, i, j, i_{k+2}, \ldots, i_{N-1}\}.$$

By arguing that the cost of this schedule is no worse than the cost of the schedule

$$\{i_0, \ldots, i_{k-1}, j, i, i_{k+2}, \ldots, i_{N-1}\},$$

obtained by interchanging i and j, it can be verified that

$$J_{k+2}\big(X_k - \{i\} - \{j\}, \tau_{ij}\big) \leq J_{k+2}\big(X_k - \{i\} - \{j\}, \tau_{ji}\big), \tag{5.1}$$

where τ_{ij} and τ_{ji} are the backlogs at machine B at time $k+2$ when i is processed before j and j is processed before i, respectively, and the backlog at time k was τ_k. A straightforward calculation shows that

$$\tau_{ij} = b_i + b_j - a_i - a_j + \max(\tau_k, a_i, a_i + a_j - b_i), \tag{5.2}$$

$$\tau_{ji} = b_j + b_i - a_j - a_i + \max(\tau_k, a_j, a_j + a_i - b_j). \tag{5.3}$$

Clearly, J_{k+2} is monotonically increasing in τ, so from Eq. (5.1) we obtain

$$\tau_{ij} \leq \tau_{ji}.$$

In view of Eqs. (5.2) and (5.3), this relation implies two possibilities. The first is

$$\tau_k \geq \max(a_i, a_i + a_j - b_i),$$

$$\tau_k \geq \max(a_j, a_j + a_i - b_j),$$

in which case $\tau_{ij} = \tau_{ji}$ and the order of i and j makes no difference. (This is the case where the backlog at time k is so large that both jobs i and j will find B working on an earlier job.) The second possibility is that

$$\max(a_i, a_i + a_j - b_i) \leq \max(a_j, a_j + a_i - b_j),$$

which can be seen to be equivalent to

$$\min(a_i, b_j) \leq \min(a_j, b_i).$$

A schedule satisfying these necessary conditions for optimality can be constructed by the following procedure:

1. Find $\min_i \min(a_i, b_i)$.
2. If the minimizing value is an a take the corresponding job first; if it is a b, take the corresponding job last.
3. Repeat the procedure with the remaining jobs until a complete schedule is constructed.

To show that this schedule is indeed optimal, we start with an optimal schedule. We consider the job i_0 that minimizes $\min(a_i, b_i)$ and by successive interchanges we move it to the same position as in the schedule constructed previously. It is seen from the preceding analysis that the resulting schedule is still optimal. Similarly, continuing through successive interchanges and maintaining optimality throughout, we can transform the optimal schedule into the schedule constructed earlier. We leave the details to the reader.

The Interchange Argument

Let us now consider the basic problem of Chapter 1 and formalize the interchange argument used in the preceding examples. The main requirement is that the problem has structure such that *there exists an open-loop policy that is optimal*, that is, a sequence of controls that performs as well as any sequence of control functions. This is certainly true in deterministic problems as discussed in Chapter 1, but it is also true in some stochastic problems such as those of Examples 5.1 and 5.2.

To apply the interchange argument, we start with an optimal sequence

$$\{u_0, \ldots, u_{k-1}, u^*, \overline{u}, u_{k+2}, \ldots, u_{N-1}\}$$

and focus attention on the controls u^* and $\overline{u}$ applied at times k and $k+1$, respectively. We then argue that if the order of u^* and $\overline{u}$ is interchanged the expected cost cannot decrease. In particular, if X_k is the set of states that can occur with positive probability starting from the given initial state x_0 and using the control subsequence $\{u_0, \ldots, u_{k-1}\}$, we must have for all $x_k \in X_k$

$$\begin{aligned} &E\{g_k(x_k, u^*, w_k) + g_{k+1}(x^*_{k+1}, \tilde{u}, w_{k+1}) + J^*_{k+2}(x^*_{k+2})\} \\ &\qquad \le E\{g_k(x_k, \tilde{u}, w_k) + g_{k+1}(\tilde{x}_{k+1}, u^*, w_{k+1}) + J^*_{k+2}(\tilde{x}_{k+2})\}, \end{aligned} \tag{5.4}$$

where x^*_{k+1} and x^*_{k+2} (or $\tilde{x}_{k+1}$ and $\tilde{x}_{k+2}$) are the states subsequent to x_k when $u_k = u^*$ and $u_{k+1} = \tilde{u}$ (or $u_k = \tilde{u}$ and $u_{k+1} = u^*$, respectively) are applied, and $J^*_{k+2}(\cdot)$ is the optimal cost-to-go function for time $k+2$.

Relation (5.4) is a *necessary* condition for optimality. It holds for every k and every optimal policy that is open-loop. There is no guarantee that this necessary condition is powerful enough to lead to an optimal solution, but it is worth considering in some specially structured problems.

4.6 NOTES, SOURCES, AND EXERCISES

The certainty equivalence principle for dynamic linear-quadratic problems was first discussed in [Sim56]. His work was preceded by [The54], who considered a single-period case, and [HMS55], who considered a deterministic case. Similar problems were independently considered in [KaK58], [JoT61], and [GuF63]. The linear-quadratic problem has received a lot of attention in control theory. See the special issue [IEE71], which contains hundreds of references.

The literature on inventory control stimulated by the pioneering paper of Arrow et al. [AHM51] is also voluminous. The 1966 survey paper

[Vei66] contains 118 references. An important work summarizing most of the research up to 1958 is [AKS58]. The proof of optimality of (s, S) policies in the case of nonzero fixed costs is due to [Sca60].

Most of the material in Section 4.3 is taken from [Mos68]. Many applications of DP in economics can be found in [Sar87] and [StL89].

The material of Section 4.4 is largely drawn from [Whi69]. Example 5.1 is given in [Ros70], Example 5.2 is given in [Ros83], and Example 5.3 is due to [WeP80]. An extensive reference on scheduling is [Pin95].

EXERCISES

4.1 (Linear-Quadratic Problems with Forecasts)

Consider the linear-quadratic problem first examined in Section 4.1 (A_k, B_k: known) for the case where at the beginning of period k we have a forecast $y_k \in \{1, 2, \ldots, n\}$ consisting of an accurate prediction that w_k will be selected in accordance with a particular probability distribution $P_{k|y_k}$ (cf. Section 1.4). The vectors w_k need not have zero mean under the distribution $P_{k|y_k}$. Show that the optimal control law is of the form

$$\mu_k(x_k, y_k) = -(B_k' K_{k+1} B_k + R_k)^{-1} B_k' K_{k+1} \big(A_k x_k + E\{w_k \mid y_k\}\big) + \alpha_k,$$

where the matrices K_k are given by the Riccati equation (1.5) and (1.6) and α_k are appropriate vectors.

4.2

Consider a scalar linear system

$$x_{k+1} = a_k x_k + b_k u_k + w_k, \qquad k = 0, 1, \ldots, N-1,$$

where $a_k, b_k \in R$, and each w_k is a Gaussian random variable with zero mean and variance σ^2. Show that the control law $\{\mu_0^*, \mu_1^*, \ldots, \mu_{N-1}^*\}$ that minimizes the cost function

$$E\left\{\exp\left[x_N^2 + \sum_{k=0}^{N-1}\big(x_k^2 + r u_k^2\big)\right]\right\}, \qquad r > 0,$$

is linear in the state variable. We assume no control constraints, independent disturbances, and that the optimal cost is finite for every x_0. Show by example that the Gaussian assumption is essential for the result to hold. (For analyses of multidimensional versions of this exercise, see [Jac73] and [Whi82].)

4.3

Consider an inventory problem similar to the problem of Section 4.2 (zero fixed cost). The only difference is that at the beginning of each period k the decision maker, in addition to knowing the current inventory level x_k, receives an accurate forecast that the demand w_k will be selected in accordance with one out of two possible probability distributions P_ℓ, P_s (large demand, small demand). The a priori probability of a large demand forecast is known (cf. Section 1.4).

(a) Obtain the optimal inventory ordering policy for the case of a single-period problem.

(b) Extend the result to the N-period case.

(c) Extend the result to the case of any finite number of possible distributions.

4.4

Consider the inventory problem of Section 4.2 (zero fixed cost), where the purchase costs c_k, $k = 0, 1, \dots, N-1$, are not initially known, but instead they are independent random variables with a priori known probability distributions. The exact value of the cost c_k, however, becomes known to the decision maker at the beginning of the kth period, so that the inventory purchasing decision at time k is made with exact knowledge of the cost c_k. Characterize the optimal ordering policy assuming that p is greater than all possible values of c_k.

4.5

Consider the inventory problem of Section 4.2 for the case where the cost has the general form

$$E\left\{\sum_{k=0}^{N} r_k(x_k)\right\}.$$

The functions r_k are convex and differentiable and

$$\lim_{x\to-\infty}\frac{dr_k(x)}{dx} = -\infty, \qquad \lim_{x\to\infty}\frac{dr_k(x)}{dx} = \infty, \qquad k = 0, \dots, N.$$

(a) Assume that the fixed cost is zero. Write the DP algorithm for this problem and show that the optimal ordering policy has the same form as the one derived in Section 4.2.

(b) Suppose there is a one-period time lag between the order and the delivery of inventory; that is, the system equation is of the form

$$x_{k+1} = x_k + u_{k-1} - w_k, \qquad k = 0, 1, \dots, N-1,$$

where u_{-1} is given. Reformulate the problem so that it has the form of the problem of part (a). *Hint*: Make a change of variables $y_k = x_k + u_{k-1}$.

(c) Repeat parts (a) and (b) for the case where the fixed cost is nonzero.

4.6 (Inventory Control for Nonzero Fixed Cost)

Consider the inventory problem of Section 4.2 (nonzero fixed cost) under the assumption that unfilled demand at each stage is not backlogged but rather is lost; that is, the system equation is $x_{k+1} = \max(0, x_k + u_k - w_k)$ instead of $x_{k+1} = x_k + u_k - w_k$. Complete the details of the following argument, which shows that a multiperiod (s, S) policy is optimal.

Abbreviated Proof: (S. Shreve) Let $J_N(x) = 0$ and for all k

$$G_k(y) = cy + E\big\{h\max(0, y - w_k) + p\max(0, w_k - y) + J_{k+1}\big(\max(0, y - w_k)\big)\big\},$$

$$J_k(x) = -cx + \min_{u\geq 0}\big[K\delta(u) + G_k(x+u)\big],$$

where $\delta(0) = 0$, $\delta(u) = 1$ for $u > 0$. The result will follow if we can show that G_k is K-convex, continuous, and $G_k(y) \to \infty$ as $|y| \to \infty$. The difficult part is to show K-convexity, since K-convexity of G_{k-1} does not imply K-convexity of $E\{J_{k+1}(\max(0, y - w))\}$. It will be sufficient to show that K-convexity of G_{k+1} implies K-convexity of

$$H(y) = p\max(0, -y) + J_{k+1}\big(\max(0, y)\big), \tag{6.1}$$

or equivalently that

$$K + H(y+z) \geq H(y) + z\left(\frac{H(y) - H(y-b)}{b}\right), \quad z \geq 0,\ b > 0,\ y \in \Re. \tag{6.2}$$

By K-convexity of G_{k+1} we have for appropriate scalars s_{k+1} and S_{k+1} such that $G_{k+1}(S_{k+1}) = \min_y G_{k+1}(y)$ and $K + G_{k+1}(S_{k+1}) = G_{k+1}(s_{k+1})$:

$$J_{k+1}(x) = \begin{cases} K + G_{k+1}(S_{k+1}) - cx & \text{if } x < s_{k+1}, \\ G_{k+1}(x) - cx & \text{if } x \geq s_{k+1}, \end{cases} \tag{6.3}$$

and J_{k+1} is K-convex by the theory of Section 4.2.

Case 1: $0 \leq y - b < y \leq y + z$. In this region, Eq. (6.2) follows from K-convexity of J_{k+1}.

Case 2: $y - b < y \leq y + z \leq 0$. In this region, H is linear and hence K-convex.

Case 3: $y - b < y \leq 0 \leq y + z$. In this region, Eq. (6.2) may be written [in view of Eq. (6.1)] as

$$K + J_{k+1}(y+z) \geq J_{k+1}(0) - p(y+z).$$

We will show that

$$K + J_{k+1}(z) \geq J_{k+1}(0) = pz, \qquad z \geq 0. \tag{6.4}$$

If $0 < s_{k+1} \leq z$, then using Eq. (6.3) and the fact $p > c$, we have

$$K + J_{k+1}(z) = K - cz + G_{k+1}(z) \geq K - pz + G_{k+1}(S_{k+1}) = J_{k+1}(0) = pz.$$

If $0 \leq z \leq s_{k+1}$, then using Eq. (6.3) and the fact $p > c$, we have

$$K + J_{k+1}(z) = 2K - cz + G_{k+1}(S_{k+1}) \geq K - pz + G_{k+1}(S_{k+1}) = J_{k+1}(0) - pz.$$

If $s_{k+1} \leq 0 \leq z$, then using Eq. (6.3), the fact $p > c$, and part (iv) of the lemma in Section 4.2, we have

$$K + J_{k+1}(z) = K - cz + G_{k+1}(z) \geq G_{k+1}(0) - pz = J_{k+1}(0) - pz.$$

Thus Eq. (6.4) is proved and Eq. (6.2) follows for the case under consideration.

Case 4: $y - b < 0 < y \leq y + z$. Then $0 < y < b$. If

$$\frac{H(y) - H(0)}{y} \geq \frac{H(y) - H(y-b)}{b}, \tag{6.5}$$

then since H agrees with J_{k+1} on $[0, \infty)$ and J_{k+1} is K-convex,

$$\begin{aligned} K + H(y+z) &\geq H(y) + z\left(\frac{H(y) - H(0)}{y}\right) \\ &\geq H(y) + z\left(\frac{H(y) - H(y-b)}{b}\right), \end{aligned}$$

where the last step follows from Eq. (6.5). If

$$\frac{H(y) - H(0)}{y} < \frac{H(y) - H(y-b)}{b},$$

then we have

$$H(y) - H(0) < \frac{y}{b}\big(H(y) - H(y-b)\big) = \frac{y}{b}\big(H(y) - H(0) + p(y-b)\big).$$

It follows that

$$\left(1 - \frac{y}{b}\right)\big(H(y) - H(0)\big) < \left(\frac{y}{b}\right)p(y-b) = -py\left(1 - \frac{y}{b}\right),$$

and since $b > y$,

$$H(y) - H(0) < -py. \tag{6.6}$$

Now we have, using the definition of H, Eqs. (6.4), and (6.6),

$$\begin{aligned} H(y) + z\frac{H(y) - H(y-b)}{b} &= H(y) + z\left(\frac{H(0) - py - H(0) + p(y-b)}{b}\right) \\ &= H(y) - pz \\ &< H(0) - p(y+z) \\ &\leq K + H(y+z). \end{aligned}$$

Hence Eq. (6.4) is proved for this case as well. **Q.E.D.**

4.7

Consider the inventory problem of Section 4.2 (zero fixed cost) with the difference that successive demands are correlated and satisfy a relation of the form

$$w_k = e_k - \gamma e_{k-1}, \qquad k = 0, 1, \ldots,$$

where γ is a given scalar, e_k are independent random variables, and e_{-1} is given.

(a) Show that this problem can be converted into an inventory problem with independent demands. *Hint*: Given $w_0, w_1, \ldots, w_{k-1}$, we can determine e_{k-1} in view of the relation

$$e_{k-1} = \gamma^k e_{-1} + \sum_{i=0}^{k-1} \gamma^i w_{k-1-i}.$$

Define $z_k = x_k + \gamma e_{k-1}$ as a new state variable.

(b) Show that the same is true when in addition there is a one-period delay in the delivery of inventory (cf. Exercise 4.5).

4.8

Consider the inventory problem of Section 4.2 (zero fixed cost), the only difference being that there is an upper bound $\overline{b}$ and a lower bound $\underline{b}$ to the allowable values of the stock x_k. This imposes the additional constraint on u_k

$$\underline{b} + d \leq u_k + x_k \leq \overline{b},$$

where $d > 0$ is the maximum value that the demand w_k can take (we assume $\underline{b} + d < \overline{b}$). Show that an optimal policy $\{\mu_0^*, \ldots, \mu_{N-1}^*\}$ is of the form

$$\mu_k^*(x_k) = \begin{cases} S_k - x_k & \text{if } x_k < S_k, \\ 0 & \text{if } x_k \geq S_k, \end{cases}$$

where $S_0, S_1, \ldots, S_{N-1}$ are some scalars.

4.9

Consider the inventory problem of Section 4.2 (nonzero fixed cost) with the difference that demand is deterministic and must be met at each time period (i.e., the shortage cost per unit is ∞). Show that it is optimal to order a positive amount at period k if and only if the stock x_k is insufficient to meet the demand w_k. Furthermore, when a positive amount is ordered, it should bring up stock to a level that will satisfy demand for an integral number of periods.

4.10 [Vei65], [Tsi84b]

Consider the inventory control problem of Section 4.2 (zero fixed cost) with the only difference that the orders u_k are constrained to be nonnegative *integers*. Let J_k be the optimal cost-to-go function. Show that:

(a) J_k is continuous.

(b) $J_k(x+1) - J_k(x)$ is a nondecreasing function of x.

(c) There exists a sequence $\{S_k\}$ of numbers such that the policy given by

$$\mu_k(x_k) = \begin{cases} n & \text{if } S_k - n \leq x_k < S_k - n + 1, \quad n = 1, 2, \ldots, \\ 0 & \text{if } x_k \geq S_k \end{cases}$$

is optimal.

4.11 (Capacity Expansion Problem)

Consider a problem of expanding over N time periods the capacity of a production facility. Let us denote by x_k the production capacity at the beginning of the kth period and by $u_k \geq 0$ the addition to capacity during the kth period. Thus capacity evolves according to

$$x_{k+1} = x_k + u_k, \qquad k = 0, 1, \ldots, N-1.$$

The demand at the kth period is denoted w_k and has a known probability distribution that does not depend on either x_k or u_k. Also, successive demands are assumed to be independent and bounded. We denote:

$C_k(u_k)$: expansion cost associated with adding capacity u_k

$P_k(x_k + u_k - w_k)$: penalty associated with capacity $x_k + u_k$ and demand w_k

$S(x_N)$: salvage value of final capacity x_N.

Thus the cost function has the form

$$\mathop{E}_{\substack{w_k \\ k=0,1,\ldots,N-1}} \left\{ -S(x_N) + \sum_{k=0}^{N-1} \big(C_k(u_k) + P_k(x_k + u_k - w_k) \big) \right\}.$$

(a) Derive the DP algorithm for this problem.

(b) Assume that S is a concave function with $\lim_{x \to \infty} dS(x)/dx = 0$, P_k are convex functions, and the expansion cost C_k is of the form

$$C_k(u) = \begin{cases} K + c_k u & \text{if } u > 0, \\ 0 & \text{if } u = 0, \end{cases}$$

where $K \geq 0$, $c_k > 0$ for all k. Show that the optimal policy is of the (s, S) type assuming $c_k y + E\big\{P_k(y - w_k)\big\} \to \infty$ as $|y| \to \infty$.

4.12

We want to use a machine to produce a certain item in quantities that meet as closely as possible a known (nonrandom) sequence of demands d_k over N periods. The machine can be in one of two states: good (G) or bad (B). The state of the machine is perfectly observed and evolves from one period to the next according to

$$P(G \mid G) = \lambda_G, \quad P(B \mid G) = 1 - \lambda_G, \quad P(B \mid B) = \lambda_B, \quad P(G \mid B) = 1 - \lambda_B,$$

where λ_G and λ_B are given probabilities. Let x_k be the stock at the beginning of the kth period. If the machine is in good state at period k, it can produce u_k, where $u_k \in [0, \overline{u}]$, and the stock evolves according to

$$x_{k+1} = x_k + u_k - d_k;$$

otherwise the stock evolves according to

$$x_{k+1} = x_k - d_k.$$

There is a cost $g(x_k)$ for having stock x_k in period k, and the terminal cost is also $g(x_N)$. We assume that the cost per stage g is a convex function such that $g(x) \to \infty$ as $|x| \to \infty$. The objective is to find a production policy that minimizes the total expected cost.

(a) Prove inductively a convexity property of the cost-to-go functions, and show that for each k there is a target stock level S_{k+1} such that if the machine is in the good state, it is optimal to produce $u_k^* \in [0, \overline{u}]$ that will bring x_{k+1} as close as possible to S_{k+1}.

(b) Generalize part (a) for the case where each demand d_k is random and takes values in an interval $[0, \overline{d}]$ with given probability distribution. The stock and the state of the machine are still perfectly observable.

4.13 (A Gambling Problem)

A gambler enters a game whereby he may at time k stake any amount $u_k \geq 0$ that does not exceed his current fortune x_k (defined to be his initial capital plus his gain or minus his loss thus far). He wins his stake back and as much more with probability p, where $\frac{1}{2} < p < 1$, and he loses his stake with probability $(1 - p)$. Show that the gambling strategy that maximizes $E\{\ln x_N\}$, where x_N denotes his fortune after N plays, is to stake at each time k an amount $u_k = (2p - 1)x_k$. *Hint*: The problem is related to the portfolio problem of Section 4.3.

4.14

Consider the dynamic portfolio problem of Section 4.3 for the case where at each period k there is a forecast that the rates of return of the risky assets for that period will be selected in accordance with a particular probability distribution as in Section 1.4. Show that a partially myopic policy is optimal.

4.15

Consider a problem involving the linear system

$$x_{k+1} = A_k x_k + B_k u_k, \qquad k = 0, 1, \ldots, N-1,$$

where the $n \times n$ matrices A_k are given, and the $n \times m$ matrices B_k are random and independent with given probability distributions that do not depend on x_k, u_k. The problem is to find a policy that maximizes $E\{U(c'x_N)\}$, where c is a given n-dimensional vector. We assume that U is a concave twice continuously differentiable utility function satisfying for all y

$$-\frac{U'(y)}{U''(y)} = a + by,$$

and that the control is unconstrained. Show that the optimal policy consists of linear functions of the current state. *Hint*: Reduce the problem to a one-dimensional problem and use the results of Section 4.3.

4.16

Suppose that a person wants to sell a house and an offer comes at the beginning of each day. We assume that successive offers are independent and an offer is w_j with probability p_j, $j = 1, \ldots, n$, where w_j are given nonnegative scalars. Any offer not immediately accepted is not lost but may be accepted at any later date. Also, a maintenance cost c is incurred for each day that the house remains unsold. The objective is to maximize the price at which the house is sold minus the maintenance costs. Consider the problem when there is a deadline to sell the house within N days and characterize the optimal policy.

4.17

Assume that we have x_0 items of a certain type that we want to sell over a period of N days. At each day we may sell at most one item. At the kth day, knowing the current number x_k of remaining unsold items, we can set the selling price u_k of a unit item to a nonnegative number of our choice; then, the probability $\lambda_k(u_k)$ of selling an item on the kth day depends on u_k as follows:

$$\lambda_k(u_k) = \alpha e^{-u_k},$$

where α is a given scalar with $0 < \alpha \leq 1$. The objective is to find the optimal price setting policy so as to maximize the total expected revenue over N days.

(a) Assuming that, for all k, the cost-to-go function $J_k(x_k)$ is monotonically nondecreasing as a function of x_k, prove that for $x_k > 0$, the optimal prices have the form

$$\mu_k^*(x_k) = 1 + J_{k+1}(x_k) - J_{k+1}(x_k - 1),$$

and that

$$J_k(x_k) = \alpha e^{-\mu_k^*(x_k)} + J_{k+1}(x_k).$$

(b) Prove simultaneously by induction that, for all k, the cost-to-go function $J_k(x_k)$ is indeed monotonically nondecreasing as a function of x_k, that the optimal price $\mu_k^*(x_k)$ is monotonically nonincreasing as a function of x_k, and that $J_k(x_k)$ is given in closed form by

$$J_k(x_k) = \begin{cases} (N-k)\alpha e^{-1} & \text{if } x_k \geq N-k, \\ \sum_{i=k}^{N-x_k} \alpha e^{-\mu_i^*(x_k)} + x_k \alpha e^{-1} & \text{if } 0 < x_k < N-k, \\ 0 & \text{if } x_k = 0. \end{cases}$$

4.18 (Optimal Termination of Sampling)

This is a classical problem, which when appropriately paraphrased, is known as the job selection, or as the secretary selection, or as the spouse selection problem. A collection of $N \geq 2$ objects is observed randomly and sequentially one at a time. The observer may either select the current object observed, in which case the selection process is terminated, or reject the object and proceed to observe the next. The observer can rank each object relative to those already observed, and the objective is to maximize the probability of selecting the "best" object according to some criterion. It is assumed that no two objects can be judged to be equal. Let r^* be the smallest positive integer r such that

$$\frac{1}{N-1} + \frac{1}{N-2} + \cdots + \frac{1}{r} \leq 1.$$

Show that an optimal policy requires that the first r^* objects be observed. If the r^*th object has rank 1 relative to the others already observed, it should be selected; otherwise, the observation process should be continued until an object of rank 1 relative to those already observed is found. *Hint*: We assume that, if the rth object has rank 1 relative to the previous $(r-1)$ objects, then the probability that it is best is r/N. For $k \geq r^*$, let $J_k(0)$ be the maximal probability of finding the best object assuming k objects have been selected and the kth object is not best relative to the previous $(k-1)$ objects. Show that

$$J_k(0) = \frac{k}{N}\left(\frac{1}{N-1} + \cdots + \frac{1}{k}\right).$$

4.19

A driver is looking for parking on the way to his destination. Each parking place is free with probability p independently of whether other parking places are free or not. The driver cannot observe whether a parking place is free until he reaches it. If he parks k places from his destination, he incurs a cost k. If he reaches the destination without having parked the cost is C.

(a) Let F_k be the minimal expected cost if he is k parking places from his destination. Show that

$$F_0 = C,$$

$$F_k = p\min(k, F_{k-1}) + qF_{k-1}, \qquad k = 1, 2, \ldots,$$

where $q = 1 - p$.

(b) Show that an optimal policy is of the form: never park if $k \geq k^*$, but take the first free place if $k < k^*$, where k is the number of parking places from the destination and k^* is the smallest integer i satisfying $q^{i-1} < (pC + q)^{-1}$.

4.20 [Whi82]

A person may go hunting for a certain type of animal on a given day or stay home. When the animal population is x, the probability of capturing one animal is $p(x)$, a known increasing function, and the probability of capturing more than one is zero. A captured animal is worth one unit and a day of hunting costs c units. Assume that x does not change due to deaths or births, that the hunter knows x at all times, that the horizon is finite, and that the terminal reward is zero. Show that it is optimal to hunt only when $p(x) \geq c$.

4.21

Consider the scalar linear system

$$x_{k+1} = ax_k + bu_k,$$

where a and b are known. At each period k we have the option of using a control u_k and incurring a cost $qx_k^2 + ru_k^2$, or else stopping and incurring a stopping cost tx_k^2. If we have not stopped by period N, the terminal cost is the stopping cost tx_N^2. We assume that $q \geq 0$, $r > 0$, $t > 0$. Show that there is a threshold value for t below which immediate stopping is optimal at every initial state, and above which continuing at every state x_k and period k is optimal.

4.22

Consider a situation involving a blackmailer and his victim. In each period the blackmailer has a choice of: a) Accepting a lump sum payment of R from the victim and promising not to blackmail again. b) Demanding a payment of u, where $u \in [0, 1]$. If blackmailed, the victim will either: 1) Comply with the demand and pay u to the blackmailer. This happens with probability $1 - u$. 2) Refuse to pay and denounce the blackmailer to the police. This happens with probability u. Once known to the police, the blackmailer cannot ask for any more money. The blackmailer wants to maximize the expected amount of money he gets over N periods by optimal choice of the payment demands u_k. (Note that there is no additional penalty for being denounced to the police.) Write a DP algorithm and find the optimal policy.

4.23 [Whi82]

The Greek adventurer Theseus is trapped in King Minos' Labyrinth maze. He can try on each day one of N passages. If he enters passage i he will escape with probability p_i, he will be killed with probability q_i, and he will determine that the passage is a dead end with probability $(1-p_i-q_i)$, in which case he will return to the point from which he started. Use an interchange argument to show that trying passages in order of decreasing p_i/q_i maximizes the probability of escape within N days.

4.24 (Hardy's Theorem)

Let $\{a_1, \ldots, a_n\}$ and $\{b_1, \ldots, b_n\}$ be monotonically nondecreasing sequences of numbers. Let us associate with each $i = 1, \ldots, n$ a distinct index j_i, and consider the expression $\sum_{i=1}^n a_i b_{j_i}$. Use an interchange argument to show that this expression is maximized when $j_i = i$ for all i, and is minimized when $j_i = n - i + 1$ for all i.

4.25

A busy professor has to complete N projects. Each project k has a deadline d_k and the time it takes the professor to complete it is t_k. The professor can work on only one project at a time and must complete it before moving on to a new project. For a given order of completion of the projects, denote by c_k the time of completion of project k, i.e.,

$$c_k = t_k + \sum_{\substack{\text{projects } i \\ \text{completed before } k}} t_i.$$

The professor wants to order the projects so as to minimize the maximum tardiness, given by

$$\max_{k \in \{1, \ldots, N\}} \max(0, c_k - d_k).$$

Use an interchange argument to show that it is optimal to complete the projects in the order of their deadlines (do the project with the closest deadline first).

5

Problems with Imperfect State Information

Contents

We have assumed so far that the controller has access to the exact value of the current state, but this assumption is often unrealistic. For example, some state variables may be inaccessible, the sensors used for measuring them may be inaccurate, or the cost of obtaining the exact value of the state may be prohibitive. We model situations of this type by assuming that at each stage the controller receives some observations about the value of the current state, which may be corrupted by stochastic uncertainty.

Problems where the controller uses observations of this type in place of the state are called problems of *imperfect state information*, and are the subject of this chapter. We will find that even though there are DP algorithms for imperfect information problems, these algorithms are far more computationally intensive than in the perfect information case. For this reason, in the absence of an analytical solution, imperfect information problems are typically solved suboptimally in practice. On the other hand, we will see that conceptually, imperfect state information problems are no different than the perfect state information problems we have been studying so far. In fact by various reformulations, we can reduce an imperfect state information problem to one with perfect state information. We will study two different reductions of this type, which will yield two different DP algorithms. The first reduction is the subject of the next section, while the second reduction will be given in Section 5.4.

5.1 REDUCTION TO THE PERFECT INFORMATION CASE

We first formulate the imperfect state information counterpart of the basic problem.

Basic Problem with Imperfect State Information

Consider the basic problem of Section 1.2 where the controller, instead of having perfect knowledge of the state, has access to observations z_k of the form

$$z_0 = h_0(x_0, v_0), \qquad z_k = h_k(x_k, u_{k-1}, v_k), \qquad k = 1, 2, ..., N-1. \quad (1.1)$$

The observation z_k belongs to a given observation space Z_k. The random observation disturbance v_k belongs to a given space V_k and is characterized by a given probability distribution

$$P_{v_k}(\cdot \mid x_k, \ldots, x_0, u_{k-1}, \ldots, u_0, w_{k-1}, \ldots, w_0, v_{k-1}, \ldots, v_0),$$

which depends on the current state and the past states, controls, and disturbances.

The initial state x_0 is also random and characterized by a given probability distribution P_{x_0}. The probability distribution $P_{w_k}(\cdot \mid x_k, u_k)$ of w_k is given, and it may depend explicitly on x_k and u_k but not on the prior disturbances $w_0, \ldots, w_{k-1}, v_0, \ldots, v_{k-1}$. The control u_k is constrained to take values from a given nonempty subset U_k of the control space C_k. It is assumed that this subset does not depend on x_k.

Let us denote by I_k the information available to the controller at time k and call it the *information vector*. We have

$$\begin{aligned} I_k &= (z_0, z_1, \ldots, z_k, u_0, u_1, \ldots, u_{k-1}), \qquad k = 1, 2, \ldots, N-1, \\ I_0 &= z_0. \end{aligned} \tag{1.2}$$

We consider the class of policies consisting of a sequence of functions $\pi = \{\mu_0, \mu_1, \ldots, \mu_{N-1}\}$, where each function μ_k maps the information vector I_k into the control space C_k and

$$\mu_k(I_k) \in U_k, \qquad \text{for all } I_k,\ k = 0, 1, \ldots, N-1.$$

Such policies are called *admissible*. We want to find an admissible policy $\pi = \{\mu_0, \mu_1, \ldots, \mu_{N-1}\}$ that minimizes the cost function

$$J_\pi = \mathop{E}_{\substack{x_0, w_k, v_k \\ k=0,\ldots,N-1}} \left\{ g_N(x_N) + \sum_{k=0}^{N-1} g_k\big(x_k, \mu_k(I_k), w_k\big) \right\} \tag{1.3}$$

subject to the system equation

$$x_{k+1} = f_k\big(x_k, \mu_k(I_k), w_k\big), \qquad k = 0, 1, \ldots, N-1,$$

and the measurement equation

$$\begin{aligned} z_0 &= h_0(x_0, v_0), \\ z_k &= h_k\big(x_k, \mu_{k-1}(I_{k-1}), v_k\big), \qquad k = 1, 2, \ldots, N-1. \end{aligned}$$

Note the difference from the perfect state information case. Whereas before we were trying to find a rule that would specify the control u_k to be applied for each state x_k and time k, now we are looking for a rule that gives the control to be applied for every possible information vector I_k (or state of information), that is, for every sequence of observations received and controls employed up to time k.

Example 1.1 (Multiaccess Communication)

Consider a collection of transmitting stations sharing a common channel, for example, a set of ground stations communicating with a satellite at a common frequency. The stations are synchronized to transmit packets of data at integer times. Each packet requires one time unit (also called a *slot*) for transmission. The total number a_k of packet arrivals during slot k is independent of prior arrivals and has a given probability distribution. The stations do not know the backlog x_k at the beginning of the kth slot (the number of packets waiting to be transmitted). Packet transmissions are scheduled using a strategy (known as *slotted Aloha*) whereby each packet residing in the system at the beginning of the kth slot is transmitted during the kth slot with probability u_k (common for all packets). If two or more packets are transmitted simultaneously, they collide and have to rejoin the backlog for retransmission at a later slot. However, the stations can observe the channel and determine whether in any one slot there was a collision (two or more packets), a success (one packet), or an idle (no packets). These observations provide information about the state of the system (the backlog x_k) and can be used to select appropriately the control (the transmission probability u_k). The objective is to keep the backlog small, so we assume a cost per stage $g_k(x_k)$, which is a monotonically increasing function of x_k.

The state of the system here is the backlog x_k and evolves according to the equation

$$x_{k+1} = x_k + a_k - t_k,$$

where a_k is the number of new arrivals and t_k is the number of packets successfully transmitted during slot k. Both a_k and t_k may be viewed as disturbances, and the distribution of t_k depends on the state x_k and the control u_k. It can be seen that $t_k = 1$ (a success) with probability $x_k u_k (1 - u_k)^{x_k - 1}$, and $t_k = 0$ (idle or collision) otherwise [the probability of any one of the x_k waiting packets being transmitted, while all the other packets are not transmitted, is $u_k(1 - u_k)^{x_k - 1}$].

If we had perfect state information (i.e., the backlog x_k were known at the beginning of slot k), the optimal policy would be to select the value of u_k that maximizes $x_k u_k (1 - u_k)^{x_k - 1}$, which is the success probability.† By setting the derivative of this probability to zero, we find the optimal (perfect state information) policy to be

$$\mu_k(x_k) = \frac{1}{x_k}, \qquad \text{for all } x_k \geq 1.$$

---

† For a more detailed derivation, note that the DP algorithm for the perfect state information problem is

$$J_k(x_k) = g_k(x_k) + \min_{0 \leq u_k \leq 1} \mathop{E}_{a_k} \big\{ p(x_k, u_k) J_{k+1}(x_k + a_k - 1) + \big(1 - p(x_k, u_k)\big) J_{k+1}(x_k + a_k) \big\},$$

where $p(x_k, u_k)$ is the success probability $x_k u_k (1 - u_k)^{x_k - 1}$. Since the cost per stage $g_k(x_k)$ is an increasing function of the backlog x_k, it is clear that each cost-to-go function $J_k(x_k)$ is an increasing function of x_k (this can also be easily proved by induction). Thus $J_{k+1}(x_k + a_k) \geq J_{k+1}(x_k + a_k - 1)$ for all x_k and a_k, based on which the DP algorithm implies that the optimal u_k maximizes $p(x_k, u_k)$ over $[0, 1]$.

In practice, however, x_k is not known (imperfect state information), and the optimal control must be chosen on the basis of the available observations (i.e., the entire channel history of successes, idles, and collisions). These observations relate to the backlog history (the past states) and the past transmission probabilities (the past controls), but are corrupted by stochastic uncertainty. Mathematically, we may write an equation $z_{k+1} = v_{k+1}$, where z_{k+1} is the observation obtained at the end of the kth slot, and the random variable v_{k+1} yields an idle with probability $(1 - u_k)^{x_k}$, a success with probability $x_k u_k (1 - u_k)^{x_k - 1}$, and a collision otherwise.

It can be seen that this is a problem that fits the given imperfect state information framework. Unfortunately, the optimal solution to this problem is very complicated and for all practical purposes cannot be computed. A suboptimal solution will be discussed in Section 6.1.

Reformulation as a Perfect State Information Problem

We now show how to effect the reduction from imperfect to perfect state information. As in the discussion of state augmentation in Section 1.4, it is intuitively clear that we should define a new system whose state at time k is the set of all variables the knowledge of which can be of benefit to the controller when making the kth decision. Thus a first candidate as the state of the new system is the information vector I_k. Indeed we will show that this choice is appropriate.

We have by the definition of the information vector [cf. Eq. (1.2)]

$$I_{k+1} = (I_k, z_{k+1}, u_k), \qquad k = 0, 1, \ldots, N-2, \qquad I_0 = z_0. \tag{1.4}$$

These equations can be viewed as describing the evolution of a system of the same nature as the one considered in the basic problem of Section 1.2. The state of the system is I_k, the control is u_k, and z_{k+1} can be viewed as a random disturbance. Furthermore, we have

$$P(z_{k+1} \mid I_k, u_k) = P(z_{k+1} \mid I_k, u_k, z_0, z_1, \ldots, z_k), \tag{1.5}$$

since $z_0, z_1, \ldots, z_k$ are part of the information vector I_k. Thus the probability distribution of z_{k+1} depends explicitly only on the state I_k and control u_k of the new system (1.4) and not on the prior "disturbances" $z_k, \ldots, z_0$.

By writing

$$E\{g_k(x_k, u_k, w_k)\} = E\left\{ \underset{x_k, w_k}{E} \{g_k(x_k, u_k, w_k) \mid I_k, u_k\} \right\},$$

we can similarly reformulate the cost function in terms of the variables of the new system. The cost per stage as a function of the new state I_k and the control u_k is

$$\tilde{g}_k(I_k, u_k) = \underset{x_k, w_k}{E} \{g_k(x_k, u_k, w_k) \mid I_k, u_k\}. \tag{1.6}$$

Thus the basic problem with imperfect state information has been reformulated as a problem with perfect state information that involves the system (1.4) and the cost per stage (1.6). By writing the DP algorithm for this latter problem and substituting the expressions (1.4) and (1.6), we obtain

$$J_{N-1}(I_{N-1}) = \min_{u_{N-1}\in U_{N-1}} \Bigg[\mathop{E}_{x_{N-1},\, w_{N-1}} \Big\{ g_N\big(f_{N-1}(x_{N-1}, u_{N-1}, w_{N-1})\big) + g_{N-1}(x_{N-1}, u_{N-1}, w_{N-1}) \mid I_{N-1}, u_{N-1} \Big\} \Bigg], \tag{1.7}$$

and for $k = 0, 1, \ldots, N-2$,

$$J_k(I_k) = \min_{u_k\in U_k} \Bigg[\mathop{E}_{x_k,\, w_k,\, z_{k+1}} \Big\{ g_k(x_k, u_k, w_k) + J_{k+1}(I_k, z_{k+1}, u_k) \mid I_k, u_k \Big\} \Bigg]. \tag{1.8}$$

These equations constitute one possible DP algorithm for the imperfect state information problem. An optimal policy $\{\mu_0^*, \mu_1^*, \ldots, \mu_{N-1}^*\}$ is obtained by first minimizing in the right-hand side of the DP Eq. (1.7) for every possible value of the information vector I_{N-1} to obtain $\mu_{N-1}^*(I_{N-1})$. Simultaneously, $J_{N-1}(I_{N-1})$ is computed and used in the computation of $J_{N-2}(I_{N-2})$ via the minimization in the DP Eq. (1.8), which is carried out for every possible value of I_{N-2}. Proceeding similarly, one obtains $J_{N-3}(I_{N-3})$ and μ_{N-3}^* and so on, until $J_0(I_0) = J_0(z_0)$ is computed. The optimal cost J^* is then obtained from

$$J^* = \mathop{E}_{z_0}\big\{J_0(z_0)\big\}. \tag{1.9}$$

Machine Repair Example

A machine can be in one of two states denoted P and $\overline{P}$. State P corresponds to a machine in proper condition (good state) and state $\overline{P}$ to a machine in improper condition (bad state). If the machine is operated for one time period, it stays in state P with probability $\frac{2}{3}$ if it started in P, and it stays in state $\overline{P}$ with probability 1 if it started in $\overline{P}$. The machine is operated for a total of three time periods and starts in state P. At the end of the first and second time periods the machine is inspected and there are two possible inspection outcomes denoted G (probably good state) and B (probably bad state). If the machine is in the good state P, the inspection outcome is G with probability $\frac{3}{4}$; if the machine is in the bad state $\overline{P}$, the inspection outcome is B with probability $\frac{3}{4}$:

$$P(G \mid x = P) = \frac{3}{4}, \qquad P(B \mid x = P) = \frac{1}{4},$$

$$P(G \mid x = \overline{P}) = \frac{1}{4}, \qquad P(B \mid x = \overline{P}) = \frac{3}{4};$$

see Fig. 5.1.1. After each inspection one of two possible actions can be taken:

C: Continue operation of the machine.

S: Stop the machine, determine its state through an accurate diagnostic test, and if it is in the bad state $\overline{P}$ bring it back to the good state P.

At each period there is a cost of 2 and 0 units for starting the period with a machine in the bad state $\overline{P}$ and the good state P, respectively. The cost for taking the stop-and-repair action S is 1 unit and the terminal cost is 0.

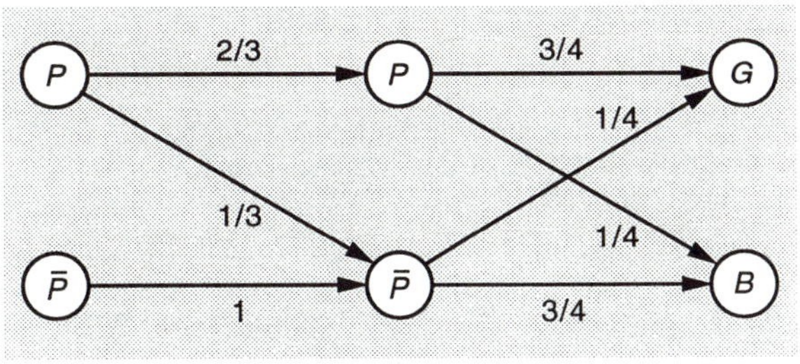

Figure 5.1.1 State transition diagram and probabilities of inspection outcomes in the machine repair example.

The problem is to determine the policy that minimizes the expected costs over the three time periods. In other words, we want to find the optimal action after the result of the first inspection is known, and after the results of the first and second inspections, as well as the action taken after the first inspection, are known.

It can be seen that this example falls within the general framework of the problem of this section. The state space consists of the two states P and $\overline{P}$,

$$\text{state space} = \{P, \overline{P}\},$$

and the control space consists of the two actions

$$\text{control space} = \{C, S\}.$$

The system evolution may be described by introducing a system equation

$$x_{k+1} = w_k, \qquad k = 0, 1,$$

where for $k = 0, 1$, the probability distribution of w_k is given by

$$P(w_k = P \mid x_k = P, u_k = C) = \frac{2}{3}, \qquad P(w_k = \overline{P} \mid x_k = P, u_k = C) = \frac{1}{3},$$

$$P(w_k = P \mid x_k = \overline{P}, u_k = C) = 0, \qquad P(w_k = \overline{P} \mid x_k = \overline{P}, u_k = C) = 1,$$

$$P(w_k = P \mid x_k = P, u_k = S) = \frac{2}{3}, \qquad P(w_k = \overline{P} \mid x_k = P, u_k = S) = \frac{1}{3},$$

$$P(w_k = P \mid x_k = \overline{P}, u_k = S) = \frac{2}{3}, \qquad P(w_k = \overline{P} \mid x_k = \overline{P}, u_k = S) = \frac{1}{3}.$$

We denote by x_0, x_1, x_2 the state of the machine at the end of the first, second, and third time period, respectively. Also we denote by u_0 the action taken after the first inspection (end of first time period) and by u_1 the action taken after the second inspection (end of second time period). The probability distribution of x_0 is

$$P(x_0 = P) = \frac{2}{3}, \qquad P(x_0 = \overline{P}) = \frac{1}{3}.$$

Note that we do not have perfect state information, since the inspections do not reveal the state of the machine with certainty. Rather the result of each inspection may be viewed as a measurement of the system state of the form

$$z_k = v_k, \qquad k = 0, 1,$$

where for $k = 0, 1$, the probability distribution of v_k is given by

$$P(v_k = G \mid x_k = P) = \frac{3}{4}, \qquad P(v_k = B \mid x_k = P) = \frac{1}{4},$$

$$P(v_k = G \mid x_k = \overline{P}) = \frac{1}{4}, \qquad P(v_k = B \mid x_k = \overline{P}) = \frac{3}{4}.$$

The cost resulting from a sequence of states x_0, x_1 and actions u_0, u_1 is

$$g(x_0, u_0) + g(x_1, u_1),$$

where continue stop

$$g(P, C) = 0, \qquad g(P, S) = 1, \qquad g(\overline{P}, C) = 2, \qquad g(\overline{P}, S) = 1.$$

The information vector at times 0 and 1 is

$$I_0 = z_0, \qquad I_1 = (z_0, z_1, u_0),$$

and we seek functions $\mu_0(I_0), \mu_1(I_1)$ that minimize

$$\begin{aligned} &\mathop{E}_{\substack{x_0, w_0, w_1\\ v_0, v_1}} \big\{ g\big(x_0, \mu_0(I_0)\big) + g\big(x_1, \mu_1(I_1)\big) \big\} \\ &\quad = \mathop{E}_{\substack{x_0, w_0, w_1\\ v_0, v_1}} \Big\{ g\big(x_0, \mu_0(z_0)\big) + g\Big(x_1, \mu_1\big(z_0, z_1, \mu_0(z_0)\big)\Big) \Big\}. \end{aligned} \tag{1.10}$$

We now apply the DP algorithm.

Last Stage: We use Eq. (1.7) to compute $J_1(I_1)$ for each of the eight possible information vectors $I_1 = (z_0, z_1, u_0)$. For each of these vectors, we shall compute the expected cost of the two possible actions, $u_1 = C$ and $u_1 = S$, and select as optimal the action with the smallest cost. We have

$$\text{cost of } C = 2 \cdot P(x_1 = \overline{P} \mid I_1), \qquad \text{cost of } S = 1,$$

and therefore

$$J_1(I_1) = \min\big[2P(x_1 = \overline{P} \mid I_1),\, 1\big].$$

The probabilities $P(x_1 = \overline{P} \mid I_1)$ can be computed by using Bayes' rule and the problem data. Some of the details will be omitted. We have:

(1) For $I_1 = (G, G, S)$

$$P(x_1 = \overline{P} \mid G, G, S) = \frac{P(x_1 = \overline{P}, G, G, S)}{P(G, G, S)}$$
$$= \frac{\frac{1}{3} \cdot \frac{1}{4} \cdot \left(\frac{2}{3} \cdot \frac{3}{4} + \frac{1}{3} \cdot \frac{1}{4}\right)}{\left(\frac{2}{3} \cdot \frac{3}{4} + \frac{1}{3} \cdot \frac{1}{4}\right)^2} = \frac{1}{7}.$$

Hence

$$J_1(G, G, S) = \frac{2}{7}, \qquad \mu_1^*(G, G, S) = C.$$

(2) For $I_1 = (B, G, S)$

$$P(x_1 = \overline{P} \mid B, G, S) = P(x_1 = \overline{P} \mid G, G, S) = \frac{1}{7},$$

$$J_1(B, G, S) = \frac{2}{7}, \qquad \mu_1^*(B, G, S) = C.$$

(3) For $I_1 = (G, B, S)$

$$P(x_1 = \overline{P} \mid G, B, S) = \frac{P(x_1 = \overline{P}, G, B, S)}{P(G, B, S)}$$
$$= \frac{\frac{1}{3} \cdot \frac{3}{4} \cdot \left(\frac{2}{3} \cdot \frac{3}{4} + \frac{1}{3} \cdot \frac{1}{4}\right)}{\left(\frac{2}{3} \cdot \frac{1}{4} + \frac{1}{3} \cdot \frac{3}{4}\right)\left(\frac{2}{3} \cdot \frac{3}{4} + \frac{1}{3} \cdot \frac{1}{4}\right)} = \frac{3}{5},$$

$$J_1(G, B, S) = 1, \qquad \mu_1^*(G, B, S) = S.$$

(4) For $I_1 = (B, B, S)$

$$P(x_1 = \overline{P} \mid B, B, S) = P(x_1 = \overline{P} \mid G, B, S) = \frac{3}{5},$$

$$J_1(B, B, S) = 1, \qquad \mu_1^*(B, B, S) = S.$$

(5) For $I_1 = (G, G, C)$

$$P(x_1 = \overline{P} \mid G, G, C) = \frac{P(x_1 = \overline{P}, G, G, C)}{P(G, G, C)} = \frac{1}{5},$$

$$J_1(G, G, C) = \frac{2}{5}, \qquad \mu_1^*(G, G, C) = C.$$

(6) For $I_1 = (B, G, C)$

$$P(x_1 = \overline{P} \mid B, G, C) = \frac{11}{23},$$

$$J_1(B, G, C) = \frac{22}{23}, \qquad \mu_1^*(B, G, C) = C.$$

(7) For $I_1 = (G, B, C)$

$$P(x_1 = \overline{P} \mid G, B, C) = \frac{9}{13},$$

$$J_1(G, B, C) = 1, \qquad \mu_1^*(G, B, C) = S.$$

(8) For $I_1 = (B, B, C)$

$$P(x_1 = \overline{P} \mid B, B, C) = \frac{33}{37},$$

$$J_1(B, B, C) = 1, \qquad \mu_1^*(B, B, C) = S.$$

Summarizing the results for the last stage, the optimal policy is to continue ($u_1 = C$) if the result of the last inspection was G, and to stop ($u_1 = S$) if the result of the last inspection was B.

First Stage: Here we use the DP Eq. (1.8) to compute $J_0(I_0)$ for each of the two possible information vectors $I_0 = (G)$, $I_0 = (B)$. We have

$$\begin{aligned} \text{cost of } C &= 2P(x_0 = \overline{P} \mid I_0, C) + \mathop{E}_{z_1}\{J_1(I_0, z_1, C) \mid I_0, C\} \\ &= 2P(x_0 = \overline{P} \mid I_0, C) + P(z_1 = G \mid I_0, C) J_1(I_0, G, C) \\ &\quad + P(z_1 = B \mid I_0, C) J_1(I_0, B, C), \end{aligned}$$

$$\begin{aligned} \text{cost of } S &= 1 + \mathop{E}_{z_1}\{J_1(I_0, z_1, S) \mid I_0, S\} \\ &= 1 + P(z_1 = G \mid I_0, S) J_1(I_0, G, S) + P(z_1 = B \mid I_0, S) J_1(I_0, B, S), \end{aligned}$$

and

$$\begin{aligned} J_0(I_0) = \min\Big[& 2P(x_0 = \overline{P} \mid I_0, C) + \mathop{E}_{z_1}\{J_1(I_0, z_1, C) \mid I_0, C\}, \\ & 1 + \mathop{E}_{z_1}\{J_1(I_0, z_1, S) \mid I_0, S\}\Big] \end{aligned}$$

(1) For $I_0 = (G)$: Direct calculation yields

$$P(z_1 = G \mid G, C) = \frac{15}{28}, \qquad P(z_1 = B \mid G, C) = \frac{13}{28},$$

$$P(z_1 = G \mid G, S) = \frac{7}{12}, \qquad P(z_1 = B \mid G, S) = \frac{5}{12},$$

$$P(x_0 = \overline{P} \mid G, C) = \frac{1}{7},$$

and hence

$$J_0(G) = \min \Big[2 \cdot \frac{1}{7} + \frac{15}{28} J_1(G, G, C) + \frac{13}{28} J_1(G, B, C), \\ 1 + \frac{7}{12} J_1(G, G, S) + \frac{5}{12} J_1(G, B, S)\Big].$$

Using the values of J_1 obtained in the previous stage

$$J_0(G) = \min \left[2 \cdot \frac{1}{7} + \frac{15}{28} \cdot \frac{2}{5} + \frac{13}{28} \cdot 1,\ 1 + \frac{7}{12} \cdot \frac{2}{7} + \frac{5}{12} \cdot 1\right]$$
$$= \min \left[\frac{27}{28}, \frac{19}{12}\right] = \frac{27}{28},$$

$$J_0(G) = \frac{27}{28}, \qquad \mu_0^*(G) = C.$$

(2) For $I_0 = (B)$: Direct calculation yields

$$P(z_1 = G \mid B, C) = \frac{23}{60}, \qquad P(z_1 = B \mid B, C) = \frac{37}{60},$$

$$P(z_1 = G \mid B, S) = \frac{7}{12}, \qquad P(z_1 = B \mid B, S) = \frac{5}{12},$$

$$P(x_0 = \overline{P} \mid B, C) = \frac{3}{5},$$

and

$$J_0(B) = \min \Big[2 \cdot \frac{3}{5} + \frac{23}{60} J_1(B, G, C) + \frac{37}{60} J_1(B, B, C), \\ 1 + \frac{7}{12} J_1(B, G, S) + \frac{5}{12} J_1(B, B, S)\Big].$$

Using the values of J_1 obtained in the previous state

$$J_0(B) = \min \left[\frac{131}{60}, \frac{19}{12}\right] = \frac{19}{12},$$

$$J_0(B) = \frac{19}{12}, \qquad \mu_0^*(B) = S.$$

Summarizing, the optimal policy for both stages is to continue if the result of the latest inspection is G, and to stop and repair otherwise.

The optimal cost is

$$J^* = P(G)J_0(G) + P(B)J_0(B).$$

We can verify that $P(G) = \frac{7}{12}$ and $P(B) = \frac{5}{12}$, so that

$$J^* = \frac{7}{12} \cdot \frac{27}{28} + \frac{5}{12} \cdot \frac{19}{12} = \frac{176}{144}.$$

In the above example, the computation of the optimal policy and the optimal cost by means of the DP algorithm (1.7) and (1.8) was possible because the problem was very simple. It is easy to see that for a more complex problem, the computational requirements of the DP algorithm can be prohibitive, particularly if the number of possible information vectors I_k is large (or infinite). Unfortunately, even if the control and observation spaces are simple (one-dimensional or finite), the space of the information vector I_k may have large dimension. This makes the application of the algorithm very difficult or computationally impossible in many cases. However, there are problems where an analytical solution can be found, and the next two sections deal with such problems.

5.2 LINEAR SYSTEMS AND QUADRATIC COST

We will show how the DP algorithm of the preceding section can be used to solve the imperfect state information analog of the linear system/quadratic cost problem of Section 4.1. We have the same linear system

$$x_{k+1} = A_k x_k + B_k u_k + w_k, \qquad k = 0, 1, \ldots, N-1, \tag{2.1}$$

and quadratic cost

$$E\left\{x_N' Q_N x_N + \sum_{k=0}^{N-1} (x_k' Q_k x_k + u_k' R_k u_k)\right\}, \tag{2.2}$$

but now the controller does not have access to the current state. Instead it receives at the beginning of each period k an observation of the form

$$z_k = C_k x_k + v_k, \qquad k = 0, 1, \ldots, N-1, \tag{2.3}$$

where $z_k \in \Re^s$, C_k is a given $s \times n$ matrix, and $v_k \in \Re^s$ is an observation noise vector with given probability distribution. Furthermore, the vectors v_k are independent, and independent from w_k and x_0 as well. We make the same assumptions as in Section 4.1 concerning the input disturbances w_k, i.e., that they are independent, zero mean, and that they have finite variance. The system matrices A_k, B_k are known; there is no analytical solution of the imperfect information counterpart of the model with random system matrices considered in Section 4.1.

From the DP Eq. (1.7) we have

$$\begin{aligned} J_{N-1}(I_{N-1}) = \min_{u_{N-1}} \Big[\mathop{E}_{x_{N-1}, w_{N-1}} & \{ x'_{N-1} Q_{N-1} x_{N-1} + u'_{N-1} R_{N-1} u_{N-1} \\ & + (A_{N-1} x_{N-1} + B_{N-1} u_{N-1} + w_{N-1})' \\ & \cdot Q_N (A_{N-1} x_{N-1} + B_{N-1} u_{N-1} + w_{N-1}) \mid I_{N-1} \} \Big] \end{aligned}$$

Since $E\{w_{N-1} \mid I_{N-1}\} = E\{w_{N-1}\} = 0$, this expression can be written as

$$\begin{aligned} J_{N-1}(I_{N-1}) = & \mathop{E}_{x_{N-1}} \{ x'_{N-1}(A'_{N-1} Q_N A_{N-1} + Q_{N-1}) x_{N-1} \mid I_{N-1} \} \\ & + \mathop{E}_{w_{N-1}} \{ w'_{N-1} Q_N w_{N-1} \} \\ & + \min_{u_{N-1}} \big[u'_{N-1}(B'_{N-1} Q_N B_{N-1} + R_{N-1}) u_{N-1} \\ & \qquad + 2E\{x_{N-1} \mid I_{N-1}\}' A'_{N-1} Q_N B_{N-1} u_{N-1} \big]. \end{aligned} \tag{2.4}$$

The minimization yields the optimal policy for the last stage:

$$\begin{aligned} u^*_{N-1} &= \mu^*_{N-1}(I_{N-1}) \\ &= -(B'_{N-1} Q_N B_{N-1} + R_{N-1})^{-1} B'_{N-1} Q_N A_{N-1} E\{x_{N-1} \mid I_{N-1}\}, \end{aligned} \tag{2.5}$$

and upon substitution in Eq. (2.4), we obtain

$$\begin{aligned} J_{N-1}(I_{N-1}) = & \mathop{E}_{x_{N-1}} \{ x'_{N-1} K_{N-1} x_{N-1} \mid I_{N-1} \} \\ & + \mathop{E}_{x_{N-1}} \big\{ (x_{N-1} - E\{x_{N-1} \mid I_{N-1}\})' \\ & \qquad \cdot P_{N-1} (x_{N-1} - E\{x_{N-1} \mid I_{N-1}\}) \mid I_{N-1} \big\} \\ & + \mathop{E}_{w_{N-1}} \{ w'_{N-1} Q_N w_{N-1} \}, \end{aligned} \tag{2.6}$$

where the matrices K_{N-1} and P_{N-1} are given by

$$P_{N-1} = A'_{N-1} Q_N B_{N-1} (R_{N-1} + B'_{N-1} Q_N B_{N-1})^{-1} B'_{N-1} Q_N A_{N-1},$$
$$K_{N-1} = A'_{N-1} Q_N A_{N-1} - P_{N-1} + Q_{N-1}.$$

Note that the optimal policy (2.5) is identical to its perfect state information counterpart except that x_{N-1} is replaced by its conditional expectation $E\{x_{N-1} \mid I_{N-1}\}$. Note also that the cost-to-go $J_{N-1}(I_{N-1})$ exhibits a corresponding similarity to its perfect state information counterpart except that $J_{N-1}(I_{N-1})$ contains an additional middle term, which is in effect a penalty for estimation error.

Now the DP equation for period $N-2$ is

$$
\begin{aligned}
J_{N-2}(I_{N-2}) = \min_{u_{N-2}} \Big[& \underset{x_{N-2}, w_{N-2}, z_{N-1}}{E} \{x'_{N-2} Q_{N-2} x_{N-2} + u'_{N-2} R_{N-2} u_{N-2} \\
& + J_{N-1}(I_{N-1}) \mid I_{N-2}, u_{N-2}\}\Big] \\
= E\{ & x'_{N-2} Q_{N-2} x_{N-2} \mid I_{N-2}\} \\
& + \min_{u_{N-2}} \big[u'_{N-2} R_{N-2} u_{N-2} + x'_{N-1} K_{N-1} x_{N-1} \mid I_{N-2}\}\big] \\
& + E\Big\{\big(x_{N-1} - E\{x_{N-1} \mid I_{N-1}\}\big)' \\
& \quad \cdot P_{N-1}\big(x_{N-1} - E\{x_{N-1} \mid I_{N-1}\}\big) \mid I_{N-2}, u_{N-2}\Big\} \\
& + \underset{w_{N-1}}{E} \{w'_{N-1} Q_N w_{N-1}\}.
\end{aligned}
\tag{2.7}
$$

Note that we have excluded the next to last term from the minimization with respect to u_{N-2}. We have done so since this term turns out to be independent of u_{N-2}. To show this fact, we need the following lemma.

The lemma says essentially that the quality of estimation as expressed by the statistics of the error $x_k - E\{x_k \mid I_k\}$ cannot be influenced by the choice of control. This is due to the linearity of both the system and the measurement equation.

Lemma 2.1: For every k, there is a function M_k such that we have

$$x_k - E\{x_k \mid I_k\} = M_k(x_0, w_0, \ldots, w_{k-1}, v_0, \ldots, v_k),$$

independently of the policy being used.

Proof: Fix a policy and consider the following two systems. In the first system there is control as determined by the policy,

$$x_{k+1} = A_k x_k + B_k u_k + w_k, \qquad z_k = C_k x_k + v_k,$$

while in the second system there is no control,

$$\overline{x}_{k+1} = A_k \overline{x}_k + \overline{w}_k, \qquad \overline{z}_k = C_k \overline{x}_k + \overline{v}_k.$$

We consider the evolution of these two systems when their initial conditions are identical,

$$x_0 = \overline{x}_0,$$

and when their system disturbance and observation noise vectors are also identical,

$$w_k = \overline{w}_k, \qquad v_k = \overline{v}_k, \qquad k = 0, 1, \ldots, N-1.$$

Consider the vectors

$$Z^k = (z_0, \ldots, z_k)', \qquad \overline{Z}^k = (\overline{z}_0, \ldots, \overline{z}_k)',$$
$$W^k = (w_0, \ldots, w_k)', \qquad V^k = (v_0, \ldots, v_k)', \qquad U^k = (u_0, \ldots, u_k)'.$$

Linearity implies the existence of matrices F_k, G_k, and H_k such that

$$x_k = F_k x_0 + G_k U^{k-1} + H_k W^{k-1},$$
$$\overline{x}_k = F_k x_0 + H_k W^{k-1}.$$

Since the vector $U^{k-1} = (u_0, \ldots, u_{k-1})'$ is part of the information vector I_k, we have $U^{k-1} = E\{U^{k-1} \mid I_k\}$, so

$$E\{x_k \mid I_k\} = F_k E\{x_0 \mid I_k\} + G_k U^{k-1} + H_k E\{W^{k-1} \mid I_k\},$$
$$E\{\overline{x}_k \mid I_k\} = F_k E\{x_0 \mid I_k\} + H_k E\{W^{k-1} \mid I_k\}.$$

We thus obtain

$$x_k - E\{x_k \mid I_k\} = \overline{x}_k - E\{\overline{x}_k \mid I_k\}.$$

From the equations for z_k and $\overline{z}_k$, we see that

$$\overline{Z}^k = Z^k - R_k U^{k-1} = S_k W^{k-1} + T_k V^k,$$

where R_k, S_k, and T_k are some matrices of appropriate dimension. Thus, the information provided by $I_k = (Z^k, U^{k-1})$ regarding $\overline{x}_k$ is summarized in $\overline{Z}^k$, and we have $E\{\overline{x}_k \mid I_k\} = E\{\overline{x}_k \mid \overline{Z}^k\}$, so that

$$x_k - E\{x_k \mid I_k\} = \overline{x}_k - E\{\overline{x}_k \mid \overline{Z}^k\}.$$

The function M_k given by

$$M_k(x_0, w_0, \ldots, w_{k-1}, v_0, \ldots, v_k) = \overline{x}_k - E\{\overline{x}_k \mid \overline{Z}^k\}$$

serves the purpose stated in the lemma. **Q.E.D.**

We can now justify excluding the term

$$E\{(x_{N-1} - E\{x_{N-1} \mid I_{N-1}\})' P_{N-1} (x_{N-1} - E\{x_{N-1} \mid I_{N-1}\}) \mid I_{N-2}, u_{N-2}\}$$

from the minimization in Eq. (2.7), as being independent of u_{N-2}. Indeed, by using the lemma, we see that

$$x_{N-1} - E\{x_{N-1} \mid I_{N-1}\} = \xi_{N-1},$$

where ξ_{N-1} is a function of $x_0, w_0, \ldots, w_{N-2}, v_0, \ldots, v_{N-1}$. We have that ξ_{N-1} as well as I_{N-2} are independent of u_{N-2}, so the conditional expectation of $\xi'_{N-1} P_{N-1} \xi_{N-1}$ satisfies

$$E\{\xi'_{N-1} P_{N-1} \xi_{N-1} \mid I_{N-2}, u_{N-2}\} = E\{\xi'_{N-1} P_{N-1} \xi_{N-1} \mid I_{N-2}\}.$$

Returning now to our problem, the minimization in Eq. (2.7) yields, using an argument similar to the one for the last stage,

$$\begin{aligned} u^*_{N-2} &= \mu^*_{N-2}(I_{N-2}) \\ &= -(R_{N-2} + B'_{N-2} K_{N-1} B_{N-2})^{-1} B'_{N-2} K_{N-1} A_{N-2} E\{x_{N-2} \mid I_{N-2}\}. \end{aligned}$$

We can proceed similarly to obtain the optimal policy for every stage:

$$\mu^*_k(I_k) = L_k E\{x_k \mid I_k\}, \tag{2.8}$$

where the matrix L_k is given by

$$L_k = -(R_k + B'_k K_{k+1} B_k)^{-1} B'_k K_{k+1} A_k, \tag{2.9}$$

with the matrices K_k given recursively by the Riccati equation

$$K_N = Q_N,$$

$$P_k = A'_k K_{k+1} B_k (R_k + B'_k K_{k+1} B_k)^{-1} B'_k K_{k+1} A_k, \tag{2.10}$$

$$K_k = A'_k K_{k+1} A_k - P_k + Q_k. \tag{2.11}$$

The key step in this derivation is that at stage k of the DP algorithm, the minimization over u_k that defines $J_k(I_k)$ involves the additional terms

$$E\big\{\big(x_s - E\{x_s \mid I_s\}\big)' P_s \big(x_s - E\{x_s \mid I_s\}\big) \mid I_k, u_k\big\},$$

where $s = k+1, \ldots, N-1$. By using the argument of the proof of the earlier lemma, it can be seen that none of these terms depends on u_k so that the presence of these terms does not affect the minimization in the DP algorithm. As a result, the optimal policy is the same as the one for the perfect information case, except that the state x_k is replaced by its conditional expectation $E\{x_k \mid I_k\}$.

It is interesting to note that the optimal controller can be decomposed into the two parts shown in Fig. 5.2.1:

(a) An *estimator*, which uses the data to generate the conditional expectation $E\{x_k \mid I_k\}$.

(b) An *actuator*, which multiplies $E\{x_k \mid I_k\}$ by the gain matrix L_k and applies the control input $u_k = L_k E\{x_k \mid I_k\}$.

Furthermore, the gain matrix L_k is independent of the statistics of the problem and is the same as the one that would be used if we were faced with the deterministic problem, where w_k and x_0 would be fixed and equal to their expected values. On the other hand, as shown in Appendix E, the estimate $\hat{x}$ of a random vector x given some information (random vector) I, which minimizes the mean squared error $E_x\{\|x - \hat{x}\|^2 \mid I\}$ is precisely the conditional expectation $E\{x \mid I\}$ (expand the quadratic form and set to zero the derivative with respect to $\hat{x}$). Thus the *estimator portion of the optimal controller is an optimal solution of the problem of estimating the state x_k assuming no control takes place, while the actuator portion is an optimal solution of the control problem assuming perfect state information prevails*. This property, which shows that the two portions of the optimal controller can be designed independently as optimal solutions of an estimation and a control problem, has been called the *separation theorem for linear systems and quadratic cost* and occupies a central position in modern automatic control theory.

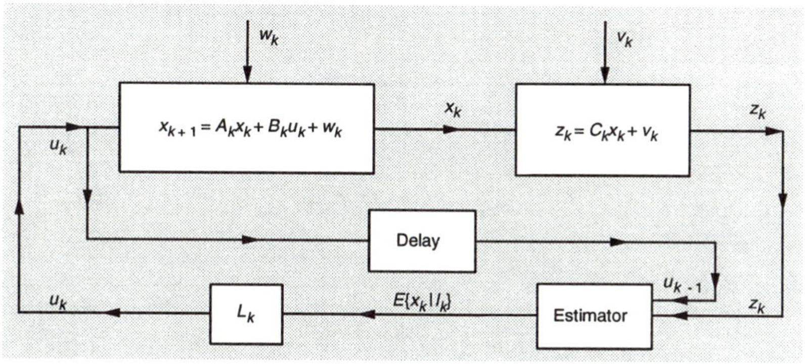

Figure 5.2.1 Structure of the optimal controller for the linear-quadratic problem. It consists of an estimator, which generates the conditional expectation $E\{x_k \mid I_k\}$, and an actuator, which multiplies $E\{x_k \mid I_k\}$ by the gain matrix L_k.

Another interesting observation is that the optimal controller applies at each time k the control that would be applied when faced with the deterministic problem of minimizing the cost-to-go

$$x_N' Q_N x_N + \sum_{i=k}^{N-1} (x_i' Q_i x_i + u_i' R_i u_i),$$

and the input disturbances $w_k, w_{k+1}, \ldots, w_{N-1}$ and current state x_k were known and fixed at their conditional expected values, which are zero and $E\{x_k \mid I_k\}$, respectively. This is another manifestation of the *certainty equivalence principle*, which was referred to in Section 4.1. A similar result holds in the case of correlated disturbances; see Exercise 5.1.

Implementation Aspects – Steady-State Controller

As explained in the perfect information case, the linear form of the actuator portion of the optimal policy is particularly attractive for implementation. In the imperfect information case, however, we need to implement an estimator that produces the conditional expectation

$$\hat{x}_k = E\{x_k \mid I_k\},$$

and this is not easy in general. Fortunately, in the important special case, where *the disturbances* w_k*,* v_k*, and the initial state* x_0 *are Gaussian random vectors*, a convenient implementation of the estimator is possible by means of the Kalman filtering algorithm, which is developed in Appendix E. This algorithm is organized recursively so that to produce $\hat{x}_{k+1}$ at time $k+1$, only the most recent measurement z_{k+1} is needed, together with $\hat{x}_k$ and u_k. In particular, we have for all $k = 0, \ldots, N-1$,

$$\hat{x}_{k+1} = A_k \hat{x}_k + B_k u_k + \Sigma_{k+1|k+1} C'_{k+1} N_{k+1}^{-1} \big(z_{k+1} - C_{k+1}(A_k \hat{x}_k + B_k u_k)\big), \tag{2.12}$$

and

$$\hat{x}_0 = E\{x_0\} + \Sigma_{0|0} C'_0 N_0^{-1} \big(z_0 - C_0 E\{x_0\}\big), \tag{2.13}$$

where the matrices $\Sigma_{k|k}$ are precomputable and are given recursively by

$$\Sigma_{k+1|k+1} = \Sigma_{k+1|k} - \Sigma_{k+1|k} C'_{k+1} (C_{k+1} \Sigma_{k+1|k} C'_{k+1} + N_{k+1})^{-1} C_{k+1} \Sigma_{k+1|k}, \tag{2.14}$$

$$\Sigma_{k+1|k} = A_k \Sigma_{k|k} A'_k + M_k, \qquad k = 0, 1, \ldots, N-1, \tag{2.15}$$

with

$$\Sigma_{0|0} = S - S C'_0 (C_0 S C'_0 + N_0)^{-1} C_0 S.$$

In these equations, M_k, N_k, and S are the covariance matrices of w_k, v_k, and x_0, respectively, and we assume that w_k and v_k have zero mean; that is

$$E\{w_k\} = E\{v_k\} = 0,$$

$$M_k = E\{w_k w'_k\}, \qquad N_k = E\{v_k v'_k\},$$

$$S = E\Big\{\big(x_0 - E\{x_0\}\big)\big(x_0 - E\{x_0\}\big)'\Big\}.$$

In addition, the matrices N_k are assumed to be positive definite.

Consider now the case where the system and measurement equations and the disturbance statistics are stationary. We can then drop subscripts from the system matrices. Assume that *the pair* (A, B) *is controllable and that the matrix* Q *can be written as* $Q = F'F$*, where* F *is a matrix such that the pair* (A, F) *is observable*. By the theory of Section 4.1, if the horizon tends to infinity, the optimal controller tends to the steady-state policy

$$\mu^*(I_k) = L\hat{x}_k, \tag{2.16}$$

where

$$L = -(R + B'KB)^{-1}B'KA, \tag{2.17}$$

and K is the unique positive semidefinite symmetric solution of the algebraic Riccati equation

$$K = A'\left(K - KB(R + B'KB)^{-1}B'K\right)A + Q. \tag{2.18}$$

By a similar argument, it can be shown (see Appendix E) that $\hat{x}_k$ can be generated in the limit as $k \to \infty$ by a steady-state Kalman filtering algorithm:

$$\hat{x}_{k+1} = (A + BL)\hat{x}_k + \overline{\Sigma}C'N^{-1}\big(z_{k+1} - C(A + BL)\hat{x}_k\big), \tag{2.19}$$

where $\overline{\Sigma}$ is given by

$$\overline{\Sigma} = \Sigma - \Sigma C'(C\Sigma C' + N)^{-1}C\Sigma, \tag{2.20}$$

and Σ is the unique positive semidefinite symmetric solution of the Riccati equation

$$\Sigma = A\big(\Sigma - \Sigma C'(C\Sigma C' + N)^{-1}C\Sigma\big)A' + M. \tag{2.21}$$

The assumptions required for this are that *the pair (A, C) is observable and that the matrix M can be written as $M = DD'$, where D is a matrix such that the pair (A, D) is controllable.* The steady-state controller of Eqs. (2.16), (2.17), and (2.19) is particularly attractive for practical implementation. Furthermore, as shown in Appendix E, it results in a stable closed-loop system, under the preceding controllability and observability assumptions.

5.3 MINIMUM VARIANCE CONTROL OF LINEAR SYSTEMS

We have considered so far the control of linear systems in state variable form as in the previous section. However, linear systems are often modeled by means of an input-output equation, which is more economical in the number of parameters needed to describe the system dynamics. In this section we consider single-input, single-output, linear, time-invariant systems, and a special type of quadratic cost function. The resulting optimal policy is particularly simple and has found wide application. We first introduce some of the basic facts regarding linear systems in input-output form. Detailed discussions may be found in [AsW84] , [AsW90], [GoS84], and [Whi63].

We consider a single-input single-output discrete-time linear system, which is specified by an equation of the form

$$y_k + a_1 y_{k-1} + \cdots + a_m y_{k-m} = b_0 u_k + b_1 u_{k-1} + \cdots + b_m u_{k-m}, \qquad (3.1)$$

where a_i, b_i are given scalars. The scalar sequences $\{u_k \mid k = 0, \pm 1, \pm 2, \ldots\}$, $\{y_k \mid k = 0, \pm 1, \pm 2, \ldots\}$ are viewed as the input and output of the system, respectively. Note that we allow time to extend to $-\infty$ as well as $+\infty$; this will be useful for describing generic properties of the system relating to stability. We will later revert to our usual convention of starting at time 0 and proceeding forward.

It is convenient to describe this type of system by means of the *backward shift operator*, denoted s, which when operating on a sequence $\{x_k \mid k = 0, \pm 1, \pm 2, \ldots\}$ shifts its index by one unit; that is,

$$s(x_k) = x_{k-1}, \qquad k = 0, \pm 1, \pm 2, \ldots$$

We denote by s^r the operator resulting from r successive applications of s:

$$s^r(x_k) = x_{k-r}, \qquad k = 0, \pm 1, \pm 2, \ldots \qquad (3.2)$$

We also write for simplicity $s^r x_k = x_{k-r}$. The *forward shift operator*, denoted s^{-1}, is the inverse of s and is defined by

$$s^{-1}(x_k) = x_{k+1}, \qquad k = 0, \pm 1, \pm 2, \ldots$$

Thus the notation (3.2) holds for all integers r. We can form linear combinations of operators of the form s^r. Thus, for example, the operator $(s + 2s^2)$ is defined by

$$(s + 2s^2)(x_k) = x_{k-1} + 2x_{k-2}, \qquad k = 0, \pm 1, \pm 2, \ldots$$

With this notation, Eq. (3.1) can be written as

$$A(s) y_k = B(s) u_k, \qquad (3.3)$$

where $A(s)$, $B(s)$ are the operators

$$A(s) = 1 + a_1 s + \cdots + a_m s^m, \qquad (3.4)$$

$$B(s) = b_0 + b_1 s + \cdots + b_m s^m. \qquad (3.5)$$

Sometimes it is convenient to write the equation $A(s) y_k = B(s) u_k$ as

$$y_k = \frac{B(s)}{A(s)} u_k$$

or

$$\frac{A(s)}{B(s)} y_k = u_k.$$

The meaning of both equations is that the sequences $\{y_k\}$ and $\{u_k\}$ are related via $A(s)y_k = B(s)u_k$. There is a certain ambiguity here in that, for a fixed $\{u_k\}$, the equation $A(s)y_k = B(s)u_k$ has an infinite number of solutions in $\{y_k\}$. For example, the equation

$$y_k + a y_{k-1} = u_k$$

for $u_k \equiv 0$ has as solutions all sequences of the form $y_k = \beta(-a)^k$, where β is any scalar; the solution becomes unique only after some boundary condition for the sequence $\{y_k\}$ is specified. As will be discussed shortly, however, for stable systems and for a *bounded* sequence $\{u_k\}$ there is a unique solution $\{y_k\}$ that is *bounded*. It is this solution that will be denoted by $\big(B(s)/A(s)\big)u_k$ in what follows. The reader who is familiar with linear dynamic system theory will note that $B(s)/A(s)$ can be viewed as a *transfer function* involving z-transforms.

We now introduce some terminology. When the sequences $\{y_k\}$ and $\{u_k\}$ satisfy $A(s)y_k = B(s)u_k$, we say that y_k is *obtained by passing u_k through the filter $B(s)/A(s)$*. This comes from engineering terminology, where linear time-invariant systems are commonly referred to as filters. We also refer to the equation $A(s)y_k = B(s)u_k$ as the filter $B(s)/A(s)$.

A filter $B(s)/A(s)$ is said to be *stable* if the polynomial $A(s)$ has all its (complex) roots strictly outside the unit circle of the complex plane; that is, $|\rho| > 1$ for all complex ρ satisfying $A(\rho) = 0$. A stable filter $B(s)/A(s)$ has the following two properties:

(a) Every solution $\{y_k\}$ of

$$A(s)y_k = 0$$

satisfies $\lim_{k\to\infty} y_k = 0$; that is, the output y_k tends to zero if the input sequence $\{u_k\}$ is identically zero.

(b) For every bounded sequence $\{\overline{u}_k\}$, the equation

$$A(s)y_k = B(s)\overline{u}_k$$

has a *unique* solution $\{\overline{y}_k\}$ within the class of bounded sequences. Furthermore, every solution $\{y_k\}$ (possibly unbounded) of the equation satisfies

$$\lim_{k\to\infty} (y_k - \overline{y}_k) = 0. \tag{3.6}$$

For example, consider the system

$$y_k - 0.5 y_{k-1} = u_k.$$

Given the bounded input sequence $\overline{u}_k = \{\ldots, 1, 1, 1, \ldots\}$, the set of all solutions is given by

$$y_k = 2 + \frac{\beta}{2^k},$$

where β is a scalar, but of these the only bounded solution is $\overline{y}_k = \{\ldots, 2, 2, 2, \ldots\}$. The solution $\{\overline{y}_k\}$ can thus be naturally associated with the input sequence $\{u_k\}$; it is also known as the *forced response* of the system due to the input $\{u_k\}$.

ARMAX Models – Reduction to State Space Form

We now consider a linear system with output y_k, which is driven by two inputs: a random noise input ϵ_k, and a control input u_k. It has the form

$$\begin{aligned} y_k + a_1 y_{k-1} + \cdots + a_m y_{k-m} = b_1 u_{k-1} + \cdots + b_m u_{k-m} \\ + \epsilon_k + c_1 \epsilon_{k-1} + \cdots + c_m \epsilon_{k-m}, \end{aligned} \tag{3.7}$$

and it is known as an ARMAX model (AutoRegressive, Moving Average, with eXogenous input). We assume throughout that the random variables ϵ_k are mutually independent. We can write the model in the shorthand form

$$A(s) y_k = B(s) u_k + C(s) \epsilon_k,$$

where the polynomials $A(s)$, $B(s)$, and $C(s)$ are given by

$$A(s) = 1 + a_1 s + \cdots + a_m s^m,$$
$$B(s) = b_1 s + \cdots + b_m s^m,$$
$$C(s) = 1 + c_1 s + \cdots + c_m s^m.$$

The ARMAX model is very common and its derivation is outlined in Appendix F, where it is shown that without loss of generality we can assume that $C(s)$ *has no roots strictly inside the unit circle.* For much of the analysis in subsequent sections, it will be necessary to exclude the critical case where $C(s)$ has roots *on* the unit circle and assume that $C(s)$ has all its roots strictly outside the unit circle. This assumption is satisfied in most practical cases.

In several situations, analysis and algorithms relating to the ARMAX model are greatly simplified if $C(s) = 1$ so that the noise terms $C(s)\epsilon_k = \epsilon_k$ are independent. However, this is typically an unrealistic assumption. To emphasize this point and see how easily the noise can be correlated, suppose that we have a first-order system

$$x_{k+1} = a x_k + w_k,$$

where we observe

$$y_k = x_k + v_k.$$

Then

$$\begin{aligned} y_{k+1} &= x_{k+1} + v_{k+1} \\ &= a x_k + w_k + v_{k+1} \\ &= a(y_k - v_k) + w_k + v_{k+1}, \end{aligned}$$

so finally

$$y_{k+1} = a y_k + v_{k+1} - a v_k + w_k.$$

However, the noise sequence $\{v_{k+1} - a v_k + w_k\}$ is correlated even if $\{v_k\}$ and $\{w_k\}$ are individually and mutually independent.

The ARMAX model (3.7) can be put into state space form. The process is based on state augmentation and can perhaps be best understood in terms of an example. Consider the system

$$y_k + a_1 y_{k-1} + a_2 y_{k-2} = b_1 u_{k-1} + b_2 u_{k-2} + \epsilon_k + c_1 \epsilon_{k-1}. \tag{3.8}$$

We have

$$\begin{pmatrix} y_{k+1} \\ y_k \\ u_k \\ \epsilon_{k+1} \end{pmatrix} = \begin{pmatrix} -a_1 & -a_2 & b_2 & c_1 \\ 1 & 0 & 0 & 0 \\ 0 & 0 & 0 & 0 \\ 0 & 0 & 0 & 0 \end{pmatrix} \begin{pmatrix} y_k \\ y_{k-1} \\ u_{k-1} \\ \epsilon_k \end{pmatrix} + \begin{pmatrix} b_1 \\ 0 \\ 1 \\ 0 \end{pmatrix} u_k + \begin{pmatrix} \epsilon_{k+1} \\ 0 \\ 0 \\ \epsilon_{k+1} \end{pmatrix}. \tag{3.9}$$

By setting

$$x_k = \begin{pmatrix} y_k \\ y_{k-1} \\ u_{k-1} \\ \epsilon_k \end{pmatrix}, \qquad w_k = \begin{pmatrix} \epsilon_{k+1} \\ 0 \\ 0 \\ \epsilon_{k+1} \end{pmatrix},$$

$$A = \begin{pmatrix} -a_1 & -a_2 & b_2 & c_1 \\ 1 & 0 & 0 & 0 \\ 0 & 0 & 0 & 0 \\ 0 & 0 & 0 & 0 \end{pmatrix}, \qquad B = \begin{pmatrix} b_1 \\ 0 \\ 1 \\ 0 \end{pmatrix},$$

we can write Eq. (3.9) as

$$x_{k+1} = A x_k + B u_k + w_k, \tag{3.10}$$

where $\{w_k\}$ is a stationary independent process. We have arrived at this state space model through state augmentation. Notice that the state x_k includes ϵ_k. Thus if the controller is assumed to know at time k only the present and past outputs $y_k, y_{k-1}, \ldots$, and past controls $u_{k-1}, u_{k-2}, \ldots$ (but not $\epsilon_k, \epsilon_{k-1}, \ldots$), we are faced with a model of imperfect state information. If $c_1 = 0$ in Eq. (3.8) then the state space model can be simplified so that

$$x_k = \begin{pmatrix} y_k \\ y_{k-1} \\ u_{k-1} \end{pmatrix},$$

in which case we have perfect state information. More generally, we have perfect state information in the ARMAX model (3.7) if $b_1 \neq 0$ and $c_1 = c_2 = \cdots = c_m = 0$.

Minimum Variance Control: Perfect State Information Case

Consider the perfect state information case of the ARMAX model (3.7):

$$y_k + a_1 y_{k-1} + \cdots + a_m y_{k-m} = b_1 u_{k-1} + \cdots + b_m u_{k-m} + \epsilon_k, \tag{3.11}$$

where $b_1 \neq 0$. A problem of interest, known as the *minimum variance control problem*, is to select u_k as a function of the present and past outputs $y_k, y_{k-1}, \ldots$, as well as the past controls $u_{k-1}, u_{k-2}, \ldots$, so as to minimize the cost

$$E\left\{\sum_{k=1}^{N} (y_k)^2\right\}. \tag{3.12}$$

There are no constraints on u_k. By transforming the system to state space form, we see that this problem can be reduced to a perfect state information linear-quadratic problem where the state x_k is

$$\big(y_k, y_{k-1}, \ldots, y_{k-m+1}, u_{k-1}, \ldots, u_{k-m+1}\big)'.$$

The problem is of the same nature as the linear-quadratic problem of Section 4.1 except that the corresponding matrices R_k in the quadratic cost function are zero here. Nonetheless, in Section 4.1 we used the invertibility of R_k only to ensure that various matrices in the optimal policy and the Riccati equation are invertible. If invertibility of these matrices can be guaranteed by other means, the same analysis applies even if R_k is positive semidefinite. This is indeed the case here. An analysis analogous to the one of Section 4.1 shows that the optimal control u_k^* at time k (given $y_k, y_{k-1}, \ldots, y_{k-m+1}$ and $u_{k-1}, \ldots, u_{k-m+1}$) is the same as the one that would be applied if all future disturbances $\epsilon_{k+1}, \ldots, \epsilon_N$ were set equal to zero, their expected value (certainty equivalence). It follows that

$$\begin{aligned}\mu_k^*(y_k, \ldots, y_{k-m+1}, u_{k-1}, \ldots, u_{k-m+1}) = \frac{1}{b_1}(&a_1 y_k + \cdots + a_m y_{k-m+1} \\ &- b_2 u_{k-1} - \cdots - b_m u_{k-m+1}),\end{aligned}$$

and $\{u_k^*\}$ is generated via the equation

$$b_1 u_k^* + b_2 u_{k-1}^* + \cdots + b_m u_{k-m+1}^* = a_1 y_k + a_2 y_{k-1} + \cdots + a_m y_{k-m+1}. \tag{3.13}$$

In other words, $\{u_k^*\}$ is generated by passing $\{y_k\}$ through the linear filter $\overline{A}(s)/\overline{B}(s)$, where

$$\overline{A}(s) = a_1 + a_2 s + \cdots + a_m s^{m-1} = s^{-1}\big(A(s) - 1\big), \tag{3.14}$$

$$\overline{B}(s) = b_1 + b_2 s + \cdots + b_m s^{m-1} = s^{-1} B(s), \tag{3.15}$$

as shown in Fig. 5.3.1. The resulting closed-loop system is

$$y_k = \epsilon_k \tag{3.16}$$

and the associated cost is

$$N\, E\{(\epsilon_k)^2\}.$$

Notice that the optimal policy, called *minimum variance* control law, is time invariant and does not depend on the horizon N.

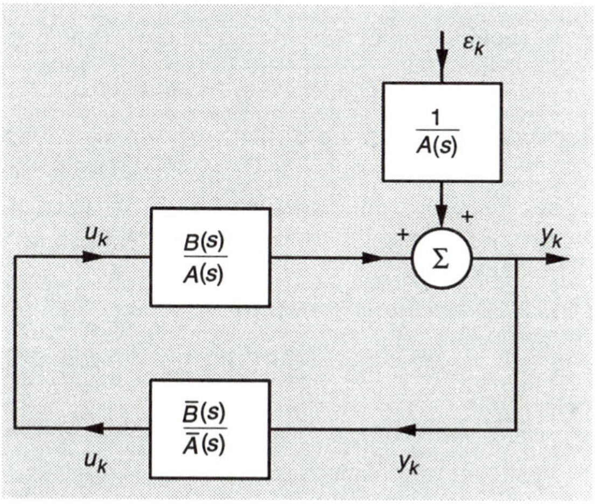

Figure 5.3.1 Minimum variance control with perfect state information. Structure of the optimal closed-loop system, where $A(s) = 1 + a_1 s + \cdots + a_m s^m$, $B(s) = b_1 s + \cdots + b_m s^m$, $\overline{A}(s) = s^{-1}\big(A(s) - 1\big)$, and $\overline{B}(s) = s^{-1}B(s)$.

Whereas the optimal closed-loop system as given by Eq. (3.16) is clearly stable, we can anticipate serious difficulties if the filter $\overline{A}(s)/\overline{B}(s)$ in the feedback loop is unstable. For if $\overline{B}(s)$ has some roots inside the unit circle, then the sequence $\{u_k\}$ will tend to be unbounded. This is illustrated by the following example.

Example 3.1 (An Optimal but Unstable Controller)

Consider the system

$$y_k + y_{k-1} = u_{k-1} - 2u_{k-2} + \epsilon_k.$$

The optimal policy is

$$u_k = 2u_{k-1} + y_k$$

and the optimal closed-loop system is

$$y_k = \epsilon_k,$$

which is a stable system. On the other hand, the last two equations yield

$$u_k = 2u_{k-1} + \epsilon_k.$$

Thus, u_k is generated by an *unstable* system, and in fact it is given by

$$u_k = \sum_{n=0}^{k} 2^n \epsilon_{k-n}.$$

Therefore, even though the output y_k stays bounded, the control u_k typically becomes unbounded.

For another view of the same difficulty, suppose that the coefficients $b_1, \ldots, b_m$ of $\overline{B}(s)$ are slightly different from the ones of the true system. We will show that if the feedback filter $\overline{A}(s)/\overline{B}(s)$ is unstable, then the closed-loop system is also unstable in the sense that both u_k and y_k become unbounded with probability one.

Assume that the system is governed by

$$A^0(s)y_k = B^0(s)u_k + \epsilon_k, \tag{3.17}$$

while the policy is calculated under the assumption that the system model is

$$A(s)y_k = B(s)u_k + \epsilon_k,$$

where the coefficients of $A(s)$ and $B(s)$ differ slightly from those of $A^0(s)$, $B^0(s)$. Define $\overline{A}^0(s)$, $\overline{B}^0(s)$ by

$$1 + s\overline{A}^0(s) = A^0(s), \tag{3.18a}$$

$$s\overline{B}^0(s) = B^0(s). \tag{3.18b}$$

Note that $\overline{A}^0(s) = \overline{A}(s)$ and $\overline{B}^0(s) = \overline{B}(s)$ if $A^0(s) = A(s), B^0(s) = B(s)$. By multiplying Eq. (3.17) with $\overline{B}(s)$ and by using the relation defining the optimal policy

$$\overline{B}(s)u_k = \overline{A}(s)y_k,$$

we obtain

$$\overline{B}(s)A^0(s)y_k = B^0(s)\overline{A}(s)y_k + \overline{B}(s)\epsilon_k.$$

If the coefficients of $\overline{A}^0(s)$ and $\overline{B}^0(s)$ are close to those of $\overline{A}(s)$, $\overline{B}(s)$, then the roots of the polynomial

$$\overline{B}(s) + s\big(\overline{B}(s)\overline{A}^0(s) - \overline{B}^0(s)\overline{A}(s)\big)$$

are close to the roots of $\overline{B}(s)$. Thus *the closed-loop system is stable only if the roots of* $\overline{B}(s)$ *are outside the unit circle*, or equivalently, if and only if the filter $\overline{A}(s)/\overline{B}(s)$ is stable. If our model is exact, the closed-loop system will be stable due to what is commonly referred to as a *pole-zero cancellation*. However, the slightest modeling discrepancy will induce instability.

The conclusion from the preceding analysis is that the minimum variance control law is advisable only if it can be realized through a stable filter [$\overline{B}(s)$ has roots outside the unit circle]. Even if $\overline{B}(s)$ has its roots outside the unit circle, but some of these roots are near the unit circle, the performance of the minimum variance policy can be very sensitive to variations in the parameters of the polynomials $A(s)$ and $B(s)$. One way to overcome this sensitivity is to change the cost to

$$E\left\{\sum_{k=1}^{N}\left((y_k)^2 + R(u_{k-1})^2\right)\right\},$$

where R is some positive parameter. This requires solution via the Riccati equation as in Section 4.1. For a detailed derivation, see [Ast83].

In some problems, the system equation includes an additional external input sequence $\{v_k\}$, the values of which can be measured by the controller as they occur. In particular, consider the scalar system

$$\begin{aligned} y_k + a_1 y_{k-1} + \cdots + a_m y_{k-m} &= b_1 u_{k-1} + \cdots + b_m u_{k-m} \\ &\quad + d_1 v_{k-1} + \cdots + d_m v_{k-m} + \epsilon_k, \end{aligned}$$

where each value v_k becomes known to the controller without error at time k. The minimum variance controller then takes the form

$$\begin{aligned} &\mu_k^*(y_k, \ldots, y_{k-m+1}, u_{k-1}, \ldots, u_{k-m+1}, v_k, \ldots, v_{k-m+1}) \\ &\qquad = \frac{1}{b_1}\big(a_1 y_k + \cdots + a_m y_{k-m+1} - d_1 v_k - \cdots - d_m v_{k-m+1} \\ &\qquad\qquad - b_2 u_{k-1} \cdots - b_m u_{k-m+1}\big), \end{aligned}$$

and the optimal controls $\{u_k^*\}$ are generated by

$$\overline{B}(s)u_k^* = \overline{A}(s)y_k - \overline{D}(s)v_k,$$

where

$$\overline{A}(s) = a_1 + a_2 s + \cdots + a_m s^{m-1},$$

$$\overline{B}(s) = b_1 + b_2 s + \cdots + b_m s^{m-1},$$

$$\overline{D}(s) = d_1 + d_2 s + \cdots + d_m s^{m-1}.$$

The closed-loop system is again $y_k = \epsilon_k$, but for practical purposes it is stable only if $\overline{B}(s)$ has its roots outside the unit circle. The process whereby external inputs are measured and used for control is commonly referred to as *feedforward control*.

Imperfect State Information Case

Consider now the general ARMAX model

$$y_k + a_1 y_{k-1} + \cdots + a_m y_{k-m} = b_M u_{k-M} + \cdots + b_m u_{k-m} + \epsilon_k + c_1 \epsilon_{k-1} + \cdots + c_m \epsilon_{k-m}$$

or, equivalently,

$$A(s) y_k = B(s) u_k + C(s) \epsilon_k,$$

where

$$A(s) = 1 + a_1 s + \cdots + a_m s^m,$$
$$B(s) = b_M s^M + \cdots + b_m s^m,$$
$$C(s) = 1 + c_1 s + \cdots + c_m s^m.$$

We assume the following:

(1) $b_M \neq 0$ and $1 \leq M \leq m$.

(2) $\{\epsilon_k\}$ is a zero mean, independent, stationary process.

(3) The polynomial $C(s)$ has all its roots outside the unit circle. (As explained in Appendix F, this assumption is not overly restrictive.)

The controller knows at each time k the past inputs and outputs. Thus the information vector at time k is

$$I_k = \big(y_k, y_{k-1}, \ldots, y_{-m+1}, u_{k-1}, u_{k-2}, \ldots, u_{-m+M}\big). \tag{3.19}$$

(We include in the information vector the control inputs $u_{-1}, \ldots, u_{-m+M}$. If control starts at time 0, these inputs will be zero.) There are no constraints on u_k. The problem is to find a policy $\{\mu_0(I_0), \ldots, \mu_{N-1}(I_{N-1})\}$ that minimizes

$$E\left\{\sum_{k=1}^{N} (y_k)^2\right\}.$$

By using state augmentation, we can cast this problem into the framework of the linear-quadratic problem of Section 5.2. The corresponding linear system in state space format involves a state x_k given by

$$x_k = \big(y_{k+M-1}, \ldots, y_{k+M-m}, u_{k-1}, \ldots, u_{k+M-m}, \epsilon_{k+M-1}, \ldots, \epsilon_{k+M-m}\big).$$

Because $y_{k+M-1}, \ldots, y_{k+1}$ and $\epsilon_{k+M-1}, \ldots, \epsilon_{k+M-m}$ are unknown to the controller, we are faced with a problem of imperfect state information.

An analysis analogous to the one of Section 5.2 shows that certainty equivalence holds; that is, the *optimal control* u_k^* *at time* k *given* I_k *is the same as the one that would be applied in the deterministic problem where the current state*

$$x_k = \big(y_{k+M-1}, \ldots, y_{k+M-m}, u_{k-1}, \ldots, u_{k+M-m}, \epsilon_{k+M-1}, \ldots, \epsilon_{k+M-m}\big)$$

is set equal to its conditional expected value given I_k, and the future disturbances $\epsilon_{k+M}, \ldots, \epsilon_N$ are set equal to zero (their expected value).

Thus the optimal control $u_k^* = \mu_k^*(I_k)$ is obtained by solving for u_k the equation

$$\begin{aligned} E\{y_{k+M} \mid u_k, I_k\} &= E\{y_{k+M} \mid y_k, y_{k-1}, \ldots, y_{-m+1}, u_k, u_{k-1}, \ldots, u_{-m+M}\} \\ &= 0. \end{aligned} \tag{3.20}$$

This leads to the problem of calculating $E\{y_{k+M} \mid I_k, u_k\}$, known as the *forecasting* or *prediction* problem, which is important in its own right. We first treat the easier case where there is no delay ($M = 1$) and then discuss the more general case where the delay can be positive.

Forecasting for ARMAX Models – No Delay ($M = 1$)

Assume that $M = 1$. We would like to generate an equation for the forecast $E\{y_{k+1} \mid I_k, u_k\}$, and then determine the optimal control $u_k^* = \mu_k^*(I_k)$ by setting this forecast to zero. Let us introduce an auxiliary sequence $\{z_k\}$ via the equation

$$z_k = y_k - \epsilon_k. \tag{3.21}$$

A key fact is that, since $\{\epsilon_k\}$ is an independent, zero-mean sequence, we have

$$E\{z_{k+1} \mid I_k, u_k\} = E\{y_{k+1} \mid I_k, u_k\}.$$

We can thus obtain the desired forecast of y_{k+1} by forecasting z_{k+1} instead. We can then obtain the optimal control u_k^* by setting $E\{z_{k+1} \mid I_k, u_k^*\} = 0$.

By using the definition $z_k = y_k - \epsilon_k$ to express y_k in terms of z_k in the ARMAX model equation for $M = 1$, we obtain

$$z_{k+1} + c_1 z_k + \cdots + c_m z_{k-m+1} = b_1 u_k + \cdots + b_m u_{k-m+1} + w_k, \tag{3.22}$$

where

$$w_k = (c_1 - a_1) y_k + \cdots + (c_m - a_m) y_{k-m+1}.$$

We note that w_k is perfectly observable by the controller; however, the scalars $z_k, \ldots, z_{k-m+1}$ are not known to the controller because the initial conditions $z_0, \ldots, z_{1-m}$ of the system (3.22) are unknown. Nonetheless, the system (3.22) is stable, since the roots of the polynomial $C(s)$ have been assumed to be outside the unit circle. As a result, the initial conditions do not matter asymptotically. In other words, if we generate a sequence $\{\hat{y}_k\}$ using the system (3.22) and zero initial conditions, i.e.,

$$\hat{y}_{k+1} + c_1 \hat{y}_k + \cdots + c_m \hat{y}_{k-m+1} = b_1 u_k + \cdots + b_m u_{k-m+1} + w_k, \tag{3.23}$$

with

$$\hat{y}_0 = 0, \qquad \hat{y}_{-1} = 0, \qquad \ldots \qquad \hat{y}_{1-m} = 0,$$

then we will have

$$\lim_{k\to\infty} \big(\hat{y}_k - z_k\big) = 0.$$

Thus, $\hat{y}_{k+1}$ is an asymptotically accurate approximation to the optimal forecast $E\{y_{k+1} \mid I_k, u_k\}$.

Minimum Variance Control: Imperfect State Information and No Delay

Based on the earlier discussion, an asymptotically accurate approximation to the minimum variance policy is obtained by setting u_k to the value that makes $\hat{y}_{k+1} = 0$; that is, by solving for u_k the equation

$$\hat{y}_{k+1} + c_1\hat{y}_k + \cdots + c_m\hat{y}_{k-m+1} = b_1u_k + \cdots + b_mu_{k-m+1} + w_k.$$

If this policy is followed, however, the earlier forecasts $\hat{y}_k, \ldots, \hat{y}_{k-m+1}$ will be equal to zero. Thus the (approximate) minimum variance policy is given by

$$\begin{aligned} u_k &= \frac{1}{b_1}\big(w_k - b_2u_{k-1} - \cdots - b_mu_{k-m+1}\big) \\ &= \frac{1}{b_1}\big((a_1 - c_1)y_k + \cdots + (a_m - c_m)y_{k-m-1} \\ &\qquad - b_2u_{k-1} - \cdots - b_mu_{k-m+1}\big). \end{aligned}$$

By substituting this policy in the ARMAX model

$$\begin{aligned} y_{k+1} + a_1y_k + \cdots + a_my_{k-m+1} = {} & b_1u_k + \cdots + b_mu_{k-m+1} \\ & + \epsilon_{k+1} + c_1\epsilon_k + \cdots + c_m\epsilon_{k-m+1}, \end{aligned}$$

we see that the closed-loop system becomes

$$y_{k+1} - \epsilon_{k+1} + c_1(y_k - \epsilon_k) + \cdots + c_m(y_{k-m+1} - \epsilon_{k-m+1}) = 0,$$

or equivalently $C(s)(y_k - \epsilon_k) = 0$. Since $C(s)$ has its roots outside the unit circle, this is a stable system, and we have

$$y_k = \epsilon_k + \gamma(k),$$

where $\gamma(k) \to 0$ as $k \to \infty$.

Forecasting: The General Case

Consider now the general case where the delay M can be greater than 1. The forecasting problem can still be nicely solved by using a certain trick to transform the ARMAX equation into a more convenient form. To this end, we first obtain polynomials $F(s)$ and $G(s)$ of the form

$$F(s) = 1 + f_1 s + \cdots + f_{M-1} s^{M-1}, \tag{3.24}$$

$$G(s) = g_0 + g_1 s + \cdots + g_{m-1} s^{m-1} \tag{3.25}$$

that satisfy

$$C(s) = A(s)F(s) + s^M G(s). \tag{3.26}$$

The coefficients of $F(s)$ and $G(s)$ are uniquely determined from those of $C(s)$ and $A(s)$ by equating coefficients of both sides of the relation

$$1 + c_1 s + \cdots + c_m s^m = (1 + a_1 s + \cdots + a_m s^m)(1 + f_1 s + \cdots + f_{M-1} s^{M-1}) \\ + s^M (g_0 + g_1 s + \cdots + g_{m-1} s^{m-1}).$$

Example 3.2

Let $m = 3$ and $M = 2$. Then the preceding equation takes the form

$$1 + c_1 s + c_2 s^2 + c_3 s^3 = (1 + a_1 s + a_2 s^2 + a_3 s^3)(1 + f_1 s) + s^2 (g_0 + g_1 s + g_2 s^2),$$

and by equating coefficients we have

$$c_1 = a_1 + f_1, \quad c_2 = a_2 + a_1 f_1 + g_0, \quad c_3 = a_3 + a_2 f_1 + g_1, \quad a_3 f_1 + g_2 = 0,$$

from which f_1, g_0, g_1, and g_2 are uniquely determined.

The ARMAX model can be written as

$$A(s) y_{k+M} = \overline{B}(s) u_k + C(s) \epsilon_{k+M}, \tag{3.27}$$

where

$$\overline{B}(s) = s^{-M} B(s) = b_M + b_{M+1} s + \cdots + b_m s^{m-M}.$$

Multiplying both sides of Eq. (3.27) with $F(s)$, we have

$$F(s)A(s) y_{k+M} = F(s)\overline{B}(s) u_k + F(s)C(s) \epsilon_{k+M},$$

and using Eq. (3.26) to express $F(s)A(s)$ as $C(s) - s^M G(s)$, we obtain

$$\big(C(s) - s^M G(s)\big) y_{k+M} = F(s)\overline{B}(s) u_k + F(s)C(s) \epsilon_{k+M},$$

or equivalently

$$C(s)\big(y_{k+M} - F(s)\epsilon_{k+M}\big) = F(s)\overline{B}(s)u_k + G(s)y_k. \tag{3.28}$$

Let us now introduce the auxiliary sequence $\{z_k\}$ via the equation

$$z_{k+M} = y_{k+M} - F(s)\epsilon_{k+M} = y_{k+M} - \epsilon_{k+M} - f_1\epsilon_{k+M-1} - \cdots - f_{M-1}\epsilon_{k+1}. \tag{3.29}$$

Note that when $M = 1$, we have $F(s) = 1$ and $z_k = y_k - \epsilon_k$, so $\{z_k\}$ is the same sequence as the one introduced earlier for the case of no delay. Again, since $\{\epsilon_k\}$ is an independent, zero-mean sequence, by taking expectations in the definition $z_{k+M} = y_{k+M} - F(s)\epsilon_{k+M}$, we obtain

$$E\{z_{k+M} \mid I_k, u_k\} = E\{y_{k+M} \mid I_k, u_k\},$$

and we can obtain the desired forecast of y_{k+M} by forecasting z_{k+M} in its place. Furthermore, by Eq. (3.28), z_{k+M} is written as

$$C(s)z_{k+M} = w_k$$

or

$$z_{k+M} + c_1 z_{k+M-1} + \cdots + c_m z_{k+M-m} = w_k, \tag{3.30}$$

where

$$w_k = F(s)\overline{B}(s)u_k + G(s)y_k. \tag{3.31}$$

Since the scalar w_k of Eq. (3.31) is available at time k (i.e., it is determined from I_k and u_k), the system (3.30) can serve as a basis for forecasting z_{k+M}. We would be able to predict exactly z_{k+M} and use it as a forecast of y_{k+M} if we knew appropriate initial conditions with which to start the equation (3.30) that generates it. We don't know such initial conditions, but because this equation represents a stable system, the choice of initial conditions does not matter asymptotically, as we proceed to explain more formally.

We consider the sequence $\hat{y}_{k+M}$ generated by

$$\hat{y}_{k+M} + c_1\hat{y}_{k+M-1} + \cdots + c_m\hat{y}_{k+M-m} = w_k \tag{3.32}$$

with initial condition

$$\hat{y}_{M-1} = \hat{y}_{M-2} = \cdots = \hat{y}_{M-m} = 0, \tag{3.33}$$

and we claim that the forecast $E\{z_{k+M} \mid I_k\}$ can be approximated by $\hat{y}_{k+M}$. To see this, note that from Eqs. (3.30) to (3.33) we have

$$z_{k+M} = \hat{y}_{k+M} + \big(\gamma_1(k)z_{M-1} + \cdots + \gamma_m(k)z_{M-m}\big) \tag{3.34}$$

and

$$E\{z_{k+M} \mid I_k, u_k\} = \hat{y}_{k+M} + \sum_{i=1}^{m} \gamma_i(k) E\{z_{M-i} \mid I_k, u_k\},$$

where $\gamma_1(k), \ldots, \gamma_m(k)$ are appropriate scalars depending on k. Since $C(s)$ has all its roots outside the unit circle, we have (compare with the discussion on stability earlier in this section)

$$\lim_{k\to\infty} \gamma_1(k) = \lim_{k\to\infty} \gamma_2(k) = \cdots = \lim_{k\to\infty} \gamma_m(k) = 0. \tag{3.35}$$

It follows that, for large values of k,

$$\hat{y}_{k+M} \simeq E\{z_{k+M} \mid I_k, u_k\} = E\{y_{k+M} \mid I_k, u_k\}.$$

(More precisely, we have $|\hat{y}_{k+M} - E\{y_{k+M} \mid I_k, u_k\}| \to 0$ as $k \to \infty$, where the convergence is in the mean-square sense.)

In conclusion, an asymptotically accurate approximation to the optimal forecast $E\{y_{k+M} \mid I_k, u_k\}$ is given by $\hat{y}_{k+M}$ and is generated by the equation

$$\hat{y}_{k+M} + c_1\hat{y}_{k+M-1} + \cdots + c_m\hat{y}_{k+M-m} = F(s)\overline{B}(s)u_k + G(s)y_k \tag{3.36}$$

with the initial condition

$$\hat{y}_{M-1} = \hat{y}_{M-2} = \cdots = \hat{y}_{M-m} = 0. \tag{3.37}$$

Minimum Variance Control: The General Case

Based on the earlier discussion, the minimum variance policy is obtained by solving for u_k the equation $E\{y_{k+M} \mid I_k, u_k\} = 0$. Thus an asymptotically accurate approximation is obtained by setting u_k to the value that makes $\hat{y}_{k+M} = 0$, that is, by solving for u_k the equation [cf. Eqs. (3.36) and (3.37)]

$$F(s)\overline{B}(s)u_k + G(s)y_k = c_1\hat{y}_{k+M-1} + \cdots + c_m\hat{y}_{k+M-m}.$$

If this policy is followed, however, the earlier forecasts $\hat{y}_{k+M-1}, \ldots, \hat{y}_{k+M-m}$ will be equal to zero. Thus the (approximate) minimum variance policy is given by

$$F(s)\overline{B}(s)u_k + G(s)y_k = 0; \tag{3.38}$$

that is, u_k^* is generated by passing y_k through the linear filter

$$-G(s)/F(s)\overline{B}(s),$$

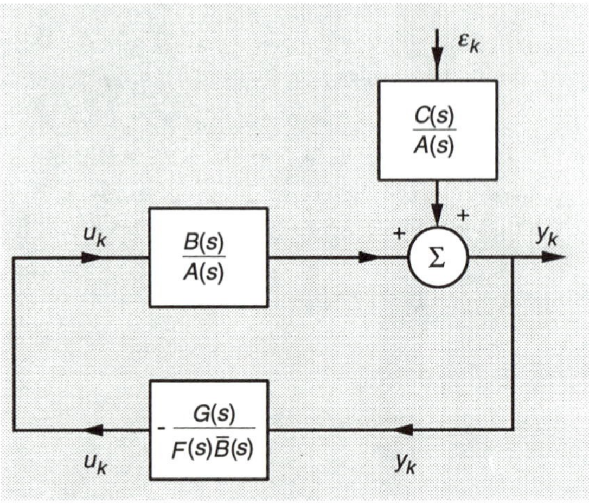

Figure 5.3.2 Minimum variance control with imperfect state information. Structure of the optimal closed-loop system.

as shown in Fig. 5.3.2.

From Eqs. (3.28) and (3.38), we obtain the equation for the closed-loop system

$$C(s)\big(y_{k+M} - F(s)\epsilon_{k+M}\big) = 0.$$

Since $C(s)$ has its roots outside the unit circle, we obtain

$$y_{k+M} = F(s)\epsilon_{k+M} + \gamma(k),$$

where $\gamma(k) \to 0$ as $k \to \infty$. So asymptotically, the closed-loop system takes the form

$$y_k = \epsilon_k + f_1\epsilon_{k-1} + \cdots + f_{M-1}\epsilon_{k-M+1}.$$

Let us consider now the stability properties of the closed-loop system when the true system parameters differ slightly from those of the assumed model. Let the true system be described by

$$A^0(s)y_k = s^M \overline{B}^0(s)u_k + C^0(s)\epsilon_k, \tag{3.39}$$

while u_k is given by the minimum variance policy

$$F(s)\overline{B}(s)u_k + G(s)y_k = 0, \tag{3.40}$$

where

$$C(s) = A(s)F(s) + s^M G(s).$$

Operating on Eq. (3.39) with $F(s)\overline{B}(s)$ and using Eq. (3.40), we obtain

$$F(s)\overline{B}(s)A^0(s)y_k = -s^M\overline{B}^0(s)G(s)y_k + F(s)\overline{B}(s)C^0(s)\epsilon_k.$$

Combining the last two equations and collecting terms, we have

$$\{F(s)\overline{B}(s)A^0(s) + \big(C(s) - A(s)F(s)\big)\overline{B}^0(s)\}y_k = F(s)\overline{B}(s)C^0(s)\epsilon_k$$

or

$$\{\overline{B}^0(s)C(s) + F(s)\big(\overline{B}(s)A^0(s) - A(s)\overline{B}^0(s)\big)\}y_k = F(s)\overline{B}(s)C^0(s)\epsilon_k.$$

If the coefficients of $A^0(s)$, $\overline{B}^0(s)$, and $C^0(s)$ are near those of $A(s)$, $\overline{B}(s)$, and $C(s)$, then the poles of the closed-loop system will be near the roots of $\overline{B}(s)C(s)$. Thus *the closed-loop system will be in effect stable only if the roots of* $\overline{B}(s)$ *are strictly outside the unit circle*, similar to the perfect state information case examined earlier.

5.4 SUFFICIENT STATISTICS AND FINITE-STATE MARKOV CHAINS

The main difficulty with the DP algorithm for imperfect state information problems is that it is carried out over a state space of expanding dimension. As a new measurement is added at each stage k, the dimension of the state (the information vector I_k) increases accordingly. This motivates an effort to reduce the data that are truly necessary for control purposes. In other words, it is of interest to look for quantities known as *sufficient statistics*, which ideally would be of smaller dimension than I_k and yet summarize all the essential content of I_k as far as control is concerned.

Consider the DP algorithm (1.7) and (1.8) restated here for convenience:

$$J_{N-1}(I_{N-1}) = \min_{u_{N-1}\in U_{N-1}} \Big[\underset{x_{N-1},w_{N-1}}{E} \{g_N\big(f_{N-1}(x_{N-1},u_{N-1},w_{N-1})\big) + g_{N-1}(x_{N-1},u_{N-1},w_{N-1}) \mid I_{N-1},u_{N-1}\}\Big], \tag{4.1}$$

$$J_k(I_k) = \min_{u_k\in U_k} \Big[\underset{x_k,w_k,z_{k+1}}{E} \{g_k(x_k,u_k,w_k) + J_{k+1}(I_k,z_{k+1},u_k) \mid I_k,u_k\}\Big]. \tag{4.2}$$

Suppose that we can find a function $S_k(I_k)$ of the information vector, such that a minimizing control in Eqs. (4.1) and (4.2) depends on I_k via $S_k(I_k)$. By this we mean that the minimization in the right-hand side of the DP algorithm (4.1) and (4.2) can be written in terms of some function H_k as

$$\min_{u_k\in U_k} H_k\big(S_k(I_k),u_k\big).$$

Such a function S_k is called a *sufficient statistic*. Its salient feature is that an optimal policy obtained by the preceding minimization can be written as

$$\mu_k^*(I_k) = \overline{\mu}_k\big(S_k(I_k)\big),$$

where $\overline{\mu}_k$ is an appropriate function. Thus, if the sufficient statistic is characterized by a set of fewer numbers than the information vector I_k, it may be easier to implement the policy in the form $u_k = \overline{\mu}_k\big(S_k(I_k)\big)$ and take advantage of the resulting data reduction.

The Conditional State Distribution as a Sufficient Statistic

There are many different functions that can serve as sufficient statistics. The identity function $S_k(I_k) = I_k$ is certainly one of them. To obtain another important sufficient statistic, we assume that *the probability distribution of the observation disturbance v_{k+1} depends explicitly only on the immediately preceding state, control, and system disturbance x_k, u_k, w_k, and not on $x_{k-1},\ldots,x_0$, $u_{k-1},\ldots,u_0$, $w_{k-1}\ldots,w_0$, $v_{k-1},\ldots,v_0$.* Under this assumption we can show that a sufficient statistic is given by the conditional probability distribution $P_{x_k|I_k}$ of the state x_k, given the information vector I_k.

Indeed, for a fixed I_k and u_k, in order to calculate the expression

$$\mathop{E}_{x_k,w_k,z_{k+1}}\big\{g_k(x_k,u_k,w_k) + J_{k+1}(I_k,z_{k+1},u_k) \mid I_k,u_k\big\}$$

in the DP equation (4.2), we need the joint distribution

$$P(x_k,w_k,z_{k+1} \mid I_k,u_k)$$

or, equivalently,

$$P\Big(x_k,w_k,h_{k+1}\big(f_k(x_k,u_k,w_k),u_k,v_{k+1}\big) \mid I_k,u_k\Big).$$

By using Bayes' rule, this distribution can be expressed in terms of $P_{x_k|I_k}$, the given distributions

$$P(w_k \mid x_k,u_k), \qquad P\big(v_{k+1} \mid f_k(x_k,u_k,w_k),u_k,w_k\big),$$

and the system equation $x_{k+1} = f_k(x_k,u_k,w_k)$. Therefore the DP algorithm can be written as

$$J_k(I_k) = \min_{u_k\in U_k} H_k\big(P_{x_k|I_k},u_k\big)$$

for a suitable function H_k, and the distribution $P_{x_k|I_k}$ is a sufficient statistic.

Note that if the conditional distribution $P_{x_k|I_k}$ is uniquely determined by another expression $S_k(I_k)$, that is, $P_{x_k|I_k} = G_k\big(S_k(I_k)\big)$ for an appropriate function G_k, then $S_k(I_k)$ is also a sufficient statistic. Thus, for example, if we can show that $P_{x_k|I_k}$ is a Gaussian distribution, then the mean and the covariance matrix corresponding to $P_{x_k|I_k}$ form a sufficient statistic.

Regardless of its computational value, the representation of the optimal policy as a sequence of functions of the conditional probability distribution $P_{x_k|I_k}$,

$$\mu_k(I_k) = \overline{\mu}_k(P_{x_k|I_k}), \qquad k = 0, 1, \ldots, N-1,$$

is conceptually very useful. It provides a decomposition of the optimal controller in two parts:

(a) An *estimator*, which uses at time k the measurement z_k and the control u_{k-1} to generate the probability distribution $P_{x_k|I_k}$.

(b) An *actuator*, which generates a control input to the system as a function of the probability distribution $P_{x_k|I_k}$ (Fig. 5.4.1).

This interpretation has formed the basis for various suboptimal control schemes that separate the controller a priori into an estimator and an actuator and attempt to design each part in a manner that seems "reasonable." Schemes of this type will be discussed in Chapter 6.

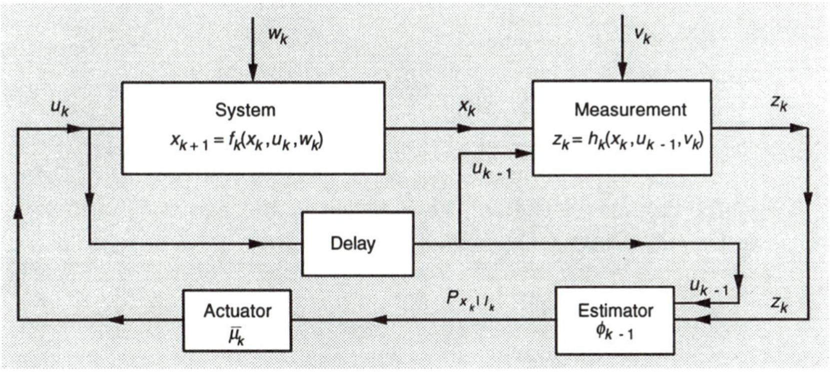

Figure 5.4.1 Conceptual separation of the optimal controller into an estimator and an actuator.

Alternative Perfect State Information Reduction

By using the sufficient statistic $P_{x_k|I_k}$ we can also write the DP algorithm in an alternative form. The key fact here is that $P_{x_k|I_k}$ is generated recursively in time and can be viewed as the state of a controlled discrete-time dynamic system. In particular, as we discuss shortly, it turns out that we can write for all k

$$P_{x_{k+1}|I_{k+1}} = \Phi_k\big(P_{x_k|I_k}, u_k, z_{k+1}\big), \tag{4.3}$$

where Φ_k is some function that can be determined from the data of the problem, u_k is the control of the system, and z_{k+1} plays the role of a random disturbance, the statistics of which are known and depend explicitly only on $P_{x_k|I_k}$ and u_k, and not on $z_k, \ldots, z_0$. Suppose now that we use the sufficient statistic property of $P_{x_k|I_k}$ to define functions $\overline{J}_k$ by

$$\overline{J}_k(P_{x_k|I_k}) = J_k(I_k) = \min_{u_k \in U_k} \Big[\mathop{E}_{x_k, w_k, z_{k+1}} \{g_k(x_k, u_k, w_k) + J_{k+1}(I_{k+1}) \mid I_k, u_k\}\Big].$$

We can then use the recursion (4.3) to write the DP algorithm for $k < N-1$ in the alternate form

$$\overline{J}_k(P_{x_k|I_k}) = \min_{u_k \in U_k} \Big[\mathop{E}_{x_k, w_k, z_{k+1}} \{g_k(x_k, u_k, w_k) + \overline{J}_{k+1}\big(\Phi_k(P_{x_k|I_k}, u_k, z_{k+1})\big) \mid I_k, u_k\}\Big]. \tag{4.4}$$

In the case where $k = N-1$, the corresponding equation is

$$\overline{J}_{N-1}(P_{x_{N-1}|I_{N-1}}) = \min_{u_{N-1} \in U_{N-1}} \Big[\mathop{E}_{x_{N-1}, w_{N-1}} \Big\{g_N\big(f_{N-1}(x_{N-1}, u_{N-1}, w_{N-1})\big) + g_{N-1}(x_{N-1}, u_{N-1}, w_{N-1}) \mid I_{N-1}, u_{N-1}\Big\}\Big]. \tag{4.5}$$

By minimizing on the right-hand side of the above equations, this alternate DP algorithm yields a policy of the form

$$\overline{\mu}_k^*(P_{x_k|I_k}), \qquad k = 0, 1, \ldots, N-1.$$

In addition, the optimal cost is given by

$$J^* = \mathop{E}_{z_0}\big\{\overline{J}_0(P_{x_0|z_0})\big\},$$

where $\overline{J}_0$ is obtained by the last step of this algorithm, and the probability distribution of z_0 is obtained from the measurement equation $z_0 = h_0(x_0, v_0)$ and the statistics of x_0 and v_0.

The preceding analysis is in effect an alternate reduction of the basic problem with imperfect state information to a problem with perfect state information. Indeed, the system

$$P_{x_{k+1}|I_{k+1}} = \Phi_k\big(P_{x_k|I_k}, u_k, z_{k+1}\big)$$

that describes the evolution of the conditional distribution $P_{x_k|I_k}$ fits the framework of the basic problem. Furthermore, the controller can calculate (at least in principle) the state $P_{x_k|I_k}$ of this system at time k, so perfect state information prevails. Thus the alternate DP algorithm (4.4) and (4.5) may be viewed as the DP algorithm of the perfect state information problem that involves the above system whose state is $P_{x_k|I_k}$ and an appropriately reformulated cost function. In the absence of perfect knowledge of the state, *the controller can be viewed as controlling the "probabilistic state"* $P_{x_k|I_k}$ *so as to minimize the expected cost-to-go conditioned on the information* I_k *available.*

The Conditional State Distribution Recursion

There remains to justify the recursion

$$P_{x_{k+1}|I_{k+1}} = \Phi_k\big(P_{x_k|I_k}, u_k, z_{k+1}\big).$$

Let us first give an example.

Example 4.1 (Search Problem)

In a classical problem of search, one has to decide at each period whether to search a site that may contain a treasure. If a treasure is present, the search reveals it with probability β, in which case the treasure is removed from the site. Here the state has two values: either a treasure is present in the site or it is not. The control u_k takes two values: search and not search. If the site is searched, the observation z_{k+1} takes two values, treasure found or not found, while if the site is not searched, the value of z_{k+1} is irrelevant.

Denote

p_k : probability a treasure is present at the beginning of period k.

This probability evolves according to the equation

$$p_{k+1} = \begin{cases} p_k & \text{if the site is not searched at time } k, \\ 0 & \text{if the site is searched and a treasure is found,} \\ \frac{p_k(1-\beta)}{p_k(1-\beta)+1-p_k} & \text{if the site is searched but no treasure is found.} \end{cases}$$

The second relation holds because the treasure is removed after a successful search. The third relation follows by application of Bayes' rule (p_{k+1} is equal to the kth period probability of a treasure being present *and* the search being unsuccessful, divided by the probability of an unsuccessful search). The preceding equation defines the desired recursion for the conditional distribution of the state.

This recursion can be used to write the DP algorithm (4.4), assuming that the treasure's worth is V, that each search costs C, and that once we decide not to search at a particular time, then we cannot search at future times. The algorithm takes the form

$$\overline{J}_k(p_k) = \max\left[0,\ -C + p_k\beta V + (1 - p_k\beta)\overline{J}_{k+1}\left(\frac{p_k(1-\beta)}{p_k(1-\beta) + 1 - p_k}\right)\right],$$

with $\overline{J}_N(p_N) = 0$. It can be shown by induction that the functions $\overline{J}_k$ satisfy $\overline{J}_k(p_k) = 0$ for all $p_k \leq C/(\beta V)$. Furthermore, it is optimal to search at period k if and only if

$$p_k \beta V \geq C,$$

that is, if and only if the expected reward from the next search is greater or equal to the cost of the search.

The general form of the recursion $P_{x_{k+1}|I_{k+1}} = \Phi_k\big(P_{x_k|I_k}, u_k, z_{k+1}\big)$ is developed in Exercise 5.7 for the case where the state, control, observation, and disturbance spaces are finite sets. In the case where these spaces are the real line and all random variables involved possess probability density functions, the conditional density $p(x_{k+1} \mid I_{k+1})$ is generated from $p(x_k \mid I_k)$, u_k, and z_{k+1} by means of the equation

$$\begin{aligned} p(x_{k+1} \mid I_{k+1}) &= p(x_{k+1} \mid I_k, u_k, z_{k+1}) \\ &= \frac{p(x_{k+1}, z_{k+1} \mid I_k, u_k)}{p(z_{k+1} \mid I_k, u_k)} \\ &= \frac{p(x_{k+1} \mid I_k, u_k)p(z_{k+1} \mid I_k, u_k, x_{k+1})}{\int_{-\infty}^{\infty} p(x_{k+1} \mid I_k, u_k)p(z_{k+1} \mid I_k, u_k, x_{k+1})dx_{k+1}}. \end{aligned}$$

In this equation all the probability densities appearing in the right-hand side may be expressed in terms of $p(x_k \mid I_k)$, u_k, and z_{k+1}. In particular, the density $p(x_{k+1} \mid I_k, u_k)$ may be expressed through $p(x_k \mid I_k)$, u_k, and the system equation $x_{k+1} = f_k(x_k, u_k, w_k)$ using the given density $p(w_k \mid x_k, u_k)$ and the relation

$$p(w_k \mid I_k, u_k) = \int_{-\infty}^{\infty} p(x_k \mid I_k)p(w_k \mid x_k, u_k)dx_k.$$

Similarly, the density $p(z_{k+1} \mid I_k, u_k, x_{k+1})$ is expressed through the measurement equation $z_{k+1} = h_{k+1}(x_{k+1}, u_k, v_{k+1})$ using the densities

$$p(x_k \mid I_k), \quad p(w_k \mid x_k, u_k), \quad p(v_{k+1} \mid x_k, u_k, w_k).$$

By substituting these expressions in the equation for $p(x_{k+1} \mid I_{k+1})$, we obtain a dynamic system equation for the conditional state distribution of the desired form. Other similar examples will be given in subsequent sections. A mathematically rigorous substantiation of the recursion $P_{x_{k+1}|I_{k+1}} = \Phi_k\big(P_{x_k|I_k}, u_k, z_{k+1}\big)$ and the alternate DP algorithm (4.4) and (4.5) can be found in [BeS78].

Finite-State Systems

When the system is a finite-state Markov chain, the conditional probability distribution $P_{x_k|I_k}$ is characterized by a finite set of numbers. This is particularly convenient, and leads to further simplification when the control and observation spaces are also finite sets. It then turns out that the cost-to-go functions $\overline{J}_k$ in the DP algorithm (4.4) and (4.5) are *piecewise linear* and *concave*. The demonstration of this fact is straightforward, but tedious, and is outlined in Exercise 5.7. The piecewise linearity of $\overline{J}_k$ is, however, an important property since it shows that $\overline{J}_k$ can be characterized by a finite set of scalars. Still, however, for fixed k, the number of these scalars can increase fast with N, and there may be no computationally efficient way to solve the problem (see [PaT88]). We will not discuss here any special procedures for computing $\overline{J}_k$ (see [Lov91], [SmS73], [Son71]). Instead we will demonstrate the DP algorithm by means of examples.

Example 4.2 (Machine Repair)

In the two-state machine repair example of Section 5.1, let us denote

$$p_1 = P(x_1 = \overline{P} \mid I_1), \qquad p_0 = P(x_0 = \overline{P} \mid I_0).$$

The equation relating p_1, p_0, u_0, z_1 [cf. Eq. (4.3)] is written as

$$p_1 = \Phi_0(p_0, u_0, z_1).$$

One may verify by straightforward calculation that Φ_0 is given by

$$p_1 = \Phi_0(p_0, u_0, z_1) = \begin{cases} \frac{1}{7} & \text{if } u_0 = S, \quad z_1 = G, \\ \frac{3}{5} & \text{if } u_0 = S, \quad z_1 = B, \\ \frac{1+2p_0}{7-4p_0} & \text{if } u_1 = C, \quad z_1 = G, \\ \frac{3+6p_0}{5+4p_0} & \text{if } u_0 = C, \quad z_1 = B. \end{cases}$$

The DP algorithm (4.4) and (4.5) may be written in terms of p_0, p_1, and Φ_0 above as

$$\overline{J}_1(p_1) = \min[2p_1, 1],$$

$$\begin{aligned} \overline{J}_0(p_0) = \min \Big[& 2p_0 + P(z_1 = G \mid I_0, C)\overline{J}_1\big(\Phi_0(p_0, C, G)\big) \\ & + P(z_1 = B \mid I_0, C)\overline{J}_1\big(\Phi_0(p_0, C, B)\big), \\ & 1 + P(z_1 = G \mid I_0, S)\overline{J}_1\big(\Phi_0(p_0, S, G)\big) \\ & + p(z_1 = B \mid I_0, S)\overline{J}_1\big(\Phi_0(p_0, S, B)\big)\Big]. \end{aligned}$$

The probabilities entering in the second equation may be expressed in terms of p_0 by straightforward calculation as

$$P(z_1 = G \mid I_0, C) = \frac{7 - 4p_0}{12}, \qquad P(z_1 = B \mid I_0, C) = \frac{5 + 4p_0}{12},$$

$$P(z_1 = G \mid I_0, S) = \frac{7}{12}, \qquad P(z_1 = B \mid I_0, S) = \frac{5}{12}.$$

Using these values we have

$$\overline{J}_0(p_0) = \min \Bigg[2p_0 + \frac{7-4p_0}{12}\overline{J}_1\left(\frac{1+2p_0}{7-4p_0}\right) + \frac{5+4p_0}{12}\overline{J}_1\left(\frac{3+6p_0}{5+4p_0}\right),$$
$$1 + \frac{7}{12}\overline{J}_1\left(\frac{1}{7}\right) + \frac{5}{12}\overline{J}_1\left(\frac{3}{5}\right)\Bigg].$$

Now by minimization in the equation defining $\overline{J}_1(p_1)$, we obtain an optimal policy for the last stage

$$\overline{\mu}_1^*(p_1) = \begin{cases} C & \text{if } p_1 \le \frac{1}{2}, \\ S & \text{if } p_1 > \frac{1}{2}. \end{cases}$$

Also by substitution of $\overline{J}_1(p_1)$ and by carrying out the straightforward calculation we obtain

$$\overline{J}_0(p_0) = \begin{cases} \frac{19}{12} & \text{if } \frac{3}{8} \le p_0 \le 1, \\ \frac{7+32p_0}{12} & \text{if } 0 \le p_0 \le \frac{3}{8}, \end{cases}$$

and an optimal policy for the first stage

$$\overline{\mu}_0^*(p_0) = \begin{cases} C & \text{if } p_0 \le \frac{3}{8}, \\ S & \text{if } p_0 > \frac{3}{8}. \end{cases}$$

Note that

$$P(x_0 = \overline{P} \mid z_0 = G) = \frac{1}{7}, \qquad P(x_0 = \overline{P} \mid z_0 = B) = \frac{3}{5},$$
$$P(z_0 = G) = \frac{7}{12}, \qquad P(z_0 = B) = \frac{5}{12},$$

so that the formula

$$J^* = \mathop{E}_{z_0}\big\{\overline{J}_0(P_{x_0 \mid z_0})\big\} = \frac{7}{12}\overline{J}_0\left(\frac{1}{7}\right) + \frac{5}{12}\overline{J}_0\left(\frac{3}{5}\right) = \frac{176}{144}$$

yields the same optimal cost as the one obtained in Section 5.1 by means of the DP algorithm (1.7) and (1.8).

A Problem of Instruction

Consider a problem of instruction where the objective is to teach a student a certain simple item. At the beginning of each period, the student may be in one of two possible states:

L: Item learned.

$\overline{L}$: Item not learned.

At the beginning of each period, the instructor must make one of two decisions:

T: Terminate the instruction.

$\overline{T}$: Continue the instruction for one period and then conduct a test that indicates whether the student has learned the item.

The test has two possible outcomes:

R: Student gives a correct answer.

$\overline{R}$: Student gives an incorrect answer.

The transition probabilities from one state to the next if instruction takes place are given by

$$P(x_{k+1} = L \mid x_k = L) = 1, \qquad P(x_{k+1} = \overline{L} \mid x_k = L) = 0,$$
$$P(x_{k+1} = L \mid x_k = \overline{L}) = t, \qquad P(x_{k+1} = \overline{L} \mid x_k = \overline{L}) = 1 - t,$$

where t is a given scalar with $0 < t < 1$.

The outcome of the test depends probabilistically on the state of knowledge of the student as follows:

$$P(z_k = R \mid x_k = L) = 1, \qquad P(z_k = \overline{R} \mid x_k = L) = 0,$$
$$P(z_k = R \mid x_k = \overline{L}) = r, \qquad P(z_k = \overline{R} \mid x_k = \overline{L}) = 1 - r,$$

where r is a given scalar with $0 < r < 1$. The probabilistic structure of the problem is illustrated in Fig. 5.4.2.

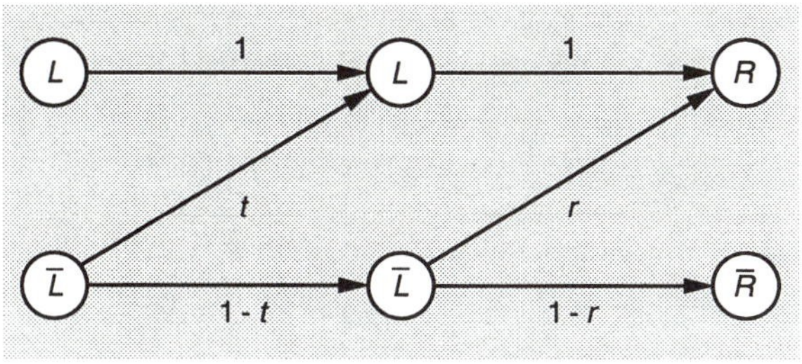

Figure 5.4.2 Probabilistic structure of the instruction problem.

The cost of instruction and testing is I per period. The cost of terminating instruction is 0 or $C > 0$ if the student has learned or has not learned the item, respectively. The objective is to find an optimal instruction-termination policy for each period k as a function of the test information accrued up to that period, assuming that there is a maximum of N periods of instruction.

It is straightforward to reformulate this problem into the framework of the basic problem with imperfect state information and to conclude that the decision whether to terminate or continue instruction at period k should depend on the conditional probability that the student has learned the item given the test results so far. This probability is denoted

$$p_k = P(x_k|I_k) = P(x_k = L \mid z_0, z_1, \ldots, z_k).$$

In addition, we can use the DP algorithm (4.4) and (4.5) defined over the space of the sufficient statistic p_k to obtain an optimal policy. However, rather than proceeding with this elaborate reformulation, we prefer to argue and obtain this DP algorithm directly.

Concerning the evolution of the conditional probability p_k (assuming instruction occurs), we have by Bayes' rule

$$p_{k+1} = P(x_{k+1} = L \mid z_0, \ldots, z_{k+1}) = \frac{P(x_{k+1} = L, z_{k+1} \mid z_0, \ldots, z_k)}{P(z_{k+1} \mid z_0, \ldots, z_k)},$$

where

$$\begin{aligned} &P(z_{k+1} \mid z_0, \ldots, z_k) \\ &\quad = P(x_{k+1} = L \mid z_0, \ldots, z_k) P(z_{k+1} \mid z_0, \ldots, z_k, x_{k+1} = L) \\ &\qquad + P(x_{k+1} = \overline{L} \mid z_0, \ldots, z_k) P(z_{k+1} \mid z_0, \ldots, z_k, x_{k+1} = \overline{L}). \end{aligned}$$

From the probabilistic descriptions given, we have

$$P(z_{k+1} \mid z_0, \ldots, z_k, x_{k+1} = L) = P(z_{k+1} \mid x_{k+1} = L) = \begin{cases} 1 & \text{if } z_{k+1} = R, \\ 0 & \text{if } z_{k+1} = \overline{R}. \end{cases}$$

$$\begin{aligned} P(z_{k+1} \mid z_0, \cdots, z_k, x_{k+1} = L) &= P(z_{k+1} \mid x_{k+1} = \overline{L}) \\ &= \begin{cases} r & \text{if } z_{k+1} = R, \\ 1 - r & \text{if } z_{k+1} = \overline{R}. \end{cases} \end{aligned}$$

$$P(x_{k+1} = L \mid z_0, \ldots, z_k) = p_k + (1 - p_k)t,$$

$$P(x_{k+1} = \overline{L} \mid z_0, \cdots, z_k) = (1 - p_k)(1 - t).$$

Combining these equations, we obtain

$$p_{k+1} = \Phi(p_k, z_{k+1}) = \begin{cases} \frac{p_k + (1-p_k)t}{p_k + (1-p_k)t + (1-p_k)(1-t)r} & \text{if } z_{k+1} = R, \\ 0 & \text{if } z_{k+1} = \overline{R}, \end{cases}$$

or equivalently

$$p_{k+1} = \Phi(p_k, z_{k+1}) = \begin{cases} \frac{1-(1-t)(1-p_k)}{1-(1-t)(1-r)(1-p_k)} & \text{if } z_{k+1} = R, \\ 0 & \text{if } z_{k+1} = \overline{R}. \end{cases} \tag{4.6}$$

This equation is a special case of the general recursive update equation for the conditional probability of the state; cf. Eq. (4.3). A cursory examination of Eq. (4.6) shows that, as expected, the conditional probability p_{k+1} that the student has learned the item increases with every correct answer and drops to zero with every incorrect answer.

We now derive the DP algorithm for the problem. At the end of the Nth period, assuming instruction has continued to that period, the expected cost is

$$\overline{J}_N(p_N) = (1 - p_N)C. \tag{4.7}$$

At the end of period $N-1$, the instructor has calculated the conditional probability p_{N-1} that the student has learned the item and wishes to decide whether to terminate instruction and incur an expected cost $(1-p_{N-1})C$ or continue the instruction and incur an expected cost $I + E\{\overline{J}_N(p_N)\}$. This leads to the following equation for the optimal expected cost-to-go:

$$\overline{J}_{N-1}(p_{N-1}) = \min\bigl[(1 - p_{N-1})C,\ I + (1-t)(1-p_{N-1})C\bigr].$$

The term $(1-p_{N-1})C$ is the cost of terminating instruction, while the term $(1-t)(1-p_{N-1})$ is the probability that the student still has not learned the item following an additional period of instruction.

Similarly, the algorithm is written for every stage k by replacing N by $k+1$:

$$\overline{J}_k(p_k) = \min\left[(1-p_k)C,\ I + \mathop{E}_{z_{k+1}}\bigl\{\overline{J}_{k+1}\bigl(\Phi(p_k, z_{k+1})\bigr)\bigr\}\right].$$

Now using expression (4.6) for the function Φ and the probabilities

$$P(z_{k+1} = \overline{R} \mid p_k) = (1-t)(1-r)(1-p_k),$$
$$P(z_{k+1} = R \mid p_k) = 1 - (1-t)(1-r)(1-p_k),$$

we have

$$\overline{J}_k(p_k) = \min\bigl[(1-p_k)C,\ I + A_k(p_k)\bigr], \tag{4.8}$$

where

$$\begin{aligned} A_k(p_k) = {} & P(z_{k+1} = R \mid I_k)\overline{J}_{k+1}\bigl(\Phi(p_k, R)\bigr) \\ & + P(z_{k+1} = \overline{R} \mid I_k)\overline{J}_{k+1}\bigl(\Phi(p_k, \overline{R})\bigr) \end{aligned}$$

or, equivalently, using Eq. (4.6),

$$\begin{aligned} A_k(p_k) = {} & \bigl(1 - (1-t)(1-r)(1-p_k)\bigr)\overline{J}_{k+1}\left(\frac{1-(1-t)(1-p_k)}{1-(1-t)(1-r)(1-p_k)}\right) \\ & + (1-t)(1-r)(1-p_k)\overline{J}_{k+1}(0). \end{aligned} \tag{4.9}$$

As shown in Fig. 5.4.3, if $I + (1-t)C \leq C$ or, equivalently, if

$$I < tC, \tag{4.10}$$

then there exists a scalar α_{N-1} with $0 < \alpha_{N-1} < 1$ that determines an optimal policy for the last period:

$$\begin{aligned} &\text{continue instruction} && \text{if } p_{N-1} \leq \alpha_{N-1}, \\ &\text{terminate instruction} && \text{if } p_{N-1} > \alpha_{N-1}. \end{aligned}$$

In the opposite case, where $I \geq tC$, the cost of instruction is so high relative to the cost of not learning that instructing the student is never optimal.

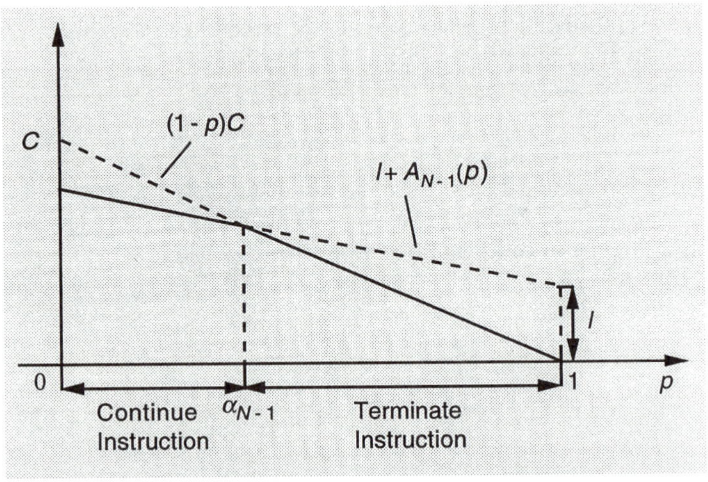

Figure 5.4.3 Determining the optimal instruction policy in the last period.

It may be shown by induction (Exercise 5.8) that if $I < tC$, the functions $A_k(p)$ are concave and piecewise linear for each k and satisfy, for all k,

$$A_k(1) = 0. \tag{4.11}$$

Furthermore, they satisfy for all k,

$$A_k(p) \geq A_k(p'), \qquad \text{for } 0 \leq p < p' \leq 1, \tag{4.12}$$

$$A_{k-1}(p) \leq A_k(p) \leq A_{k+1}(p), \qquad \text{for all } p \in [0,1]. \tag{4.13}$$

Thus the functions $(1-p_k)C$ and $I + A_k(p_k)$ intersect at a single point, and from the DP algorithm (4.8), we obtain that the optimal policy for each period is determined by the unique scalars α_k, which are such that

$$(1-\alpha_k)C = I + A_k(\alpha_k), \qquad k = 0, 1, \ldots, N-1.$$

An optimal policy for period k is given by

$$\begin{aligned} &\text{continue instruction} \quad &&\text{if } p_k \leq \alpha_k, \\ &\text{terminate instruction} \quad &&\text{if } p_k > \alpha_k. \end{aligned}$$

Since the functions $A_k(p)$ are monotonically nondecreasing with respect to k, it follows from Fig. 5.4.4 that

$$\alpha_{N-1} \leq \alpha_{N-2} \leq \cdots \leq \alpha_k \leq \cdots \leq 1 - \frac{1}{C},$$

and therefore the sequence $\{\alpha_k\}$ converges to some scalar $\overline{\alpha}$ as $k \to -\infty$. Thus, as the horizon gets longer, the optimal policy (at least for the initial stages) can be approximated by the stationary policy

$$\begin{aligned} &\text{continue instruction} \quad &&\text{if } p_k \leq \overline{\alpha}_k, \\ &\text{terminate instruction} \quad &&\text{if } p_k > \overline{\alpha}_k. \end{aligned} \tag{4.14}$$

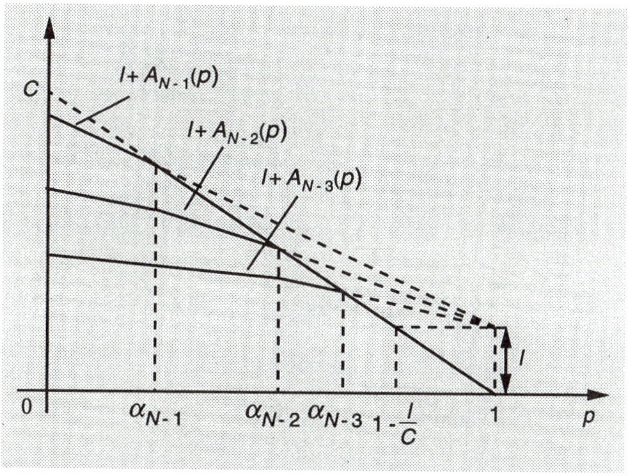

Figure 5.4.4 Demonstrating that the instruction thresholds are decreasing with time.

It turns out that this stationary policy has a convenient implementation that does not require the calculation of the conditional probability at each stage. From Eq. (4.6) we have that p_{k+1} increases over p_k, if a correct answer R is given and drops to zero if an incorrect answer $\overline{R}$ is given. For $m = 1, 2, \ldots$, define recursively the probability π_m of getting m successive correct answers following an incorrect answer:

$$\pi_1 = \Phi(0, R), \quad \pi_2 = \Phi(\pi_1, R), \quad \ldots \quad , \pi_{k+1} = \Phi(\pi_k, R), \ldots$$

Let n be the smallest integer for which $\pi_n > \overline{\alpha}$. Then the stationary policy (4.14) can be implemented by terminating instruction if and only if n successive correct answers have been received.

5.5 SEQUENTIAL HYPOTHESIS TESTING

In this section we consider a hypothesis testing problem that is typical of statistical sequential analysis. A decision maker can make observations, at a cost C each, relating to two hypotheses. Given a new observation, he can either accept one of the hypotheses or delay the decision for one more period, pay the cost C, and obtain a new observation. At issue is trading off the cost of observation with the higher probability of accepting the wrong hypothesis.

Let $z_0, z_1, \ldots, z_{N-1}$ be the sequence of observations. We assume that they are independent, identically distributed random variables taking values on a finite set Z. Suppose we know that the probability distribution of the z_k's is either f_0 or f_1 and that we are trying to decide on one of these. Here, for any element $z \in Z$, $f_0(z)$ and $f_1(z)$ denote the probabilities of z when f_0 and f_1 are the true distributions, respectively. At time k after observing $z_0, \ldots, z_k$, we may either stop observing and accept either f_0 or f_1, or we may take an additional observation at a cost $C > 0$. If we stop observing and make a choice, then we incur zero cost if our choice is correct, and costs L_0 and L_1 if we choose incorrectly f_0 and f_1, respectively. We are given the a priori probability p that the true distribution is f_0, and we assume that at most N observations are possible.

It can be seen that we are faced with an imperfect state information problem involving the two states:

$$x^0 : \text{true distribution is } f_0,$$

$$x^1 : \text{true distribution is } f_1.$$

The alternate DP algorithm (4.4) and (4.5) is defined over the interval $[0, 1]$ of possible values of the conditional probability

$$p_k = P(x_k = x^0 \mid z_0, \ldots, z_k).$$

Similar to the previous section, we will obtain this algorithm directly.

The conditional probability p_k is generated recursively according to the following equation [assuming $f_0(z) > 0$, $f_1(z) > 0$ for all $z \in Z$]:

$$p_0 = \frac{p f_0(z_0)}{p f_0(z_0) + (1-p) f_1(z_0)}, \tag{5.1}$$

$$p_{k+1} = \frac{p_k f_0(z_{k+1})}{p_k f_0(z_{k+1}) + (1-p_k) f_1(z_{k+1})}, \qquad k = 0, 1, \ldots, N-2, \tag{5.2}$$

where p is the a priori probability that the true distribution is f_0. The optimal expected cost for the last period is

$$\overline{J}_{N-1}(p_{N-1}) = \min\big[(1-p_{N-1})L_0,\ p_{N-1}L_1\big], \tag{5.3}$$

where $(1-p_{N-1})L_0$ is the expected cost for accepting f_0 and $p_{N-1}L_1$ is the expected cost for accepting f_1. Taking into account Eqs. (5.1) and (5.2), we can obtain the optimal expected cost-to-go for the kth period as

$$\overline{J}_k(p_k) = \min\Bigg[(1-p_k)L_0,\ p_kL_1,$$
$$C + \mathop{E}_{z_{k+1}}\left\{\overline{J}_{k+1}\left(\frac{p_kf_0(z_{k+1})}{p_kf_0(z_{k+1}) + (1-p_k)f_1(z_{k+1})}\right)\right\}\Bigg],$$

where the expectation over z_{k+1} is taken with respect to the probability distribution

$$p(z_{k+1}) = p_kf_0(z_{k+1}) + (1-p_k)f_1(z_{k+1}), \qquad z_{k+1} \in Z.$$

Equivalently, for $k = 0, 1, \ldots, N-2$,

$$\overline{J}_k(p_k) = \min\big[(1-p_k)L_0,\ p_kL_1,\ C + A_k(p_k)\big], \tag{5.4}$$

where

$$A_k(p_k) = \mathop{E}_{z_{k+1}}\left\{\overline{J}_{k+1}\left(\frac{p_kf_0(z_{k+1})}{p_kf_0(z_{k+1}) + (1-p_k)f_1(z_{k+1})}\right)\right\}. \tag{5.5}$$

An optimal policy for the last period (see Fig. 5.5.1) is obtained from the minimization indicated in Eq. (5.3):

$$\text{accept } f_0 \qquad \text{if } p_{N-1} \geq \gamma,$$
$$\text{accept } f_1 \qquad \text{if } p_{N-1} < \gamma,$$

where γ is determined from the relation $(1-\gamma)L_0 = \gamma L_1$ or equivalently

$$\gamma = \frac{L_0}{L_0 + L_1}.$$

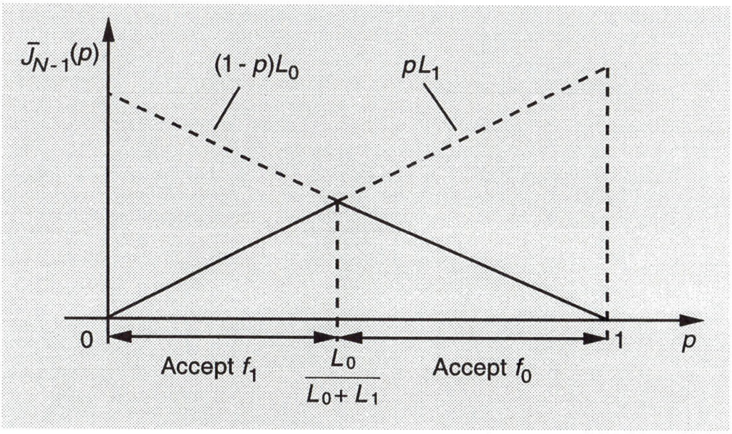

Figure 5.5.1 Determining the optimal policy for the last period.

We now prove the following lemma.

Lemma 5.1: The functions $A_k : [0,1] \to R$ of Eq. (5.5) are concave, and satisfy for all k and $p \in [0,1]$

$$A_k(0) = A_k(1) = 0,$$

$$A_{k-1}(p) \le A_k(p).$$

Proof: We have for all $p \in [0,1]$

$$\overline{J}_{N-2}(p) \le \min\big[(1-p)L_0,\, pL_0\big] = \overline{J}_{N-1}(p).$$

By making use of the stationarity of the system and the monotonicity property of DP (Exercise 1.23 in Chapter 1), we obtain

$$\overline{J}_k(p) \le \overline{J}_{k+1}(p)$$

for all k and $p \in [0,1]$. Using Eq. (5.5), we obtain $A_{k+1}(p) \le A_k(p)$ for all k and $p \in [0,1]$.

To prove concavity of A_k in view of Eqs. (5.3) and (5.4), it is sufficient to show that concavity of $\overline{J}_{k+1}$ implies concavity of A_k through relation (5.5). Indeed, assume that $\overline{J}_{k+1}$ is concave over $[0,1]$. Let $z^1, z^2, \ldots, z^n$ denote the elements of the observation space Z. We have from Eq. (5.5) that

$$A_k(p) = \sum_{i=1}^{n} \big(pf_0(z^i) + (1-p)f_1(z^i)\big)\overline{J}_{k+1}\left(\frac{pf_0(z^i)}{pf_0(z^i) + (1-p)f_1(z^i)}\right).$$

Hence it is sufficient to show that concavity of $\overline{J}_{k+1}$ implies concavity of each of the functions

$$H_i(p) = \big(pf_0(z^i) + (1-p)f_1(z^i)\big)\overline{J}_{k+1}\left(\frac{pf_0(z^i)}{pf_0(z^i) + (1-p)f_1(z^i)}\right).$$

To show concavity of H_i, we must show that for every $\lambda \in [0,1]$, p_1, $p_2 \in [0,1]$ we have

$$\lambda H_i(p_1) + (1-\lambda)H_i(p_2) \le H_i\big(\lambda p_1 + (1-\lambda)p_2\big).$$

Using the notation

$$\xi_1 = p_1 f_0(z^i) + (1-p_1)f_1(z^i), \qquad \xi_2 = p_2 f_0(z^i) + (1-p_2)f_1(z^i),$$

this inequality is equivalent to

$$\frac{\lambda\xi_1}{\lambda\xi_1+(1-\lambda)\xi_2}\overline{J}_{k+1}\left(\frac{p_1 f_0(z^i)}{\xi_1}\right)+\frac{(1-\lambda)\xi_2}{\lambda\xi_1+(1-\lambda)\xi_2}\overline{J}_{k+1}\left(\frac{p_2 f_0(z^i)}{\xi_2}\right)$$
$$\leq \overline{J}_{k+1}\left(\frac{\big(\lambda p_1+(1-\lambda)p_2\big)f_0(z^i)}{\lambda\xi_1+(1-\lambda)\xi_2}\right).$$

This relation, however, is implied by the concavity of $\overline{J}_{k+1}$. **Q.E.D.**

Using Lemma 5.1, we obtain (see Fig. 5.5.2) that if

$$C + A_{N-2}\left(\frac{L_0}{L_0+L_1}\right) < \frac{L_0L_1}{L_0+L_1},$$

then an optimal policy for each period k is of the form

$$\text{accept } f_0 \text{ if } p_k \geq \alpha_k,$$

$$\text{accept } f_1 \text{ if } p_k \leq \beta_k,$$

$$\text{continue the observations if } \beta_k < p_k < \alpha_k,$$

where the scalars α_k, β_k are determined from the relations

$$\beta_k L_1 = C + A_k(\beta_k),$$
$$(1-\alpha_k)L_0 = C + A_k(\alpha_k).$$

Furthermore, we have

$$\cdots \leq \alpha_{k+1} \leq \alpha_k \leq \alpha_{k-1} \leq \cdots \leq 1 - \frac{C}{L_0},$$

$$\cdots \geq \beta_{k+1} \geq \beta_k \geq \beta_{k-1} \geq \cdots \geq \frac{C}{L_1}.$$

Hence as $N \to \infty$ the sequences $\{\alpha_{N-i}\}$, $\{\beta_{N-i}\}$ converge to scalars $\overline{\alpha}$, $\overline{\beta}$, respectively, and the optimal policy is approximated by the stationary policy

$$\text{accept } f_0 \text{ if } p_k \geq \overline{\alpha}, \tag{5.6a}$$

$$\text{accept } f_1 \text{ if } p_k \leq \overline{\beta}, \tag{5.6b}$$

$$\text{continue the observations if } \overline{\beta} < p_k < \overline{\alpha}. \tag{5.6c}$$

Now the conditional probability p_k is given by

$$p_k = \frac{pf_0(z_0)\cdots f_0(z_k)}{pf_0(z_0)\cdots f_0(z_k)+(1-p)f_1(z_0)\cdots f_1(z_k)}, \tag{5.7}$$

where p is the a priori probability that f_0 is the true hypothesis. Using Eq. (5.7), the stationary policy (5.6) can be written in the form

$$\text{accept } f_0 \text{ if } R_k \geq A, \tag{5.8a}$$

$$\text{accept } f_1 \text{ if } R_k \leq B, \tag{5.8b}$$

$$\text{continue the observations if } B < R_k < A, \tag{5.8c}$$

where

$$A = \frac{(1-p)\overline{\alpha}}{p(1-\overline{\alpha})},$$

$$B = \frac{(1-p)\overline{\beta}}{p(1-\overline{\beta})},$$

and

$$R_k = \frac{f_0(z_0)\cdots f_0(z_k)}{f_1(z_0)\cdots f_1(z_k)}.$$

Note that R_k can be easily generated by means of the recursive equation

$$R_{k+1} = \frac{f_0(z_{k+1})}{f_1(z_{k+1})} R_k.$$

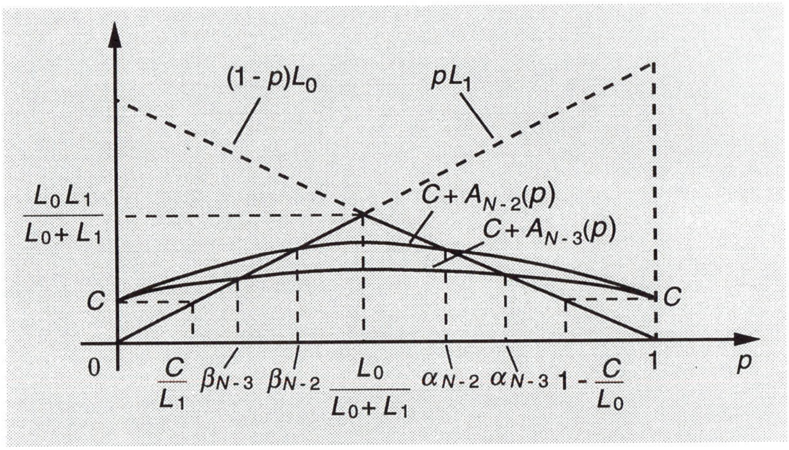

Figure 5.5.2 Determining the optimal hypothesis testing policy.

The policy (5.8) is known as the *sequential probability ratio test*, and was among the first formal methods studied in statistical sequential analysis [Wal47]. The optimality of this policy for the infinite horizon version of the problem will be shown in Vol. II, Chapter 3.

5.6 NOTES, SOURCES, AND EXERCISES

For literature on linear-quadratic problems with imperfect state information, see the references quoted for Section 4.1 and Witsenhausen's survey paper [Wit71]. Detailed discussions of the Kalman filtering algorithm can be found in many textbooks [AnM79], [Jaz70], [LjS83], [Med69], [Men73], [Nah69]. For linear-quadratic problems with Gaussian uncertainties and observation cost in the spirit of Exercise 5.6, see [AoL69]. Exercise 5.1, which indicates the form of the certainty equivalence principle when the random disturbances are correlated, is based on an unpublished report by the author [Ber70]. The minimum variance approach is also described in [AsW84] and [Whi63]. Imperfect state information problems with exponential cost functions are discussed in [JBE94] and [FeM94].

The idea of data reduction via a sufficient statistic gained wide attention following the 1965 paper by Striebel [Str65]; see also [Shi64] and [Shi66]. For the analog of the sufficient statistic idea in sequential minimax problems, see [BeR73].

Sufficient statistics have been used for analysis of finite-state problems with imperfect state information in [Eck68], [SmS73], and [Son71]. The proof of piecewise linearity of the cost-to-go functions, and an algorithm for their computation are given in [SmS73] and [Son71]. For further material on finite-state problems with imperfect state information, see [ABF93], [Lov91], and [WhS89]. The instruction model described in Section 5.4 has been considered (with some variations) by a number of authors [ABC65], [GrA66], [KaD66], [Sma71].

For a discussion of the sequential probability ratio test and related subjects, see [Che72], [DeG70], [Whi82], and the references quoted therein. The treatment given here stems from [ABG49].

EXERCISES

5.1

Consider the linear system and measurement equation of Section 5.2 and consider the problem of finding a policy $\{\mu_0^*(I_0), \dots, \mu_{N-1}^*(I_{N-1})\}$ that minimizes the quadratic cost

$$E\left\{x_N' Q x_N + \sum_{k=0}^{N-1} u_k' R_k u_k\right\}.$$

Assume, however, that the random vectors $x_0, w_0, \ldots, w_{N-1}, v_0, \ldots, v_{N-1}$ are correlated and have a given joint probability distribution, and finite first and second moments. Show that the optimal policy is given by

$$\mu_k^*(I_k) = L_k E\{y_k \mid I_k\},$$

where the gain matrices L_k are obtained from the algorithm

$$L_k = -(B_k' K_{k+1} B_k + R_k)^{-1} B_k' K_{k+1} A_k,$$

$$K_N = Q,$$

$$K_k = A_k' \big(K_{k+1} - K_{k+1} B_k (B_k' K_{k+1} B_k + R_k)^{-1} B_k' K_{k+1}\big) A_k,$$

and the vectors y_k are given by

$$y_k = x_k + A_k^{-1} w_k + A_k^{-1} A_{k+1}^{-1} w_{k+1} + \cdots + A_k^{-1} \cdots A_{N-1}^{-1} w_{N-1}$$

(assuming the matrices $A_0, A_1, \ldots, A_{N-1}$ are invertible). *Hint*: Show that the cost can be written as

$$E\left\{ y_0' K_k y_0 + \sum_{k=0}^{N-1} (u_k - L_k y_k)' P_k (u_k - L_k y_k) \right\},$$

where

$$P_k = B_k' K_{k+1} B_k + R_k.$$

5.2

Consider the scalar system

$$x_{k+1} = x_k + u_k + w_k,$$

$$z_k = x_k + v_k,$$

where we assume that the initial condition x_0, and the disturbances w_k and v_k are all independent. Let the cost be

$$E\left\{ x_N^2 + \sum_{k=0}^{N-1} \left(x_k^2 + u_k^2\right) \right\},$$

and let the given probability distributions be

$$p(x_0 = 2) = \frac{1}{2}, \qquad p(w_k = 1) = \frac{1}{2}, \qquad p\left(v_k = \frac{1}{4}\right) = \frac{1}{2},$$

$$p(x_0 = -2) = \frac{1}{2}, \qquad p(w_k = -1) = \frac{1}{2}, \qquad p\left(v_k = -\frac{1}{4}\right) = \frac{1}{2}.$$

(a) Determine the optimal policy. *Hint*: For this problem, $E\{x_k \mid I_k\}$ can be determined from $E\{x_{k-1} \mid I_{k-1}\}$, u_{k-1}, and z_k.

(b) Determine the policy that is optimal within the class of all policies that are linear functions of the measurements.

(c) Determine the asymptotic form of the policies in parts (a) and (b) as $N \to \infty$.

5.3

A linear system with Gaussian disturbances

$$x_{k+1} = Ax_k + Bu_k + w_k,$$

is to be controlled so as to minimize a quadratic cost similar to that in Section 5.2. The difference is that the controller has the option of choosing at each time k one of two types of measurements for the next stage $(k+1)$:

$$\text{first type:} \qquad z_{k+1} = C^1 x_{k+1} + v^1_{k+1}$$

$$\text{second type:} \qquad z_{k+1} = C^2 x_{k+1} + v^2_{k+1}.$$

Here C^1 and C^2 are given matrices of appropriate dimension, and $\{v^1_k\}$ and $\{v^2_k\}$ are zero-mean, independent, random sequences with given finite covariances that do not depend on x_0 and $\{w_k\}$. There is a cost g_1 (or g_2) each time a measurement of type 1 (or type 2) is taken. The problem is to find the optimal control and measurement selection policy that minimizes the expected value of the sum of the quadratic cost

$$x'_N Q x_N + \sum_{k=0}^{N-1} (x'_k Q x_k + u'_k R u_k)$$

and the total measurement cost. Assume for convenience that $N = 2$ and that the first measurement z_0 is of type 1. Show that the optimal measurement selection at $k = 0$ and $k = 1$ does not depend on the value of the information vectors I_0 and I_1, and can be determined a priori. Describe the nature of the optimal policy.

5.4

Consider a scalar single-input, single-output system given by

$$y_k + a_1 y_{k-1} + \cdots + a_m y_{k-m} = b_M u_{k-M} + \cdots + b_m u_{k-m}$$
$$+ \epsilon_k + c_1 \epsilon_{k-1} + \cdots + c_m \epsilon_{k-m} + v_{k-n},$$

where $1 \le M \le m$, $0 \le n \le m$, and v_k is generated by an equation of the form

$$v_k + d_1 v_{k-1} + \cdots + d_m v_{k-m} = \xi_k + \epsilon_1 \xi_{k-1} + \cdots + \epsilon_m \xi_{k-m},$$

and the polynomials $(1 + c_1 s + \cdots + c_m s^m)$, $(1 + d_1 s + \cdots + d_m s^m)$, and $(1 + \epsilon_1 s + \cdots + \epsilon_m s^m)$ have roots strictly outside the unit circle. The value of the scalar v_k is observed by the controller at time k together with y_k. The sequences $\{\epsilon_k\}$ and $\{\xi_k\}$ are zero mean independent identically distributed with finite variances. Find an easily implementable approximation to the minimum variance controller minimizing $E\{\sum_{k=0}^N (y_k)^2\}$. Discuss the stability properties of the closed-loop system.

5.5

(a) Within the framework of the basic problem with imperfect state information, consider the case where the system and the observations are linear:

$$x_{k+1} = A_k x_k + B_k u_k + w_k,$$

$$z_k = C_k x_k + v_k.$$

The initial state x_0 and the disturbances w_k and v_k are assumed Gaussian and mutually independent. Their covariances are given, and w_k and v_k have zero mean. Show that $E\{x_0 \mid I_0\}, \ldots, E\{x_{N-1} \mid I_{N-1}\}$ constitute a sufficient statistic for this problem.

(b) Use the result of part (a) to obtain an optimal policy for the special case of the single-stage problem involving the scalar system and observation

$$x_1 = x_0 + u_0,$$

$$z_0 = x_0 + v_0,$$

and the cost function $E\big\{|x_k|\big\}$.

(c) Generalize part (b) for the case of the scalar system

$$x_{k+1} = a x_k + u_k,$$

$$z_k = c x_k + v_k,$$

and the cost function $E\big\{\sum_{k=1}^{N} |x_k|\big\}$. The scalars a and c are given. *Note*: You may find useful the following "differentiation of an integral" formula:

$$\frac{d}{dy} \int_{\alpha(y)}^{\beta(y)} f(y, \xi) d\xi = \int_{\alpha(y)}^{\beta(y)} \frac{df(y, \xi)}{dy} d\xi$$
$$+ f\big(y, \beta(y)\big) \frac{d\beta(y)}{dy} - f\big(y, \alpha(y)\big) \frac{d\alpha(y)}{dy}.$$

5.6

Consider a machine that can be in one of two states, good or bad. Suppose that the machine produces an item at the end of each period. The item produced is either good or bad depending on whether the machine is in a good or bad state at the beginning of the corresponding period, respectively. We suppose that once the machine is in a bad state it remains in that state until it is replaced. If the machine is in a good state at the beginning of a certain period, then with probability t it will be in the bad state at the end of the period. Once an item is produced, we may inspect the item at a cost I or not inspect. If an inspected item is found to be bad, the machine is replaced with a machine in good state at a cost R. The cost for producing a bad item is $C > 0$. Write a DP algorithm for obtaining an optimal inspection policy assuming a machine initially in good state and a horizon of N periods. Solve the problem for $t = 0.2$, $I = 1$, $R = 3$, $C = 2$, and $N = 8$. (The optimal policy is to inspect at the end of the third period and not inspect in any other period.)

5.7 (Finite-State Systems – Imperfect State Information)

Consider a system that at any time can be in any one of a finite number of states $1, 2, \ldots, n$. When a control u is applied, the system moves from state i to state j with probability $p_{ij}(u)$. The control u is chosen from a finite collection $u^1, u^2, \ldots, u^m$. Following each state transition, an observation is made by the controller. There is a finite number of possible observation outcomes $z^1, z^2, \ldots, z^q$. The probability of occurrence of z^θ, given that the current state is j and the preceding control was u, is denoted by $r_j(u, \theta)$, $\theta = 1, \ldots, q$.

(a) Consider the column vector of conditional probabilities

$$P_k = [p_k^1, \ldots, p_k^n]',$$

where

$$p_k^j = P(x_k = j \mid z_0, \ldots, z_k, u_0, \ldots, u_{k-1}), \qquad j = 1, \ldots, n,$$

and show that it can be updated according to

$$p_{k+1}^j = \frac{\sum_{i=1}^n p_k^i p_{ij}(u_k) r_j(u_k, z_{k+1})}{\sum_{s=1}^n \sum_{i=1}^n p_k^i p_{is}(u_k) r_s(u_k, z_{k+1})}, \qquad j = 1, \ldots, n.$$

Write this equation in the compact form

$$P_{k+1} = \frac{\big[r(u_k, z_{k+1})\big] * \big[P(u_k)' P_k\big]}{r(u_k, z_{k+1})' P(u_k)' P_k},$$

where prime denotes transposition and

$P(u_k)$ is the $n \times n$ transition probability matrix with ijth element $p_{ij}(u_k)$,

$r(u_k, z_{k+1})$ is the column vector with jth coordinate $r_j(u_k, z_{k+1})$,

$\big[P(u_k)'P_k\big]_j$ is the jth coordinate of the vector $P(u_k)'P_k$,

$\big[r(u_k, z_{k+1})\big] * \big[P(u_k)'P_k\big]$ is the vector whose jth coordinate is the scalar $r_j(u_k, z_{k+1}) \big[P(u_k)'P_k\big]_j$.

(b) Assume that there is a cost for each stage k denoted $g_k(i, u, j)$ and associated with the control u and a transition from i to j. There is no terminal cost. Consider the problem of finding a policy that minimizes the sum of expected costs per stage over N periods. Show that the corresponding DP algorithm is given by

$$\overline{J}_{N-1}(P_{N-1}) = \min_{u\in\{u^1,\ldots,u^m\}} P'_{N-1}G_{N-1}(u),$$

$$\overline{J}_k(P_k) = \min_{u\in\{u^1,\ldots,u^m\}} \Bigg[P'_kG_k(u) + \sum_{\theta=1}^{q} r(u,\theta)'P(u)'P_k\overline{J}_{k+1}\left(\frac{\big[r(u,\theta)\big] * \big[P(u)'P_k\big]}{r(u,\theta)'P(u)'P_k}\right)\Bigg],$$

where $G_k(u)$ is the vector of expected kth stage costs given by

$$G_k(u) = \begin{pmatrix} \sum_{j=1}^{n} p_{1j}(u)g_k(1,u,j) \\ \vdots \\ \sum_{j=1}^{n} p_{nj}(u)g_k(n,u,j) \end{pmatrix}.$$

(c) Show that, for all k, $\overline{J}_k$ when viewed as a function on the set of vectors with nonnegative coordinates, is *positively homogeneous*; that is,

$$\overline{J}_k(\lambda P_k) = \lambda\overline{J}_k(P_k), \qquad \lambda > 0.$$

Use this fact to write the DP algorithm in the simpler form

$$\overline{J}_k(P_k) = \min_{u\in\{u^1,\ldots,u^m\}} \left[P'_kG_k(u) + \sum_{\theta=1}^{q} \overline{J}_{k+1}\Big(\big[r(u,\theta)\big] * \big[P(u)'P_k\big]\Big)\right].$$

(d) Show by induction that, for all k, $\overline{J}_k$ has the form

$$\overline{J}_k(P_k) = \min\big[P'_k\alpha_k^1,\ldots,P'_k\alpha_k^{m_k}\big],$$

where $\alpha_k^1,\ldots,\alpha_k^{m_k}$ are some vectors in $\Re^n$.

5.8

Consider the functions $\overline{J}_k(p_k)$ in the instruction problem. Show inductively that each of these functions is piecewise linear, concave, and of the form

$$\overline{J}_k(p_k) = \min\big[\alpha_k^1 + \beta_k^1 p_k, \alpha_k^2 + \beta_k^2 p_k, \ldots, \alpha_k^{m_k} + \beta_k^{m_k} p_k\big],$$

where $\alpha_k^1,\ldots,\alpha_k^{m_k}$, $\beta_k^1,\ldots,\beta_k^{m_k}$ are suitable scalars.

5.9

A person is offered N free plays to be distributed as he pleases between two slot machines A and B. Machine A pays α dollars with known probability s and nothing with probability $(1-s)$. Machine B pays β dollars with probability p and nothing with probability $(1-p)$. The person does not know p but instead has an a priori probability distribution $F(p)$ for p. The problem is to find a playing policy that maximizes expected profit. Let $(m+n)$ denote the number of plays in machine B after k free plays $(m+n \leq k)$, and let m denote the number of successes and n the number of failures. Show that a DP algorithm for this problem is given for $m+n \leq k$ by

$$\overline{J}_{N-1}(m,n) = \max\bigl[s\alpha,\, p(m,n)\beta\bigr],$$

$$\begin{aligned}\overline{J}_k(m,n) = \max\Bigl[& s\bigl(\alpha + \overline{J}_{k+1}(m,n)\bigr) + (1-s)\overline{J}_{k+1}(m,n),\\ & p(m,n)\bigl(\beta + \overline{J}_{k+1}(m+1,n)\bigr) + \bigl(1-p(m,n)\bigr)\overline{J}_{k+1}(m,n+1)\Bigr]\end{aligned}$$

where

$$p(m,n) = \frac{\int_0^1 p^{m+1}(1-p)^n dF(p)}{\int_0^1 p^m(1-p)^n dF(p)}.$$

Solve the problem for $N=6$, $\alpha=\beta=1$, $s=0.6$, $dF(p)/dp = 1$ for $0 \leq p \leq 1$. [The answer is to play machine B for the following pairs (m,n) : $(0,0)$, $(1,0)$, $(2,0)$, $(3,0)$, $(4,0)$, $(5,0)$, $(2,1)$, $(3,1)$, $(4,1)$. Otherwise, machine A should be played.]

5.10

A person is offered 2 to 1 odds in a coin-tossing game where he wins whenever a tail occurs. However, he suspects that the coin is biased and has an a priori probability distribution $F(p)$ for the probability p that a head occurs at each toss. The problem is to find an optimal policy of deciding whether to continue or stop participating in the game given the outcomes of the game so far. A maximum of N tossings is allowed. Indicate how such a policy can be found by means of DP.

5.11

Consider the ARMAX model of Section 5.3 where instead of $E\left\{\sum_{k=1}^{N}(y_k)^2\right\}$, the cost is

$$E\left\{\sum_{k=1}^{N}(y_k - \overline{y})^2\right\},$$

where $\overline{y}$ is a given scalar. Generalize the minimum variance policy for this case.

5.12

Consider the ARMAX model

$$y_k + a y_{k-1} = u_{k-M} + \epsilon_k,$$

where $M \geq 1$. Show that the minimum variance controller is

$$\mu_k(I_k) = a u_{k-1} - a^2 u_{k-2} + \cdots - (-1)^{M-1} a^{M-1} u_{k-M+1} - (-1)^M a^M y_k,$$

that the resulting closed-loop system is

$$y_k = \epsilon_k - a\epsilon_{k-1} + a^2 \epsilon_{k-2} - \cdots + (-1)^{M-1} a^{M-1} \epsilon_{k-M+1},$$

and that the long-term output variance is

$$E\{(y_k)^2\} = \frac{1 - a^{2M}}{1 - a^2} E\{(\epsilon_k)^2\}.$$

Discuss the qualitative difference between the cases $|a| < 1$ and $|a| > 1$, and relate it to the stability properties of the uncontrolled system $y_k + a y_{k-1} = \epsilon_k$ and the size of the delay M.

6

Suboptimal and Adaptive Control

Contents

We have seen that it is sometimes possible to solve the DP algorithm and to obtain an optimal policy in closed form. However, this tends to be the exception. In most cases, one has to resort to a numerical solution. The computational requirements for this are often overwhelming, and for many problems a complete solution of the problem by DP is impossible. The reason lies in what Bellman has called the "curse of dimensionality."

Consider for example a problem where the state, control, and disturbance spaces are the Euclidean spaces $\Re^n$, $\Re^m$, and $\Re^r$, respectively. In a straightforward numerical approach, these spaces are discretized. Taking d discretization points per state axis results in a state space grid with d^n points. For each of these points the minimization in the right-hand side of the DP equation must be carried out numerically, which involves comparison of as many as d^m numbers. To calculate each of these numbers, one must calculate an expected value over the disturbance, which is the weighted sum of as many as d^r numbers. Finally, the calculations must be done for each of N stages. Thus as a first approximation, the number of computational operations is at least of the order of Nd^n and can be of the order of Nd^{n+m+r}. It follows that for perfect state information problems with Euclidean state and control spaces, DP can be applied numerically only if the dimensions of the spaces are relatively small. Based on the analysis of the preceding chapter, we can also conclude that for problems of imperfect state information the situation is hopeless, except for very simple or very special cases.

Thus, in practice one often has to settle for a suboptimal control scheme that strikes a reasonable balance between convenient implementation and adequate performance. In this chapter we discuss some general approaches for suboptimal control.

6.1 CERTAINTY EQUIVALENT AND ADAPTIVE CONTROL

The *certainty equivalent controller* (CEC) is a suboptimal control scheme that is motivated by linear-quadratic control theory. It applies at each stage the control that would be optimal if some or all of the uncertain quantities were fixed at their expected values; that is, it acts as if a form of the certainty equivalence principle were holding.

We take as our model the basic problem with imperfect state information of Section 5.1. Let us assume that the probability distributions of the disturbances w_k do not depend on x_k and u_k, and that the state spaces and disturbance spaces are convex subsets of Euclidean spaces, so that the expected values

$$\overline{x}_k = E\{x_k \mid I_k\}, \qquad \overline{w}_k = E\{w_k\},$$

belong to the corresponding spaces. Example 1.2 describes a modification to handle the imperfectly observed Markov chain case.

The control input $\tilde{\mu}_k(I_k)$ applied by the CEC at each time k is determined by the following rule:

(1) Given the information vector I_k, compute

$$\overline{x}_k = E\{x_k \mid I_k\}.$$

(2) Find a control sequence $\{\tilde{u}_k, \tilde{u}_{k+1}, \ldots, \tilde{u}_{N-1}\}$ that solves the deterministic problem obtained by fixing the uncertain quantities x_k and $w_k, \ldots, w_{N-1}$ at their typical values, that is,

$$\text{minimize} \;\; g_N(x_N) + \sum_{i=k}^{N-1} g_i(x_i, u_i, \overline{w}_i)$$

subject to the initial condition $x_k = \overline{x}_k$ and the constraints

$$u_i \in U_i, \qquad x_{i+1} = f_i(x_i, u_i, \overline{w}_i), \qquad i = k, k+1, \ldots, N-1.$$

(3) Apply the control input

$$\tilde{\mu}_k(I_k) = \tilde{u}_k.$$

Note that step (1) is unnecessary if we have perfect state information. The deterministic optimization problem in step (2) must be solved at each time k, once the initial state $\overline{x}_k = E\{x_k \mid I_k\}$ becomes known by means of an estimation (or perfect observation) procedure. (In practice, one often uses a suboptimal estimation scheme that generates an approximation to $\overline{x}_k$.) A total of N such problems must be solved by the CEC at every system run. Each of these problems, however, is deterministic. In many cases of interest, these problems can be solved by powerful numerical techniques such as steepest descent, conjugate gradient, and Newton's method, see e.g. [Ber82b], [Lue84], and [Ber95]. Furthermore, the implementation of the CEC requires no storage of the type required for the optimal feedback controller.

An alternative to solving N optimal control problems in an "on-line" fashion is to solve these problems a priori. This is accomplished by computing an optimal feedback controller for the deterministic optimal control problem obtained from the original problem by replacing all uncertain quantities by their expected values. It is easy to verify, based on the equivalence of open-loop and feedback implementation of optimal controllers for deterministic problems, that the implementation of the CEC given earlier is equivalent to the following.

Let $\{\mu_0^d(x_0),\dots,\mu_{N-1}^d(x_{N-1})\}$ be an optimal controller obtained from the DP algorithm for the deterministic problem

$$\text{minimize } g_N(x_N)+\sum_{k=0}^{N-1} g_k\big(x_k,\mu_k(x_k),\overline{w}_k\big)$$

$$\text{subject to } x_{k+1}=f_k\big(x_k,\mu_k(x_k),\overline{w}_k\big),\quad \mu_k(x_k)\in U_k,\quad k=0,\dots,N-1.$$

Then the control input $\tilde{\mu}_k(I_k)$ applied by the CEC at time k is given by

$$\tilde{\mu}_k(I_k)=\mu_k^d\big(E\{x_k\mid I_k\}\big)=\mu_k^d(\overline{x}_k)$$

as shown in Fig. 6.1.1.

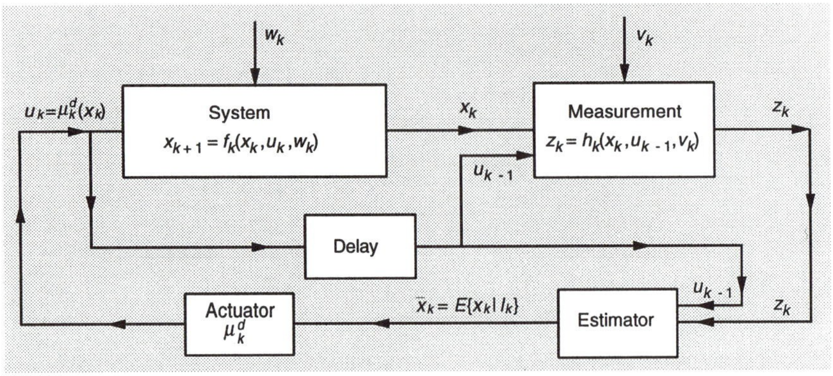

Figure 6.1.1 Structure of the certainty equivalent controller when implemented in feedback form.

In other words, an alternative implementation of the CEC consists of finding a feedback controller $\{\mu_0^d,\mu_1^d,\dots,\mu_{N-1}^d\}$ that is optimal for a corresponding deterministic problem, and subsequently using this controller for control of the uncertain system (modulo substitution of the state by its expected value). Either one of the definitions given for the CEC can serve as a basis for its implementation. Depending on the nature of the problem, one method may be preferable to the other.

The preceding description of the CEC sets all future disturbances w_i to their expected values $\overline{w}_i$. A useful variation for some imperfect state information problems is to take into account the stochastic nature of these disturbances, and to treat the problem as one of perfect state information, using an estimate $\overline{x}_k$ of x_k as if it were exact. In particular, if $\{\mu_0^p(x_0),\dots,\mu_{N-1}^p(x_{N-1})\}$ is an optimal policy obtained from the DP algorithm for the stochastic *perfect state information* problem

$$\text{minimize} \underset{\substack{w_k\\ k=0,1,\dots,N-1}}{E}\left\{g_N(x_N)+\sum_{k=0}^{N-1} g_k\big(x_k,\mu_k(x_k),w_k\big)\right\}$$

$$\text{subject to } x_{k+1}=f_k\big(x_k,\mu_k(x_k),w_k\big),\quad \mu_k(x_k)\in U_k,\quad k=0,\dots,N-1,$$

then the control input $\tilde{\mu}_k(I_k)$ applied by this variant of CEC at time k is given by

$$\tilde{\mu}_k(I_k) = \mu_k^p(\overline{x}_k).$$

Generally, there are several variants of the CEC, where the stochastic uncertainty about some of the unknown quantities is explicitly dealt with, while all other unknown quantities are replaced by estimates obtained in a variety of ways.

The CEC approach often performs well in practice and yields near-optimal policies. In fact, for the linear-quadratic problems of Sections 4.1 and 5.2, the CEC is identical to the optimal controller (certainty equivalence principle). It is possible, however, that a CEC performs strictly worse than the optimal open-loop controller (see Exercise 6.2). We now provide some examples.

Example 1.1 (Multiaccess Communication)

Consider the slotted Aloha system described in Example 1.1 of Section 5.1. It is very difficult to obtain an optimal policy for this problem, primarily because there is no simple characterization of the conditional distribution of the state (the system backlog), given the channel transmission history. We therefore resort to a suboptimal policy. As discussed in Section 5.1, the perfect state information version of the problem admits a simple optimal policy:

$$\mu_k(x_k) = \frac{1}{x_k}, \qquad \text{for all } x_k \geq 1.$$

As a result, there is a natural CEC,

$$\tilde{\mu}_k(I_k) = \min\left[1, \frac{1}{\overline{x}_k}\right],$$

where $\overline{x}_k$ is an estimate of the current packet backlog based on the entire past channel history of successes, idles, and collisions. Recursive estimators for generating $\overline{x}_k$ are discussed in [Mik79], [HaL82], [Riv85], [BeG92], and [Tsi87].

Example 1.2 (A Modified CEC for Finite-State Markov Chains)

The versions of the CEC discussed above apply to the case where the state and disturbance spaces are Euclidean, so that $\overline{x}_k$ and $\overline{w}_k$ are well defined as averages. In the case where the system is a finite-state Markov chain under imperfect state information, these averages have no meaning, but the CEC approach can be applied in the modified form described earlier. The idea is to solve the corresponding problem of perfect state information, and then use the controller thus obtained for control of the imperfectly observed system, modulo substitution of the exact state by an estimate obtained via the Viterbi algorithm described in Section 2.2.2. In particular, suppose that

$\{\mu_0^p, \ldots, \mu_{N-1}^p\}$ is an optimal policy for the corresponding problem where the state is perfectly observed. Then the modified version of the CEC, given the information vector I_k, uses the Viterbi algorithm to obtain (in real time) an estimate $\hat{x}(I_k)$ of the current state x_k, and then applies the control

$$\tilde{\mu}_k(I_k) = \mu_k^p\big(\hat{x}_k(I_k)\big).$$

Example 1.3 (Control of Systems with Unknown Parameters)

We have been dealing so far with systems having a known state equation. In practice, however, there are many cases where the system parameters are not known exactly or change over time. One possible approach is to estimate the unknown parameters from input-output records of the system by using system identification techniques (see [KuV86], [LjS83], [Lju86]). This, however, can be time consuming. Furthermore, the estimation must be repeated if the parameters change.

The alternative is to formulate the stochastic control problem so that unknown parameters are dealt with directly. It can be shown that problems involving unknown system parameters can be embedded within the framework of our basic problem with imperfect state information by using state augmentation. Indeed, let the system equation be of the form

$$x_{k+1} = f_k(x_k, \theta, u_k, w_k),$$

where θ is a vector of unknown parameters with a given a priori probability distribution. We introduce an additional state variable $y_k = \theta$ and obtain a system equation of the form

$$\begin{pmatrix} x_{k+1} \\ y_{k+1} \end{pmatrix} = \begin{pmatrix} f_k(x_k, y_k, u_k, w_k) \\ y_k \end{pmatrix}.$$

This equation can be written compactly as

$$\tilde{x}_{k+1} = \tilde{f}_k(\tilde{x}_k, u_k, w_k),$$

where $\tilde{x}_k = (x_k, y_k)$ is the new state, and $\tilde{f}_k$ is an appropriate function. The initial state is

$$\tilde{x}_0 = (x_0, \theta).$$

With a suitable reformulation of the cost function, the resulting problem becomes one that fits our usual framework.

Unfortunately, however, since $y_k = \theta$ is unobservable, we are faced with a problem of imperfect state information even if the controller knows the state x_k exactly. Thus, typically an optimal solution cannot be found. Nonetheless, a certainty equivalent controller is often convenient. In particular, suppose that for a fixed parameter vector θ, we can compute the corresponding optimal policy

$$\big\{\mu_0^*(I_0, \theta), \ldots, \mu_{N-1}^*(I_{N-1}, \theta)\big\};$$

this is true for example if for a fixed θ, the problem is linear-quadratic of the type considered in Sections 4.1 and 5.2. Then a version of the CEC takes the form

$$\tilde{\mu}_k(I_k) = \mu_k^*(I_k, \hat{\theta}_k),$$

where $\hat{\theta}_k$ is some estimate of θ based on the information vector I_k. Thus, in this approach, the system is identified while it is being controlled. However, the estimates of the unknown parameters are used as if they were exact.

The approach of the preceding example is one of the principal methods of *adaptive control*, that is, control that adapts itself to changing values of system parameters. In the remainder of this section, we discuss some of the associated issues.

6.1.1 Caution, Probing, and Dual Control

Suboptimal control is often guided by the qualitative nature of optimal control. It is therefore important to try to understand some of the characteristic features of the latter in the case where some of the system parameters are unknown. One of these is the need for balance between "caution" and "probing," two notions that are best explained by means of an example; see also [Bar81].

Example 1.4 [Kum83]

Consider the linear scalar system

$$x_{k+1} = x_k + bu_k + w_k, \qquad k = 0, 1, \ldots, N-1,$$

and the quadratic terminal cost $E\{(x_N)^2\}$. Here everything is as in Section 4.1 (perfect state information) except that the control coefficient b is unknown. Instead, it is known that the a priori probability distribution of b is Gaussian with mean and variance

$$\bar{b} = E\{b\} > 0, \qquad \sigma_b^2 = E\{(b - \bar{b})^2\}.$$

Furthermore, w_k is zero mean Gaussian with variance σ_w^2 for each k.

Consider first the case where $N = 1$, so the cost is calculated to be

$$E\{(x_1)^2\} = E\{(x_0 + bu_0 + w_0)^2\} = x_0^2 + 2\bar{b}x_0u_0 + (\bar{b}^2 + \sigma_b^2)u_0^2 + \sigma_w^2.$$

The minimum over u_0 is attained at

$$u_0 = -\frac{\bar{b}}{\bar{b}^2 + \sigma_b^2}x_0,$$

and the optimal cost is verified by straightforward calculation to be

$$\frac{\sigma_b^2}{\bar{b}^2 + \sigma_b^2} x_0^2 + \sigma_w^2.$$

Therefore, the optimal control here is *cautious* in that the optimum $|u_0|$ decreases as the uncertainty in b (i.e., σ_b^2) increases.

Consider next the case where $N = 2$. The optimal cost-to-go at stage 1 is obtained by the preceding calculation:

$$J_1(I_1) = \frac{\sigma_b^2(1)}{\big(\bar{b}(1)\big)^2 + \sigma_b^2(1)} x_1^2 + \sigma_w^2, \tag{1.1}$$

where $I_1 = (x_0, u_0, x_1)$ is the information vector and

$$\bar{b}(1) = E\{b \mid I_1\}, \qquad \sigma_b^2(1) = E\Big\{\big(b - \bar{b}(1)\big)^2 \mid I_1\Big\}.$$

Let us focus on the term $\sigma_b^2(1)$ in the expression (1.1) for $J_1(I_1)$. We can obtain $\sigma_b^2(1)$ from the equation $x_1 = x_0 + bu_0 + w_0$ (which we view as a noise-corrupted measurement of b) and least-squares estimation theory (see Appendix E). The formula for $\sigma_b^2(1)$ will be of no further use to us, so we just state it without going into the calculation:

$$\sigma_b^2(1) = \frac{\sigma_b^2 \sigma_w^2}{u_0^2 \sigma_b^2 + \sigma_w^2}. \tag{1.2}$$

The salient feature of this equation is that $\sigma_b^2(1)$ is affected by the control u_0. Basically, if $|u_0|$ is small, the measurement $x_1 = x_0 + bu_0 + w_0$ is dominated by w_0 and the "signal-to-noise ratio" is small. Thus to achieve small error variance $\sigma_b^2(1)$ [which is desirable in view of Eq. (1.1)], we must apply a control u_0 that is large in absolute value. A choice of large control to enhance parameter identification is often referred to as *probing*. On the other hand, if $|u_0|$ is large, $|x_1|$ will also be large, and this is not desirable in view of Eq. (1.1). Therefore, in choosing u_0 we must strike a balance between caution (choosing a small value to keep x_1 reasonably small) and probing (choosing a large value to improve the signal-to-noise ratio and enhance estimation of b).

The tradeoff between the control objective and the parameter estimation objective is commonly referred to as *dual control*. It manifests itself often when system parameters are unknown, but unfortunately it cannot be quantified precisely in most cases.

6.1.2 Two-Phase Control and Identifiability

An apparently reasonable form of suboptimal control in the presence of unknown parameters is to separate the control process into two phases, a *parameter identification phase* and a *control phase*. In the first phase the unknown parameters are identified, while the control takes no account of the interim results of identification. The final parameter estimates from the first phase are then used to implement an optimal control law in the second phase. This alternation of identification and control phases may be repeated several times during any system run in order to take into account subsequent changes of the parameters.

One drawback of this approach is that information gathered during the identification phase is not used to adjust the control law until the beginning of the second phase. Furthermore, it is not always easy to determine when to terminate one phase and start the other.

A second difficulty, of a more fundamental nature, is due to the fact that the control process may make some of the unknown parameters invisible to the identification process. This is the problem of parameter *identifiability* [Lju86], which is best explained by means of an example.

Example 1.5

Consider the scalar system

$$x_{k+1} = ax_k + bu_k + w_k, \qquad k = 0, 1, \ldots, N-1,$$

with the quadratic cost

$$E\left\{\sum_{k=1}^{N} (x_k)^2\right\}.$$

We assume perfect state information, so if the parameters a and b are known, this is a minimum variance control problem (cf. Section 5.3), and the optimal control law is

$$\mu_k^*(x_k) = -\frac{a}{b}x_k.$$

Assume now that the parameters a and b are unknown and consider the two-phase method. During the first phase the control law

$$\tilde{\mu}_k(x_k) = \gamma x_k \tag{1.3}$$

is used (γ is some scalar; for example, $\gamma = -\overline{a}/\overline{b}$, where $\overline{a}$ and $\overline{b}$ are a priori estimates of a and b, respectively). At the end of the first phase, the control law is changed to

$$\tilde{\mu}_k(x_k) = -\frac{\hat{a}}{\hat{b}}x_k,$$

where $\hat{a}$ and $\hat{b}$ are the estimates obtained from the identification process. However, with the control law (1.3), the closed-loop system is

$$x_{k+1} = (a + b\gamma)x_k + w_k,$$

so the identification process can at best identify the value of $(a + b\gamma)$ but not the values of both a and b. In other words, the identification process cannot discriminate between pairs of values (a_1, b_1) and (a_2, b_2) such that $a_1 + b_1\gamma = a_2 + b_2\gamma$. Therefore, a and b are not identifiable when feedback control of the form (1.3) is applied.

One way to correct the difficulty is to add an additional known input δ_k to the control law (1.3); that is, use

$$\overline{\mu}_k(x_k) = \gamma x_k + \delta_k.$$

Then the closed-loop system becomes

$$x_{k+1} = (a + b\gamma)x_k + b\delta_k + w_k,$$

and the knowledge of $\{x_k\}$ and $\{\delta_k\}$ makes it possible to identify $(a+b\gamma)$ and b. Given γ, one can then obtain estimates of a and b. Actually, to guarantee this in a more general context where the system is of higher dimension, the sequence $\{\delta_k\}$ must satisfy certain conditions: it must be "persistently exciting" (see [LjS83]).

A second possibility to bypass the identifiability problem is to change the structure of the system by artificially introducing a one-unit delay in the control feedback. Thus, instead of considering control laws of the form $\tilde{\mu}_k(x_k) = \gamma x_k$, we consider controls of the form

$$u_k = \hat{\mu}_k(x_{k-1}) = \gamma x_{k-1}.$$

The closed-loop system then becomes

$$x_{k+1} = ax_k + b\gamma x_{k-1} + w_k,$$

and given γ, it is possible to identify both parameters a and b. This technique can be generalized for systems of arbitrary order, but artificially introducing a control delay makes the system less responsive to control.

6.1.3 Certainty Equivalent Control and Identifiability

At the opposite extreme of the two-phase method we have the certainty equivalent control approach, where the parameter estimates are incorporated into the control law as they are generated, and they are treated as if they were true values. In terms of the system

$$x_{k+1} = f_k(x_k, \theta, u_k, w_k)$$

considered in Example 1.3, suppose that, for each possible value of θ, the control law $\pi^*(\theta) = \{\mu_0^*(\cdot, \theta), \ldots, \mu_{N-1}^*(\cdot, \theta)\}$ is optimal with respect to a certain cost $J_\pi(x_0, \theta)$. Then the (suboptimal) control used at time k is

$$\hat{\mu}_k(I_k) = \mu_k^*(x_k, \hat{\theta}_k),$$

where $\hat{\theta}_k$ is an estimate of θ based on the information

$$I_k = \{x_0, x_1, \dots, x_k, u_0, u_1, \dots, u_{k-1}\}$$

available at time k; for example,

$$\hat{\theta}_k = E\{\theta \mid I_k\}$$

or, more likely in practice, an estimate obtained via an on-line system identification method (see [KuV86], [LjS83], [Lju86]).

One would hope that when the horizon is very long, the parameter estimates $\hat{\theta}_k$ will converge to the true value θ, so the certainty equivalent controller will become asymptotically optimal. Unfortunately, we will see that difficulties related to identifiability arise here as well.

Suppose for simplicity that the system is stationary with a priori known transition probabilities $P\{x_{k+1} \mid x_k, u_k, \theta\}$ and that the control law used is also stationary:

$$\hat{\mu}_k(I_k) = \mu^*(x_k, \hat{\theta}_k), \qquad k = 0, 1, \dots$$

There are three systems of interest here (cf. Fig. 6.1.2):

(a) The system (perhaps falsely) believed by the controller to be true, which evolves probabilistically according to

$$P\big\{x_{k+1} \mid x_k, \mu^*(x_k, \hat{\theta}_k), \hat{\theta}_k\big\}.$$

(b) The true closed-loop system, which evolves probabilistically according to

$$P\big\{x_{k+1} \mid x_k, \mu^*(x_k, \hat{\theta}_k), \theta\big\}.$$

(c) The optimal closed-loop system that corresponds to the true value of the parameter, which evolves probabilistically according to

$$P\big\{x_{k+1} \mid x_k, \mu^*(x_k, \theta), \theta\big\}.$$

For asymptotic optimality, we would like the last two systems to be equal asymptotically. This will certainly be true if $\hat{\theta}_k \to \theta$. However, it is quite possible that either

(1) $\hat{\theta}_k$ does not converge to anything, or that

(2) $\hat{\theta}_k$ converges to a parameter $\hat{\theta} \neq \theta$.

There is not much we can say about the first case, so we concentrate on the second. To see how the parameter estimates can converge to a wrong value, assume that for some $\hat{\theta} \neq \theta$ and all x_{k+1}, x_k, we have

$$P\big\{x_{k+1} \mid x_k, \mu^*(x_k, \hat{\theta}), \hat{\theta}\big\} = P\big\{x_{k+1} \mid x_k, \mu^*(x_k, \hat{\theta}), \theta\big\}. \tag{1.4}$$

In words, *there is a false value of parameter for which the system under closed-loop control looks exactly as if the false value were true*. Then, if the controller estimates at some time the parameter to be $\hat{\theta}$, subsequent data will tend to reinforce this erroneous estimate. As a result, a situation may develop where the identification procedure locks onto a wrong parameter value, regardless of how long information is collected. This is a difficulty with identifiability of the type discussed earlier in connection with two-phase control.

On the other hand, if the parameter estimates converge to some (possibly wrong) value, we can argue intuitively that the first two systems (believed and true) typically become equal in the limit as $k \to \infty$, since, generally, parameter estimate convergence in identification methods implies that the data obtained are asymptotically consistent with the view of the system one has based on the current estimates. However, the believed and true systems may or may not become asymptotically equal to the optimal closed-loop system. We first present two examples that illustrate how, even when the parameter estimates converge, the true closed-loop system can differ asymptotically from the optimal, thereby resulting in a certainty equivalent controller that is strictly suboptimal. We then discuss the special case of the self-tuning regulator for ARMAX models with unknown parameters, where, remarkably, it turns out that all three of the above systems are typically equal in the limit, even though the parameter estimates typically converge to false values.

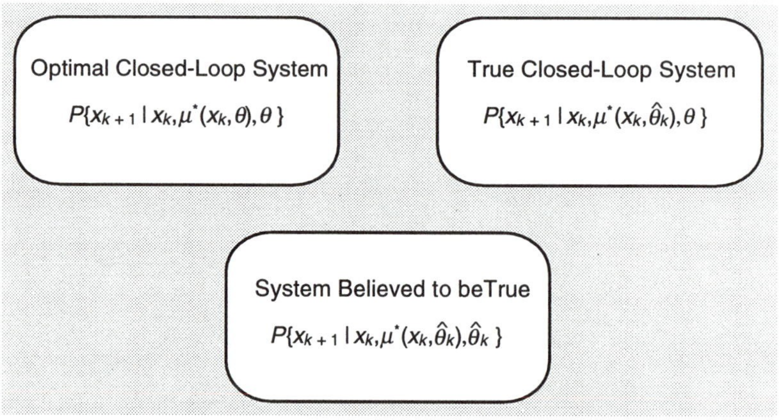

Figure 6.1.2 The three systems involved in certainty equivalent control, where θ is the true parameter and $\hat{\theta}_k$ is the parameter estimate at time k. Loss of optimality occurs when the true system differs asymptotically from the optimal closed-loop system. If the parameter estimates converge to some value $\hat{\theta}$, the true system typically becomes asymptotically equal to the system believed to be true. However, the parameter estimates need not converge, and even if they do, both systems may be different asymptotically from the optimal.

Example 1.6 [BoV79]

Consider a two-state system with two controls u^1 and u^2. The transition probabilities depend on the control applied as well as a parameter θ, which is known to take one of two values θ^* and $\hat{\theta}$. They are as shown in Fig. 6.1.3. There is zero cost for a transition from state 1 to itself and a unit cost for all other transitions. Therefore, the optimal control at state 1 is the one that maximizes the probability of the state remaining at 1. Assume that the true parameter is θ^* and that

$$p_{11}(u^1, \hat{\theta}) > p_{11}(u^2, \hat{\theta}), \qquad p_{11}(u^1, \theta^*) < p_{11}(u^2, \theta^*).$$

Then the optimal control is u^2, but if the controller *thinks* that the true parameter is $\hat{\theta}$, it will apply u^1. Suppose also that

$$p_{11}(u^1, \hat{\theta}) = p_{11}(u^1, \theta^*).$$

Then, under u^1 the system looks identical for both values of the parameter, so if the controller estimates the parameter to be $\hat{\theta}$ and applies u^1, subsequent data will tend to reinforce the controller's belief that the true parameter is indeed $\hat{\theta}$.

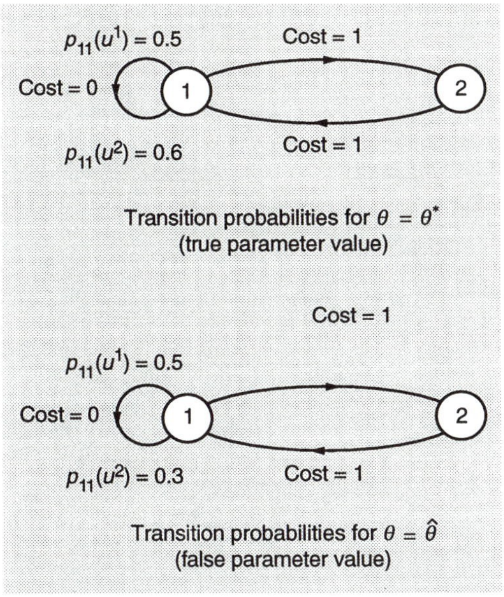

Figure 6.1.3 Transition probabilities for the two-state system of Example 1.6. Under the nonoptimal control u^1, the system looks identical under the true and the false values of the parameter θ.

More precisely, suppose that we estimate θ by selecting at each time k the value that maximizes

$$P\{\theta \mid I_k\} = \frac{P\{I_k \mid \theta\} P(\theta)}{P(I_k)},$$

where $P(\theta)$ is the a priori probability that the true parameter is θ (this is a popular estimation method). Then if $P(\hat{\theta}) > P(\theta^*)$, it can be seen, by using induction, that at each time k, the controller will estimate falsely θ to be $\hat{\theta}$ and apply the incorrect control u^1. To avoid the difficulty illustrated in this example, it has been suggested to occasionally deviate from the certainty equivalent control, applying other controls that enhance the identification of the unknown parameter (see [DoS80], [KuL82]). For example, by making sure that the control u^2 is used infrequently but infinitely often, we can guarantee that the correct parameter value will be identified by the preceding estimation scheme.

Example 1.7 [Kum83]

Consider the linear scalar system

$$x_{k+1} = ax_k + bu_k + w_k,$$

where we know that the parameters are either $(a, b) = (1, 1)$ or $(a, b) = (0, -1)$. The sequence $\{w_k\}$ is independent, stationary, zero mean, and Gaussian. The cost is quadratic of the form

$$\sum_{k=0}^{N-1} \big((x_k)^2 + 2(u_k)^2\big),$$

where N is very large, so the stationary form of the optimal control law is used (see Section 4.1). This control law can be calculated via the Riccati equation to be

$$\mu^*(x_k) = \begin{cases} -\frac{x_k}{2} & \text{if } (a, b) = (1, 1), \\ 0 & \text{if } (a, b) = (0, -1). \end{cases}$$

To estimate (a, b), we use a least-squares identification method. The value of the least-squares criterion at time k is given by

$$V_k(1, 1) = \sum_{i=0}^{k-1} (x_{i+1} - x_i - u_i)^2, \qquad \text{for } (a, b) = (1, 1), \tag{1.5a}$$

$$V_k(0, -1) = \sum_{i=0}^{k-1} (x_{i+1} + u_i)^2, \qquad \text{for } (a, b) = (0, -1). \tag{1.5b}$$

The control applied at time k is

$$u_k = \tilde{\mu}_k(I_k) = \begin{cases} -\frac{x_k}{2} & \text{if } V_k(1, 1) < V_k(0, -1), \\ 0 & \text{if } V_k(1, 1) > V_k(0, -1). \end{cases}$$

Suppose the true parameters are $\theta = (0, -1)$. Then the true system evolves according to

$$x_{k+1} = -u_k + w_k. \tag{1.6}$$

If at time k the controller estimates incorrectly the parameters to be $\hat{\theta} = (1, 1)$, because $V_k(\hat{\theta}) < V_k(\theta)$, the control applied will be $u_k = -x_k/2$ and the *true* closed-loop system will evolve according to

$$x_{k+1} = \frac{x_k}{2} + w_k. \tag{1.7}$$

On the other hand, the controller *thinks* (given the estimate $\hat{\theta}$) that the closed-loop system will evolve according to

$$x_{k+1} = x_k + u_k + w_k = x_k - \frac{x_k}{2} + w_k = \frac{x_k}{2} + w_k, \tag{1.8}$$

so from Eqs. (1.6) and (1.7) we see that *under the control law $u_k = -x_k/2$, the closed-loop system evolves identically for both the true and the false values of the parameters* [cf. Eq. (1.4)].

To see what can go wrong, note that if $V_k(\hat{\theta}) < V_k(\theta)$ for some k we will have, from Eqs. (1.5)-(1.8),

$$x_{k+1} + u_k = x_{k+1} - x_k - u_k,$$

so from Eqs. (1.5a) and (1.5b) we obtain

$$V_{k+1}(\hat{\theta}) < V_{k+1}(\theta).$$

Therefore, if $V_1(\hat{\theta}) < V_1(\theta)$, the least-squares identification method will yield the wrong estimate $\hat{\theta}$ *for every k*. To see that this can happen with positive probability, note that, since the true system is $x_{k+1} = -u_k + w_k$, we have

$$V_1(\hat{\theta}) = (x_1 - x_0 - u_0)^2 = (w_0 - x_0 - 2u_0)^2,$$

$$V_1(\theta) = (x_1 + u_0)^2 = w_0^2.$$

Therefore, the inequality $V_1(\hat{\theta}) < V_1(\theta)$ is equivalent to

$$(x_0 + 2u_0)^2 < 2w_0(x_0 + 2u_0),$$

which will occur with positive probability since w_0 is Gaussian.

The preceding examples illustrate that loss of identifiability is a serious problem that frequently arises in the context of certainty equivalent control.

6.1.4 Self-Tuning Regulators

We described earlier the nature of the identifiability issue in certainty equivalent control: under closed-loop control, incorrect parameter estimates can make the system behave as if these estimates were correct [cf. Eq. (1.4)]. As a result, the identification scheme may lock onto false parameter values. This is not necessarily bad, however, since it may happen that the control law implemented on the basis of the false parameter values is near optimal. Indeed, through a fortuitous coincidence, it turns out that *in the practically important minimum variance control formulation (Section 5.3), when the parameter estimates converge, they typically converge to false values, but the resulting control law typically converges to the optimal.* We can get an idea about this phenomenon by means of an example.

Example 1.8

Consider the simplest ARMAX model:

$$y_{k+1} + ay_k = bu_k + \epsilon_{k+1}.$$

The minimum variance control law when a and b are known is

$$u_k = \mu_k(I_k) = \frac{a}{b}y_k.$$

Suppose now that a and b are not known but are identified on-line by means of some scheme. The control applied is

$$u_k = \frac{\hat{a}_k}{\hat{b}_k}y_k, \tag{1.9}$$

where $\hat{a}_k$ and $\hat{b}_k$ are the estimates obtained at time k. Then the difficulty with identifiability occurs when

$$\hat{a}_k \to \hat{a}, \qquad \hat{b}_k \to \hat{b},$$

where $\hat{a}$ and $\hat{b}$ are such that the true closed-loop system given by

$$y_{k+1} + ay_k = \frac{b\hat{a}}{\hat{b}}y_k + \epsilon_{k+1}$$

coincides with the closed-loop system that the controller thinks is true on the basis of the estimates $\hat{a}$ and $\hat{b}$. This latter system is

$$y_{k+1} = \epsilon_{k+1}.$$

For these two systems to be identical, we must have

$$\frac{a}{b} = \frac{\hat{a}}{\hat{b}},$$

which means that the control law (1.9) asymptotically becomes optimal despite the fact that the asymptotic estimates $\hat{a}$ and $\hat{b}$ may be incorrect.

Example 1.8 can be extended to the general ARMAX model of Section 5.3 with no delay:

$$y_k + \sum_{i=1}^{m} a_i y_{k-i} = \sum_{i=1}^{m} b_i u_{k-i} + \epsilon_k + \sum_{i=1}^{m} c_i \epsilon_{k-i}. \tag{1.10}$$

If the parameter estimates converge (regardless of the identification method used and regardless of whether the limit values are correct), then a minimum variance controller *thinks* that the closed-loop system is asymptotically

$$y_k = \epsilon_k. \tag{1.11}$$

Furthermore, parameter estimate convergence intuitively means that the true closed-loop system is also asymptotically $y_k = \epsilon_k$, and this is clearly the optimal closed-loop system. Results of this type have been proved in the literature in connection with several popular methods for parameter estimation. In fact, surprisingly, in some of these results, the model adopted by the controller is allowed to be incorrect to some extent.

One issue that we have not discussed is whether the parameter estimates indeed converge. A complete analysis of this issue is quite difficult. We refer to the survey paper [Kum85], and the textbooks [GoS84], [KuV86], and [AsW90] for a discussion and sources on this subject. However, extensive simulations have shown that with proper implementation, these estimates typically converge for the type of systems likely to arise in many applications. For this reason, self-tuning regulator schemes are presently used widely in practice as general purpose process control algorithms.

6.2 OPEN-LOOP FEEDBACK CONTROL

The *open-loop feedback controller* (OLFC) is similar to the CEC except that it takes into account the uncertainty about $x_k, w_k, \ldots, w_{N-1}$ when calculating the control $\overline{\mu}_k(I_k)$ to be applied at time k. This control is determined as follows:

(1) Given the information vector I_k, compute the conditional probability distribution $P_{x_k|I_k}$.

(2) Find a control sequence $\{\overline{u}_k, \overline{u}_{k+1}, \ldots, \overline{u}_{N-1}\}$ that minimizes

$$\mathop{E}_{x_k, w_k, \ldots, w_{N-1}} \left\{ g_N(x_N) + \sum_{i=k}^{N-1} g_i(x_i, u_i, w_i) \;\Big|\; I_k \right\}$$

subject to the constraints

$$x_{i+1} = f_i(x_i, u_i, w_i), \qquad u_i \in U_i, \qquad i = k, k+1, \ldots, N-1.$$

(3) Apply the control input

$$\overline{\mu}_k(I_i) = \overline{u}_k.$$

Thus the OLFC uses at time k the new measurement z_k to calculate the conditional probability distribution $P_{x_k|I_k}$. However, it selects the control input as if no further measurements will be received in the future.

Similar to the CEC, the OLFC requires the solution of N optimal control problems. Each problem may again be solved by deterministic optimal control techniques. The computations, however, may be more complicated than those for the CEC, since now the cost involves an expectation with respect to the uncertain quantities. The main difficulty in the implementation of the OLFC is the computation of $P_{x_k|I_k}$. In many cases one cannot compute $P_{x_k|I_k}$ exactly, in which case some "reasonable" approximation scheme must be used. Of course, if we have perfect state information, this difficulty does not arise.

In any suboptimal control scheme, one would like to be assured that measurements are advantageously used. By this we mean that the scheme performs at least as well as any open-loop policy that uses a sequence of controls that is independent of the values of the measurements received. An optimal open-loop policy can be obtained by finding a sequence $\{u_0^*, u_1^*, \ldots, u_{N-1}^*\}$ that minimizes

$$\overline{J}(u_0, u_1, \ldots, u_{N-1}) = \underset{\substack{x_0, w_k \\ k=0,1,\ldots,N-1}}{E} \left\{ g_N(x_N) + \sum_{k=0}^{N-1} g_k(x_k, u_k, w_k) \right\}$$

subject to the constraints $x_{k+1} = f_k(x_k, u_k, w_k)$, and $u_k \in U_k$, $k = 0, 1, \ldots, N-1$. A nice property of the OLFC is that it performs at least as well as an optimal open-loop policy, as shown by the following proposition. By contrast, the CEC does not have this property (see Exercise 6.2).

Proposition 2.1: The cost $J_{\overline{\pi}}$ corresponding to an OLFC satisfies

$$J_{\overline{\pi}} \leq J_0^*, \tag{2.1}$$

where J_0^* is the cost corresponding to an optimal open-loop policy.

Proof: We assume throughout the proof that all expected values appearing are well defined and finite, and that the minimum in the following Eq. (2.5) is attained for every I_k. Let $\overline{\pi} = \{\overline{\mu}_0, \overline{\mu}_1, \ldots, \overline{\mu}_{N-1}\}$ be the OLFC. Its cost is given by

$$J_{\overline{\pi}} = \underset{z_0}{E}\{\overline{J}_0(I_0)\} = \underset{z_0}{E}\{\overline{J}_0(z_0)\}, \tag{2.2}$$

where $\overline{J}_0$ is obtained from the recursive algorithm

$$\begin{aligned} \overline{J}_{N-1}(I_{N-1}) = \underset{x_{N-1}, w_{N-1}}{E} \big\{ & g_N\big(f_{N-1}(x_{N-1}, \overline{\mu}_{N-1}(I_{N-1}), w_{N-1})\big) \\ & + g_{N-1}\big(x_{N-1}, \overline{\mu}_{N-1}(I_{N-1}), w_{N-1}\big) \mid I_{N-1} \big\}, \end{aligned} \tag{2.3}$$

$$\overline{J}_k(I_k) = \underset{x_k, w_k, v_{k+1}}{E} \big\{ g_k\big(x_k, \overline{\mu}_k(I_k), w_k\big) + \overline{J}_{k+1}\big(I_k, h_{k+1}\big(f_k\big(x_k, \overline{\mu}_k(I_k), w_k\big), \overline{\mu}_k(I_k), v_{k+1}\big), \overline{\mu}_k(I_k)\big) \mid I_k \big\}, \quad k = 0, \ldots, N-1, \tag{2.4}$$

where h_k is the function involved in the measurement equation as in the basic problem with imperfect state information of Section 5.1.

Consider the functions $J_k^c(I_k)$, $k = 0, 1, \ldots, N-1$, defined by

$$J_k^c(I_k) = \min_{\substack{u_i \in U_i \\ i=k,\ldots,N-1}} \underset{\substack{x_k, w_i \\ x_{i+1} = f_i(x_i, u_i, w_i) \\ i=k,\ldots,N-1}}{E} \left\{ g_N(x_N) + \sum_{i=k}^{N-1} g_i(x_i, u_i, w_i) \;\middle|\; I_k \right\}. \tag{2.5}$$

The minimization problem in this equation is the one that must be solved at time k in order to calculate the control $\overline{\mu}_k(I_k)$ of the OLFC. Clearly, $J_k^c(I_k)$ can be interpreted as the optimal open-loop cost from time k to time N when the current information vector is I_k. It can be seen that

$$\underset{z_0}{E}\big\{J_0^c(z_0)\big\} \leq J_0^*. \tag{2.6}$$

We will prove that

$$\overline{J}_k(I_k) \leq J_k^c(I_k), \qquad \text{for all } I_k \text{ and } k. \tag{2.7}$$

Then from Eqs. (2.2), (2.6), and (2.7), it will follow that

$$J_{\overline{\pi}} \leq J_0^*,$$

which is the relation to be proved. We show Eq. (2.7) by induction.

By the definition of the OLFC and Eq. (2.5), we have

$$\overline{J}_{N-1}(I_{N-1}) = J_{N-1}^c(I_{N-1}), \qquad \text{for all } I_{N-1},$$

and hence Eq. (2.7) holds for $k = N-1$. Assume that

$$\overline{J}_{k+1}(I_{k+1}) \leq J_{k+1}^c(I_{k+1}), \qquad \text{for all } I_{k+1}. \tag{2.8}$$

Then from Eqs. (2.4), (2.5), and (2.8), we have

$$\begin{aligned}
\overline{J}_k(I_k) = {} & \underset{x_k, w_k, v_{k+1}}{E} \big\{ g_k\big(x_k, \overline{\mu}_k(I_k), w_k\big) \\
& + \overline{J}_{k+1}\big(I_k, h_{k+1}\big(f_k\big(x_k, \overline{\mu}_k(I_k), w_k\big), \overline{\mu}_k(I_k), v_{k+1}\big), \overline{\mu}_k(I_k)\big) \mid I_k \big\} \\
\leq {} & \underset{x_k, w_k, v_{k+1}}{E} \big\{ g_k\big(x_k, \overline{\mu}_k(I_k), w_k\big) \\
& + J_{k+1}^c\big(I_k, h_{k+1}\big(f_k\big(x_k, \overline{\mu}_k(I_k), w_k\big), \overline{\mu}_k(I_k), v_{k+1}\big), \overline{\mu}_k(I_k)\big) \mid I_k \big\}
\end{aligned}$$

$$
\begin{aligned}
&= \underset{x_k, w_k, v_{k+1}}{E} \Bigg\{ \min_{\substack{u_i \in U_i \\ i=k+1,\dots,N-1}} \underset{\substack{x_{k+1}, w_i \\ x_{i+1}=f_i(x_i,u_i,w_i) \\ i=k+1,\dots,N-1}}{E} \Big\{ g_k\big(x_k, \overline{\mu}_k(I_k), w_k\big) \\
&\quad + \sum_{i=k+1}^{N-1} g_i(x_i, u_i, w_i) + g_N(x_N) \mid I_{k+1} \Big\} \,\Big|\, I_k \Bigg\} \\
&\leq \min_{\substack{u_i \in U_i \\ i=k+1,\dots,N-1}} \underset{\substack{x_k, w_k, w_i \\ x_{i+1}=f_i(x_i,u_i,w_i) \\ i=k+1,\dots,N-1 \\ x_{k+1}=f_k(x_k,\overline{\mu}_k(I_k),w_k)}}{E} \Big\{ g_N(x_N) + g_k\big(x_k, \overline{\mu}_k(I_k), w_k\big) \\
&\quad + \sum_{i=k+1}^{N-1} g_i(x_i, u_i, w_i) \,\Big|\, I_k \Big\} \\
&= J_k^c(I_k).
\end{aligned}
$$

The second inequality follows by interchanging expectation and minimization (since we generally have $E\{\min[\cdot]\} \leq \min[E\{\cdot\}]$) and by "integrating out" v_{k+1}. The last equality follows from the definition of OLFC. Thus Eq. (2.7) is proved for all k and the desired result is shown. **Q.E.D.**

It is worth noting that by Eq. (2.7), $J_k^c(I_k)$, which is the calculated open-loop optimal cost from time k to time N, provides a readily obtainable performance bound for the OLFC.

The preceding proposition shows that the OLFC uses the measurements advantageously even though it selects at each period the present control input as if no further measurements will be taken in the future. Of course, this says nothing about how closely the resulting cost approximates the optimal. It appears, however, that the OLFC is a fairly satisfactory mode of control for many problems.

Partial Open-Loop Feedback Control

A form of suboptimal control that is intermediate between the optimal feedback controller and the OLFC is provided by what we call the *partial open-loop feedback controller* (POLFC for short). This controller uses past measurements to compute $P_{x|I_k}$, but calculates the control input on the basis that *some* (but not all) of the measurements will in fact be taken in the future, and the other measurements will not be taken.

This method often allows one to deal with those measurements that are troublesome and complicate the solution. As an example consider an inventory problem such as the one considered in Section 4.2, where forecasts in the form of a probability distribution of each of the future demands become available over time. A POLFC calculates at each stage an optimal (s, S) policy based on the current forecast of future demands and follows this policy until a new forecast becomes available. When this happens, the current policy is abandoned in favor of a new one that is calculated on the basis of the new probability distribution of future demands, etc. Thus the complications due to the forecasts are bypassed at the expense of suboptimality of the policy obtained.

We note that an analog of Prop. 2.1 can be shown for the POLFC (see [Ber76]). In fact the corresponding error bound is potentially much better than the bound (2.1), reflecting the fact that the POLFC takes into account the future availability of some of the measurements.

6.3 LIMITED LOOKAHEAD POLICIES AND APPLICATIONS

A practical way to reduce the computation required by DP is to truncate the time horizon and use at each stage a decision based on lookahead of a small number of stages. The simplest possibility is to use a *one-step lookahead policy* whereby at stage k and state x_k one uses the control $\overline{\mu}_k(x_k)$, which attains the minimum in the expression

$$\min_{u_k \in U_k(x_k)} E\big\{g_k(x_k, u_k, w_k) + \tilde{J}_{k+1}\big(f_k(x_k, u_k, w_k)\big)\big\},$$

where $\tilde{J}_{k+1}$ is some approximation of the true cost-to-go function J_{k+1}. Similarly, a *two-step lookahead policy* applies at time k and state x_k, the control $\tilde{\mu}_k(x_k)$ attaining the minimum in the preceding equation, where now $\tilde{J}_{k+1}$ is obtained itself on the basis of a one-step lookahead approximation. In other words, for all possible states

$$x_{k+1} = f_k(x_k, u_k, w_k),$$

we have

$$\tilde{J}_{k+1}(x_{k+1}) = \min_{u_{k+1} \in U_{k+1}(x_{k+1})} E\big\{g_{k+1}(x_{k+1}, u_{k+1}, w_{k+1}) + \tilde{J}_{k+2}\big(f_{k+1}(x_{k+1}, u_{k+1}, w_{k+1})\big)\big\},$$

where $\tilde{J}_{k+2}$ is some approximation of the cost-to-go function J_{k+2}. Policies with lookahead of more than two stages are similarly defined.

The computational savings of this approach are evident. For a one-step lookahead policy, only a single minimization problem has to be solved per stage, while in a two-step policy the number of states at which the DP equation has to be solved at stage k is one plus the number of all possible next states x_{k+1} that can be generated from the current state x_k. Note also that the fixed lookahead approach can be combined with the certainty equivalent and open-loop-feedback control approaches of Sections 6.1 and 6.2 to simplify even further the calculations.

Note that the limited lookahead approach can be used equally well with infinite horizon problems. One simply uses as the terminal cost-to-go function an approximation to the optimal cost of the infinite horizon problem that starts at the end of the lookahead. Thus the following discussion applies to both finite and infinite horizon problems.

Solving Limited Lookahead Subproblems

An important fact that facilitates the application of limited lookahead policies is that a perfect state information problem with a small number of stages can often be solved as a nonlinear programming problem with a relatively small number of variables. In particular, suppose that the control space is the Euclidean space $\Re^m$, and the disturbance can take a finite number of values, say r. Then it can be shown that for a given initial state, an N-stage perfect state information problem can be formulated as a nonlinear programming problem with $m\big(1+r^{N-1}\big)$ variables. We illustrate this by means of an important example and then discuss the general case.

Example 3.1 (Two-Stage Stochastic Programming)

Here we want to find an optimal two-stage decision rule for the following situation: In the first stage we will choose a vector u_0 from a subset $U_0 \subset \Re^m$ with cost $g_0(u_0)$. Then an uncertain event represented by a random variable w will occur, where w will take one of the values $w^1, \ldots, w^r$ with corresponding probabilities $p^1, \ldots, p^r$. We will know the value w^j once it occurs, and we must then choose a vector u_1^j from a subset $U_1 \subset \Re^m$ at a cost $g_1(u_1^j, w^j)$. The objective is to minimize the expected cost

$$g_0(u_0) + \sum_{j=1}^{r} p^j g_1(u_1^j, w^j)$$

subject to

$$u_0 \in U_0, \qquad u_1^j \in U_1, \quad j = 1, \ldots, r.$$

This is a nonlinear programming problem of dimension $m(1+r)$ (the optimization variables are $u_0, u_1^1, \ldots, u_1^r$). It can also be viewed as a two-stage perfect state information problem, where $x_1 = w_0$ is the state equation, w_0 can take the values $w^1, \ldots, w^r$ with probabilities $p^1, \ldots, p^r$, the cost of the first stage is $g_0(u_0)$, and the cost of the second stage is $g_1(x_1, u_1)$.

The preceding example can be generalized. Consider the basic problem of Chapter 1 for the case where there are only two stages ($N = 2$) and the disturbances w_0 and w_1 can independently take one of the r values

$w^1, \ldots, w^r$ with corresponding probabilities $p^1, \ldots, p^r$. The optimal cost function $J_0(x_0)$ is given by the two-stage DP algorithm

$$J_0(x_0) = \min_{u_0 \in U_0(x_0)} \Bigg[\sum_{j=1}^{r} p^j \Bigg\{ g_0(x_0, u_0, w^j) + \min_{u_1^j \in U_1(f_0(x_0,u_0,w^j))} \Bigg[\sum_{i=1}^{r} p^i \Big\{ g_1\big(f_0(x_0, u_0, w^j), u_1^j, w^i\big) + g_2\Big(f_1\big(f_0(x_0, u_0, w^j), u_1^j, w^i\big)\Big) \Big\} \Bigg] \Bigg\} \Bigg].$$

This DP algorithm is equivalent to solving the nonlinear programming problem

$$\text{minimize} \sum_{j=1}^{r} p^j \Bigg\{ g_0(x_0, u_0, w^j) + \sum_{i=1}^{r} p^i \Big\{ g_1\big(f_0(x_0, u_0, w^j), u_1^j, w^i\big) + g_2\Big(f_1\big(f_0(x_0, u_0, w^j), u_1^j, w^i\big)\Big) \Big\} \Bigg\}$$

$$\text{subject to } u_0 \in U_0(x_0), \qquad u_1^j \in U_1\big(f_0(x_0, u_0, w^j)\big), \quad j = 1, \ldots, r.$$

If the controls u_0 and u_1 are elements of $\Re^m$, the number of variables in the above problem is $m(1+r)$. More generally, for an N-stage perfect state information problem a similar reformulation as a nonlinear programming problem requires $m\big(1 + r^{N-1}\big)$ variables. Thus if the number of lookahead stages is relatively small, a nonlinear programming approach may be the preferred option in calculating suboptimal limited lookahead policies.

Terminal Cost-to-Go Approximations

A key issue in implementing a limited lookahead policy is the selection of the cost-to-go approximation at the final step. It may appear important at first that the true cost-to-go function should be approximated well over the range of relevant states; however, this is not necessarily true. What is important is that the *cost-to-go differentials (or relative values) be approximated well*; that is, for an n-step lookahead policy it is important to have

$$\tilde{J}_{k+n}(x) - \tilde{J}_{k+n}(x') \approx J_{k+n}(x) - J_{k+n}(x'),$$

for any two states x and x' that can be generated n steps ahead from the current state. For example, if equality were to hold above for all x, x', then $\tilde{J}_{k+n}(x)$ and $J_{k+n}(x)$ would differ by the same constant for each relevant x and the n-step lookahead policy would be optimal.

The manner in which the cost-to-go approximation is selected depends very much on the problem being solved. There is a wide variety of possibilities here. For example, in some games like chess, the approximate cost-to-go in a certain position (state) involves a heuristic incorporation of certain features of the position into a score for the position. In other problems, a cost-to-go approximation may be based on solution of a simpler problem that is tractable computationally or analytically. We illustrate this possibility with an example; see also the manufacturing problem in Section 6.3.1.

Example 3.2

Consider the problem of the unscrupulous innkeeper who charges one of m different rates $r_1, \ldots, r_m$ for a room as the day progresses, depending on whether he has many or few vacancies, so as to maximize his expected total income during the day (Exercise 1.24 in Chapter 1). A quote of a rate r_i is accepted with probability p_i and is rejected with probability $1 - p_i$, in which case the customer departs, never to return during that day. When the number y of customers that will ask for a room during the rest of the day (including the customer currently asking for a room) is known and the number of vacancies is x, the optimal expected income $J(x, y)$ of the innkeeper is given by the DP recursion

$$J(x,y) = \max_{i=1,\ldots,m} \left[p_i\big(r_i + J(x-1, y-1)\big) + (1-p_i)J(x, y-1)\right],$$

for all $x \geq 1$ and $y \geq 1$, with initial conditions

$$J(x, 0) = J(0, y) = 0, \qquad \text{for all } x \text{ and } y.$$

On the other hand, when the innkeeper does not know y at the times of decision, but instead only has a probability distribution for y, it can be seen that the problem becomes a difficult imperfect state information problem. Yet a reasonable one-step lookahead policy is based on approximating the optimal cost-to-go of subsequent decisions with $J(x-1, \overline{y}-1)$ or $J(x, \overline{y}-1)$, where the function J is calculated by the above recursion and $\overline{y}$ is the closest integer to the expected value of y. In particular, according to this one-step lookahead policy, when the innkeeper has a number of vacancies $x \geq 1$, he quotes to the current customer the rate that maximizes $p_i\big(r_i + J(x-1, \overline{y}-1) - J(x, \overline{y}-1)\big)$.

Another approach, is to base a cost-to-go approximation on a suboptimal policy such as a certainty equivalent or open-loop-feedback controller. Generally, if a reasonably good suboptimal policy is known, it can be used to evaluate a cost-to-go approximation. For example, one may use simulation to evaluate the cost-to-go for some representative initial states and then use interpolation for the remaining states. Finally, one may start with a given cost-to-go approximation, obtain a corresponding policy by single or multiple stage lookahead, use that policy to obtain an improved cost-to-go approximation, and iterate. It is reasonable to assume that one can trade off accuracy of the cost-to-go approximation with size of lookahead; that is, by looking additional stages ahead, we can ameliorate the effect of errors in the cost-to-go approximation. The following subsections illustrate some of these approaches.

6.3.1 Flexible Manufacturing

Flexible manufacturing systems (FMS) provide a popular approach for increasing productivity in manufacturing small batches of related parts. There are several workstations in an FMS, and each is capable of carrying out a variety of operations. This allows the simultaneous manufacturing of more than one part type, reduces idle time, and allows production to continue even when a workstation is out of service because of failure or maintenance.

Consider a work center in which n part types are produced. Denote

u_k^i: the amount of part i produced in period k.

d_k^i: a known demand for part i in period k.

x_k^i: the cumulative difference of amount of part i produced and demanded up to period k.

Let us denote also by x_k, u_k, d_k the n-dimensional vectors with coordinates x_k^i, u_k^i, d_k^i, respectively. We then have

$$x_{k+1} = x_k + u_k - d_k. \tag{3.1}$$

The work center consists of a number of workstations that fail and get repaired in random fashion, thereby affecting the productive capacity of the system (i.e., the constraints on u_k). Roughly, our problem is to schedule part production so that x_k is kept around zero to the extent possible.

The productive capacity of the system depends on a random variable α_k that reflects the status of the workstations. In particular, we assume that the production vector u_k must belong to a constraint set $U(\alpha_k)$. We model the evolution of α_k by a Markov chain with known transition probabilities $P(\alpha_{k+1} \mid \alpha_k)$. In practice, these probabilities must be estimated from individual station failure and repair rates, but we will not go into the matter further. Note also that in practice these probabilities may depend on u_k. This dependence is ignored for the purpose of development of a cost-to-go approximation [cf. the following Eq. (3.7)]. It may be taken into account when the actual suboptimal control is computed [cf. the minimization in the following Eq. (3.8)].

We select as system state the pair (x_k, α_k), where x_k evolves according to Eq. (3.1) and α_k evolves according to the Markov chain described earlier. The problem is to find for every state (x_k, α_k) a production vector $u_k \in$

$U(\alpha_k)$ such that a cost function of the form

$$J_\pi(x_0) = E\left\{\sum_{k=0}^{N-1}\sum_{i=1}^{n} g^i(x_k^i)\right\}$$

is minimized. The function g^i expresses the desire to keep the current backlog or surplus of part i near zero. Two examples are $g^i(x^i) = \beta_i|x^i|$ or $g^i(x^i) = \beta_i(x^i)^2$, where $\beta_i > 0$.

The DP algorithm for this problem is

$$J_k(x_k,\alpha_k) = \sum_{i=1}^{n} g^i(x_k^i) + \min_{u_k\in U(\alpha_k)} \underset{\alpha_{k+1}}{E}\left\{J_{k+1}(x_k+u_k-d_k,\alpha_{k+1}) \mid \alpha_k\right\}. \tag{3.2}$$

If there is only one part type ($n = 1$), the optimal policy can be fairly easily determined from this algorithm (see Exercise 6.7). However, in general, the algorithm requires a prohibitive amount of calculation for an FMS of realistic size (say for $n > 10$ part types). We thus consider a one-step lookahead policy with the cost-to-go J_{k+1} replaced by an approximation $\tilde{J}_{k+1}$ that exploits the nearly separable structure of our problem.

In particular, we note that the problem can to a large extent be decomposed with respect to individual part types. Indeed, the system equation (3.1) and the cost per stage are decoupled, and the only coupling between parts comes from the constraint $u_k \in U(\alpha_k)$. Suppose we approximate $U(\alpha_k)$ by inner and outer approximating hypercubes $\underline{U}(\alpha_k)$ and $\overline{U}(\alpha_k)$ of the form

$$\underline{U}(\alpha_k) = \left\{u_k^i \mid 0 \le u_k^i \le \underline{B}_i(\alpha_k)\right\}, \tag{3.3a}$$

$$\overline{U}(\alpha_k) = \left\{u_k^i \mid 0 \le u_k^i \le \overline{B}_i(\alpha_k)\right\}, \tag{3.3b}$$

$$\underline{U}(\alpha_k) \subset U(\alpha_k) \subset \overline{U}(\alpha_k), \tag{3.4}$$

as shown in Fig. 6.3.1. If $U(\alpha_k)$ is replaced for each α_k by either $\overline{U}(\alpha_k)$ or $\underline{U}(\alpha_k)$, then the problem is decomposed completely with respect to part types. For every part i the DP algorithm for the outer approximation is given by

$$\overline{J}_k^i(x_k^i,\alpha_k) = g^i(x_k^i) + \min_{0\le u_k^i\le \overline{B}_i(\alpha_k)} \underset{\alpha_{k+1}}{E}\left\{\overline{J}_k^i(x_k^i+u_k^i-d_k^i,\alpha_{k+1}) \mid \alpha_k\right\}, \tag{3.5}$$

and for the inner approximation it is given by

$$\underline{J}_k^i(x_k^i,\alpha_k) = g^i(x_k^i) + \min_{0\le u_k^i\le \underline{B}_i(\alpha_k)} \underset{\alpha_{k+1}}{E}\left\{\underline{J}_{k+1}^i(x_k^i+u_k^i-d_k^i,\alpha_{k+1}) \mid \alpha_k\right\}. \tag{3.6}$$

Furthermore, since $\underline{U}(\alpha_k) \subset U(\alpha_k) \subset \overline{U}(\alpha_k)$, the cost-to-go functions $\overline{J}_k^i$ and $\underline{J}_k^i$ provide lower and upper bounds to the true cost-to-go function J_k,

$$\sum_{i=1}^{n} \overline{J}_k^i(x_k^i, \alpha_k) \le J_k(x_k, \alpha_k) \le \sum_{i=1}^{n} \underline{J}_k^i(x_k^i, \alpha_k), \tag{3.7}$$

and can be used to construct approximations to J_k that are suitable for a one-step lookahead policy. A simple possibility is to adopt the averaging approximation

$$\tilde{J}_k(x_k, \alpha_k) = \frac{1}{2}\sum_{i=1}^{n}\big(\overline{J}_k^i(x_k^i, \alpha_k) + \underline{J}_k^i(x_k^i, \alpha_k)\big)$$

and use at state (x_k, α_k) the control $\tilde{u}_k$ that minimizes [cf. Eq. (3.2)]

$$E\left\{\sum_{i=1}^{n}\big(\overline{J}_{k+1}^i(x_k^i + u_k^i - d_k^i, \alpha_{k+1}) + \underline{J}_{k+1}^i(x_k^i + u_k^i - d_k^i, \alpha_{k+1})\big) \;\Big|\; \alpha_k\right\} \tag{3.8}$$

over all $u_k \in U(\alpha_k)$. Multiple upper bound approximations, based on multiple choices of $\underline{B}_i(\alpha_k)$, can also be used.

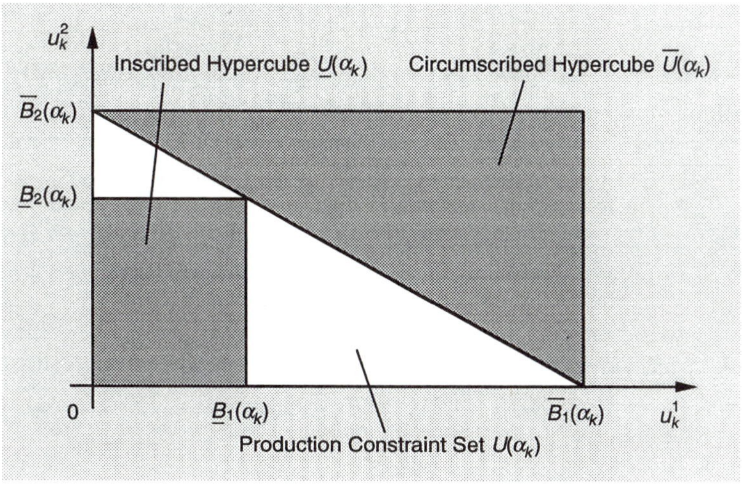

Figure 6.3.1 Inner and outer approximations of the production capacity constraint set by hypercubes.

To implement this scheme, it is necessary to compute and store the approximate cost-to-go functions $\overline{J}_k^i$ and $\underline{J}_k^i$ in tables, so that they can be used in the real-time computation of the suboptimal control via the minimization of expression (3.8). The corresponding calculations [cf. the DP algorithms (3.5) and (3.6)] are nontrivial, but they can be carried out off-line, and in any case they are much less than what would be required to compute the optimal controller. The feasibility and the benefits of the overall approach have been demonstrated by simulation for FMS of realistic size in [Kim82]. See also [KGB82] and [Tsi84a].

6.3.2 Computer Chess

Chess-playing computer programs are one of the more visible successes of artificial intelligence. Their underlying methodology relies on DP and provides an interesting case study in the use of suboptimal control. It involves the idea of limited lookahead, but also illustrates some DP ideas that we have not had much opportunity to look at in detail. These are the idea of a *forward search*, an important memory-saving technique that was discussed in Section 2.3 in the context of label correcting methods, and the idea of *alpha-beta pruning*, which is an effective method for reducing the amount of calculation required to find optimal game strategies.

The fundamental paper on which all computer chess programs are based was written by one of the most illustrious modern-day applied mathematicians, C. Shannon [Sha50]. It was argued by Shannon that whether the starting chess position is a win, loss, or draw is a question that can be answered in principle, but the answer will probably never be known. He estimated that, based on the chess rule requiring a pawn advance or a capture every 50 moves (otherwise a draw is declared), there are on the order of 10^{120} different possible sequences of moves in a chess game. He concluded that to examine these and select the best initial move for White would require 10^{90} years of a "fast" computer's time (fast here relates to the standards of the '50s). As an alternative, Shannon proposed a *limited lookahead* of a few moves and *evaluating the end positions by means of a scoring function*. The scoring function may involve, for example, the calculation of a numerical value for each of a set of major features of a position (such as material balance, mobility, pawn structure, and other positional factors), together with a method to combine these numerical values into a single score. The convention here is that White is favored in positions with high score, while Black is favored in positions with low score.

Consider first a *one-move lookahead strategy* for selecting the first move in a given position P. Let $M_1, \ldots, M_r$ be all the legal moves that can be made in position P by the side to move. Denote the resulting positions by $M_1P, \ldots, M_rP$, and let $S(M_1P), \ldots, S(M_rP)$ be the corresponding scores. Then the move selected by White (Black) in position P is the move with maximum (minimum) score. This is known as the *backed-up score* of P and is given by

$$BS(P) = \begin{cases} \max\{S(M_1P), \ldots, S(M_rP)\} & \text{if White is to move in } P, \\ \min\{S(M_1P), \ldots, S(M_rP)\} & \text{if Black is to move in } P. \end{cases}$$

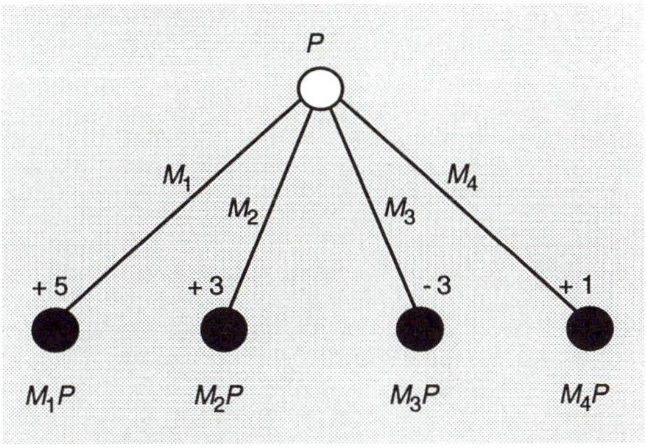

Figure 6.3.2 A one-move lookahead tree. If White moves at position P, the best move is M_1 and the backed-up score is $+5$. If Black moves in position P, the best move is M_3, and the backed-up score of P is -3.

This process is illustrated in Fig. 6.3.2.

Consider next a *two-move lookahead strategy* in a given position P. Assume for concreteness that White moves, and let the legal moves be $M_1, \ldots, M_r$ and the corresponding positions be $M_1P, \ldots, M_rP$. Then in each of the positions M_iP, $i = 1, \ldots, r$, apply the one-move lookahead strategy with Black to move. This gives a best move and a backed-up score $BS(M_iP)$ for Black in each of the positions M_iP, $i = 1, \ldots, r$. Finally, based on the backed-up scores $BS(M_1P), \ldots, BS(M_rP)$, apply a one-move lookahead strategy for White, thereby obtaining the best move at position P and a backed-up score for position P of

$$BS(P) = \max\{BS(M_1P), \ldots, BS(M_rP)\}.$$

The sequence of best moves is known as the *principal continuation*. The process is illustrated in Fig. 6.3.3. It is clear that Shannon's method as just described can be generalized for an arbitrary number of lookahead moves (see Fig. 6.3.4).

Generally, to evaluate the best move at a given position P and the corresponding backed-up score using lookahead of n moves, one can use the following DP-like procedure:

1. Evaluate the scores of all possible positions that are n moves ahead from the given position P.

2. Using the scores of the terminal positions just evaluated, calculate the backed-up scores of all possible positions that are $n-1$ moves ahead from P.

3. For $k = 1, \ldots, n-1$, using the backed-up scores of all possible positions that are $n-k$ moves ahead from P, calculate the backed-up scores of all positions that are $n-k-1$ moves ahead from P.

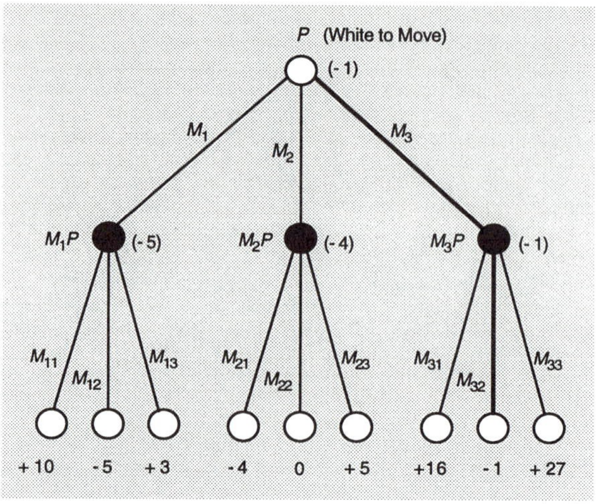

Figure 6.3.3 A two-move lookahead tree with White to move. The backed-up scores are shown in parentheses. The best initial move is M_3 and the principal continuation is (M_3, M_{32}).

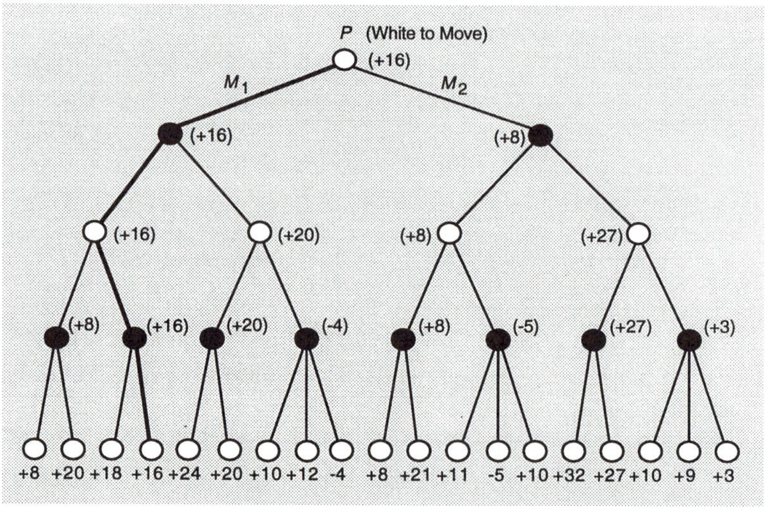

Figure 6.3.4 A four-move lookahead tree with White to move. The backed-up scores are shown in parentheses. The best initial move is M_1. The principal continuation is heavily shaded.

The above procedure requires a lot of memory storage even for a modest number of lookahead moves. Shannon pointed out that with an alternative but mathematically equivalent calculation, the amount of memory required grows only *linearly* with the depth of lookahead, thereby allowing

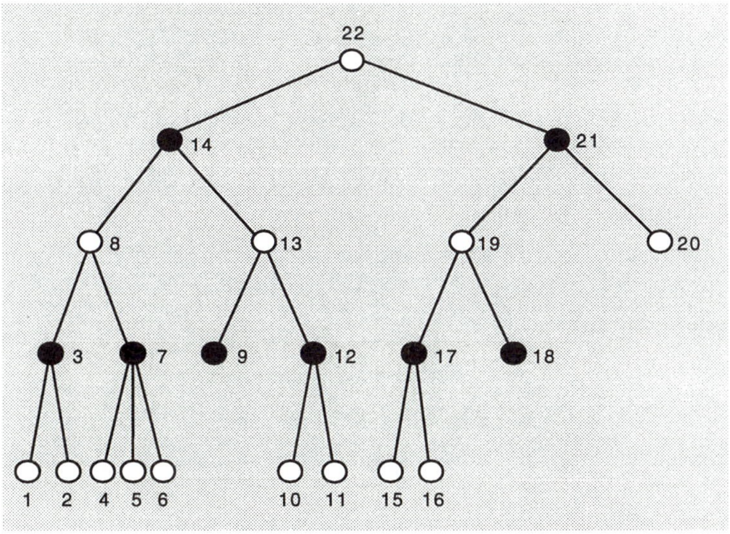

Figure 6.3.5 Traversing a tree in depth-first fashion. The numbers indicate the order in which the scores of the terminal positions and the backed-up scores of the intermediate positions are evaluated.

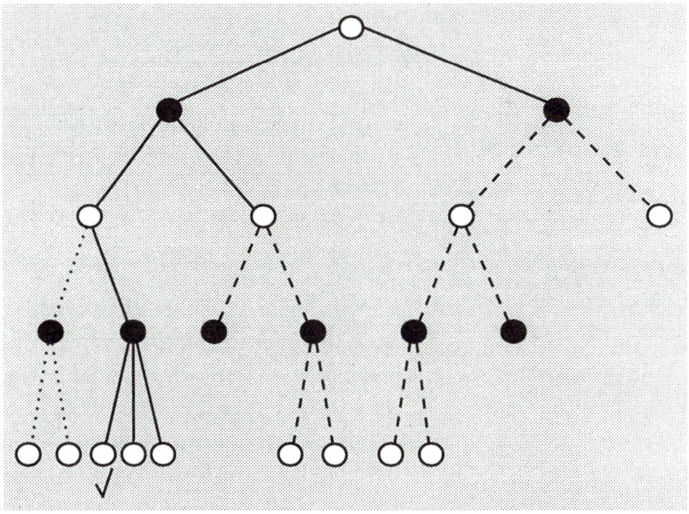

Figure 6.3.6 Storage requirements of the depth-first version of the minimax algorithm for the tree of Fig. 6.3.5. At the time that the terminal position marked by a checkmark is scored, only the solid-line moves are stored in memory. The dotted-line moves have been generated and purged from memory. The broken-line moves have not been generated as yet. The memory required grows linearly with the depth of the lookahead.

chess programs to operate in limited-memory microprocessor systems. This is accomplished by searching the tree of moves in *depth-first fashion*, and by generating new moves only when needed, as illustrated in Figs. 6.3.5 and 6.3.6. It is only necessary to store the *one* move sequence under current examination together with one list of legal moves at each level of the search tree. The precise algorithm is described by the following routine, which calls itself recursively.

Minimax Algorithm

To determine the backed-up score $BS(n)$ of position n, do the following:

1. If n is a terminal position return its score. Otherwise:
2. Generate the list of legal moves at position n and let the corresponding positions be $n_1, \ldots, n_r$. Set the tentative backed-up score $TBS(n)$ of position n to ∞ if it is White's turn to move at n and to $-\infty$ if it is Black's turn to move at n.
3. For $i = 1, \ldots, r$, do:
   a. Determine the backed-up score $BS(n_i)$ of position n_i.
   b. If it is White's move at position n, set
   $$TBS(n) := \max\{TBS(n),\, BS(n_i)\};$$
   if it is Black's move at position n, set
   $$TBS(n) := \min\{TBS(n),\, BS(n_i)\}.$$
4. Return $BS(n) = TBS(n)$.

The idea of solving one-step lookahead problems with a terminal cost (or backed-up score) that summarizes future costs is of course central in the DP algorithm. Indeed, it can be seen that the minimax algorithm just described is nothing but the DP algorithm for minimax problems (see Exercise 1.5 in Chapter 1). Here, positions and moves can be identified with states and controls, respectively, there are only terminal costs (the scores of the terminal positions), and the backed-up score of a position is nothing but the optimal cost-to-go at the corresponding state.

The minimax algorithm is also known as the *type A strategy*. Shannon argued that with this strategy, one could not expect a computer to seriously challenge human players of even moderate strength. In a typical chess position there are around 30 to 35 legal moves. It follows that for an n-move lookahead there will be around 30^n to 35^n terminal positions to be scored.

Thus the number of terminal positions grows exponentially with the size of lookahead, practically limiting n to no more than roughly 10 with present computers. Unfortunately, in some chess positions it is essential to look a substantially larger number of moves ahead. In particular, in dynamic positions involving many captures and countercaptures, the necessary size of lookahead can be very large.

These considerations led Shannon to consider another strategy, called *type B*, whereby the depth of the search tree is variable. He suggested that at each position the computer give all legal moves a preliminary examination and discard those that are "obviously bad." A scoring function together with some heuristic strategy can be used for this purpose. Similarly, he suggested that some positions that are dynamic, such as those involving many captures or checkmate threats, be explored further beyond the nominal depth of the search.

Chess-playing computer programs typically use a variation of Shannon's type B strategy. These programs use scoring functions, the forms of which have evolved by trial-and-error, but they also use sophisticated heuristics to evaluate dynamic terminal positions in detail. In particular, an effective algorithm, known as *swapoff*, is used to quickly analyze long sequences of captures and countercaptures, thereby making it possible to score realistically complex, dynamic positions (see [Lev84] for a description). One may view such heuristics as either defining a sophisticated scoring function, or as implementing a type B strategy.

Alpha-Beta Pruning

The efficiency of the minimax algorithm can be substantially improved by using the *alpha-beta pruning procedure* (denoted α-β for short), which can be used to forego some calculations involving positions that cannot affect the selection of the best move. To understand the α-β procedure, consider a chess player pondering the next move at position P. Suppose that the player has already exhaustively analyzed one relatively good move M_1 with corresponding score $BS(M_1P)$ and proceeds to examine the next move M_2. Suppose that as the opponent's replies are examined, a particularly strong response is found, which assures that the score of M_2 will be worse than that of M_1. Such a response, called a *refutation* of move M_2, makes further consideration of move M_2 unnecessary (i.e., the portion of the search tree that descends from move M_2 can be discarded). An example is shown in Fig. 6.3.7.

The α-β procedure can be generalized to trees of arbitrary or irregular depth and can be incorporated very simply into the minimax algorithm. Generally, if in the process of updating the backed-up score of a given position (step 3b) this score crosses a certain bound, then no further calculation is needed regarding that position. The cutoff bounds are adjusted dynamically as follows:

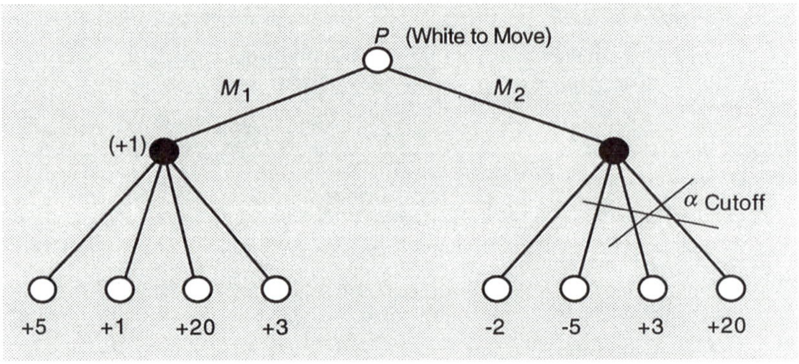

Figure 6.3.7 The α-β procedure. White has evaluated move M_1 to have backed-up score (+1), and starts evaluating move M_2. The first reply of Black is a refutation of M_2, since it leads to a temporary score of -2, less than the backed-up score of M_1. Since the backed-up score of M_2 will be -2 or less, M_2 will be inferior to M_1. Therefore, it is not necessary to evaluate move M_2 further.

1. The cutoff bound in position n, where Black has to move, is denoted α and equals the highest current score of all ancestor positions of n where White has to move. The exploration of position n can be terminated as soon as its temporary backed-up score equals or falls below α.

2. The cutoff bound in position n, where White has to move, is denoted β and equals the lowest current value of all ancestor positions of n where Black has the move. The exploration of position n can be terminated as soon as its temporary backed-up score rises above β.

The process is illustrated in Fig. 6.3.8. *It can be shown that the backed-up score and optimal move at the starting position are unaffected by the incorporation of the α-β procedure in the minimax algorithm.* We leave the verification of this fact to the reader (Exercise 6.8). It can also be seen that *the α-β procedure will be more effective if the best moves in each position are explored first.* This tends to keep the α bounds high and the β bounds low, thus saving a maximum amount of calculation. Current chess programs use sophisticated techniques for ordering moves so as to maximize the effectiveness of the α-β procedure. We discuss briefly two of these techniques: *iterative deepening* and the *killer heuristic.*

Iterative deepening, in its pure form, consists of first conducting a search based on lookahead of one move; then carrying out (from scratch) a search based on lookahead of two moves; then carrying out a search based on lookahead of three moves and so on. This process is continued either up to a fixed level of lookahead or until some limit on computation time is exceeded. At each iteration associated with a certain level of lookahead, one obtains a best move at the starting position, which is examined first in

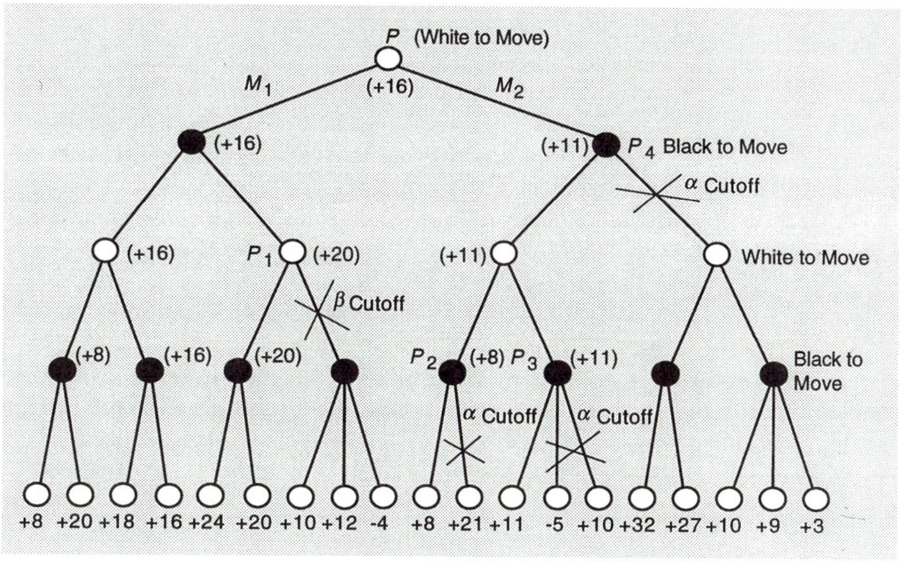

Figure 6.3.8 The α-β procedure applied to the tree of Fig. 6.3.4. For example, the β-cutoff in position P_1 is due to the fact that its temporary score $(+20)$ exceeds its current β-bound $(+16)$. The α-cutoffs in positions P_2, P_3, and P_4 are due to the fact that the corresponding temporary scores, $+8$, $+11$, and $+11$, have fallen below the current α-bound, which is $+16$, the current temporary score in position P.

the subsequent iteration that requires one extra move of lookahead. This enhances the power of the α-β procedure, thereby more than making up for the extra computation involved in doing a short lookahead search before doing a longer one. (Actually, given that the number of terminal positions increases on the average by more than a factor of about 30 with each additional level of lookahead, the extra computation is relatively small.) An additional benefit of this method is that a best move is maintained throughout the search and can be produced at any time as needed. This comes in handy in commercial programs that incorporate a feature whereby the computer is forced to move either upon exhausting a given time allocation or upon command by a human opponent. An improvement of the method is to obtain a thoroughly sorted list of moves at the starting position via a one-move lookahead, and then use the improved ordering in subsequent iterations to enhance the performance of the α-β procedure.

The killer heuristic is similar to iterative deepening in that it aims at examining first the most powerful moves at each position, thereby enhancing the pruning power of the α-β procedure. To understand the idea, suppose that in some position, White selects the first move M_1 from a candidate list $\{M_1, M_2, M_3, \ldots\}$, and upon examining Black's responses to M_1 finds that a particular move, which we will refer to as the *killer move*, is by far Black's best. Then it is often true that the killer move is also Black's best response to the second and subsequent moves $M_2, M_3, \ldots$ in White's list. It is therefore a good idea from the point of view of α-β pruning to consider the killer move first as a potential response to the remaining moves $M_2, M_3, \ldots$ Of course, this does not always work as hoped, in which case it is advisable to change the killer move depending on subsequent results of the computation. In fact, some programs maintain lists of more than one killer move at each level of lookahead.

The α-β procedure is safe in the sense that searching a game tree with and without it will produce the same result. Some computer chess programs use more drastic tree-pruning procedures, which usually require less computation for a given level of lookahead, but may miss on occasion the strongest move. There is some debate at present regarding the merits of such procedures. References [Lev84] and [New75] consider this subject, and provide a broader discussion of the limitations of computer chess programs.

6.4 APPROXIMATIONS IN DYNAMIC PROGRAMMING

Another general method to deal with the "curse of dimensionality" is based on low-dimensional approximations of the DP algorithm. There are two general approaches here:

(a) Approximating the original problem by a computationally "simpler" problem.

(b) Approximating the DP algorithm (for the original problem) by a computationally "simpler" algorithm.

In this section, we will provide an informal discussion of these two approaches.

6.4.1 Discretization of Optimal Control Problems

A typical example of approximation arises in problems with continuous state, control, or disturbance spaces, which are discretized in order to execute the DP algorithm. Here each of the continuous spaces of the problem is replaced by a space with a finite number of elements, and the system equation is appropriately modified. The approximate problem involves a finite number of states, and the system equation is essentially replaced by a set of transition probabilities between these states. Once the discretization is done, the DP algorithm is executed to yield the optimal cost-to-go functions and an optimal policy for the discrete approximating problem. The optimal cost function and/or the optimal policy for the discrete problem may then be extended to an approximate cost function or a suboptimal policy for the original continuous problem through some form of interpolation.

A prerequisite for success of this type of discretization is *consistency*. By this we mean that the optimal cost of the original problem should be achieved in the limit as the discretization becomes finer and finer. Consistency is typically guaranteed if there is a "sufficient amount of continuity" in the problem; for example, if the cost-to-go functions and the optimal policy of the original problem are continuous functions of the state. This in turn can be guaranteed through appropriate continuity assumptions on the original problem data (see the references).

Continuity of the cost-to-go functions may be sufficient to guarantee consistency, even if the optimal policy is discontinuous in the state. What may happen here is that for some states there may be a large discrepancy between the optimal policy of the continuous problem and the optimal policy of its discretized version, but this discrepancy may occur over a portion of the state space that diminishes as the discretization becomes finer. As an example consider the inventory control problem of Section 4.2 with nonzero fixed cost. We obtained an optimal policy of the (s, S) type

$$\mu_k^*(x_k) = \begin{cases} S_k - x_k & \text{if } x_k < s_k, \\ 0 & \text{if } x_k \geq s_k, \end{cases}$$

which is discontinuous at the lower threshold s_k. The optimal policy obtained from the discretized problem may not approximate well the optimal around the point of discontinuity s_k, but it is intuitively clear that the discrepancy has a diminishing effect on the optimal cost as the discretization becomes finer.

In the case of a continuous-time problem, the time must also be discretized, so that the original continuous-time problem is replaced by a discrete-time problem. The issue of consistency becomes now considerably more complex, because the time discretization interval affects not only the system equation but also the control constraint set. Furthermore, the control constraint set may change considerably as we pass to the appropriate discrete-time approximation. As an example, consider the two-dimensional system

$$\dot{x}_1(t) = u_1(t), \qquad \dot{x}_2(t) = u_2(t),$$

with the control constraint $\|u(t)\| = 1$. It can be seen then that the state at time $t + \Delta t$ can be anywhere within the sphere centered at $x(t)$ of radius Δt (note that the effect of zero control can be obtained in the continuous-time system by "chattering" between two control vectors that have norm equal to 1 and are opposite to each other). Thus, given Δt, the appropriate discrete-time approximation of the control constraint set should involve a discretization of the entire sphere of radius Δt and not just its surface. An example that illustrates some of the pitfalls associated with the discretization process is given in Exercise 6.10.

There is an interesting general method to address the discretization issues of continuous-time optimal control. The main idea in this method is to discretize, in addition to time, the state space using some kind of grid, and then to approximate the cost-to-go of nongrid states by linear interpolation of the cost-to-go values of the nearby grid states. By this we mean that if a nongrid state x is expressed as

$$x = \sum_{i=1}^{m} \xi^i x^i$$

in terms of the grid states $x^1, \ldots, x^m$, where the positive weights ξ^i add to to 1, then the cost-to-go of x is approximated by

$$\sum_{i=1}^{m} \xi^i J(x^i),$$

where $J(x^i)$ is the cost-to-go of x^i. When this is worked out, one ends up with a *stochastic* optimal control problem having as states the finite number of grid states, and transition probabilities that are determined from the weights ξ^i above. If the original continuous-time optimal control problem has fixed terminal time, the resulting stochastic control approximation has finite horizon. If the terminal time of the original problem is free and subject to optimization, the stochastic control approximation is of the stochastic shortest path type to be discussed in Section 7.2. We refer to the papers [GoR85] and [Fal87], the survey [Kus90], and the monograph [KuD92] for a description and analysis of the associated issues. An important special case is the continuous-space shortest path problem (see Exercise 6.10). For this problem, a finitely terminating adaptation of the Dijkstra shortest path algorithm has been developed [Tsi93] (see also [PBT95]).

6.4.2 Cost-to-Go Approximation

Let us now consider the approach of approximating the DP algorithm by a computationally simpler algorithm. Such an algorithm need not involve an explicit discretization of the state space. The main idea is to calculate (possibly approximate) values for the cost-to-go (and/or the optimal policy) at a finite set of state-time pairs, and then to make a "least-squares fit" of these values with a function of a given type, such as a polynomial function, a Fourier expansion, a wavelet expansion, or the output of a neural network.

In particular, suppose that we have calculated the correct value of the optimal cost-to-go $J_{N-1}(x^i)$ at the next to last stage for m states $x^1, \ldots, x^m$ through the DP formula

$$J_{N-1}(x) = \min_{u \in U(x)} \mathop{E}_{w}\bigl\{g(x,u,w) + J_N\bigl(f(x,u,w)\bigr)\bigr\}, \tag{4.1}$$

and the given terminal cost function J_N. (For simplicity in this and the following formulas, we drop the subscripts of g, x, u, and w, assuming a stationary problem, but this is not essential for the methodology described.) We can then approximate the entire function $J_{N-1}(x)$ by a function of some given form

$$\tilde{J}_{N-1}(x, r_{N-1}),$$

where r_{N-1} is a vector of parameters, which can be obtained by solving the problem

$$\min_r \sum_{i=1}^{m} \big| J_{N-1}(x^i) - \tilde{J}_{N-1}(x^i, r) \big|^2. \tag{4.2}$$

For example if x is an n-dimensional vector and $\tilde{J}_{N-1}$ is specified to be linear, the vector r consists of an n-dimensional vector α and a scalar β, and

$$\tilde{J}_{N-1}(x, r) = \alpha' x + \beta. \tag{4.3}$$

The least squares problem (4.2) is

$$\min_{\alpha,\, \beta} \sum_{i=1}^{m} \big| J_{N-1}(x^i) - \alpha' x^i - \beta \big|^2 \tag{4.4}$$

and can be solved in closed form. Actually, a linear approximating function is seldom adequate, but there are other possibilities that can be tailored to the problem at hand.

Note that this approximation procedure can be enhanced if we have additional information on the true cost-to-go function $J_{N-1}(x)$. For example, if we know that $J_{N-1}(x) \geq 0$ for all x, we may first compute the approximation $\tilde{J}_{N-1}(x, r_{N-1})$ by solving the least-squares problem (4.2) and then replace this approximation by $\max\{0, \tilde{J}_{N-1}(x, r_{N-1})\}$. This idea applies to all the approximation procedures to be presented in this section and later.

Once an approximating function $\tilde{J}_{N-1}(x, r_{N-1})$ for the next to last stage has been obtained, it can be used to similarly obtain an approximating function $\tilde{J}_{N-2}(x, r_{N-2})$. In particular, (approximate) cost-to-go function values $\hat{J}_{N-2}(x^i)$ are obtained for m states $x^1, \ldots, x^m$ through the (approximate) DP formula

$$\hat{J}_{N-2}(x) = \min_{u \in U(x)} \mathop{E}_{w} \big\{ g(x, u, w) + \tilde{J}_{N-1}\big(f(x, u, w), r_{N-1}\big) \big\}. \tag{4.5}$$

These values are then used to approximate the cost-to-go function $J_{N-2}(x)$ by a function of some given form

$$\tilde{J}_{N-2}(x, r_{N-2}),$$

where r_{N-2} is a vector of parameters, which is obtained by solving the problem

$$\min_r \sum_{i=1}^{m} \big| \hat{J}_{N-2}(x^i) - \tilde{J}_{N-2}(x^i, r) \big|^2. \tag{4.6}$$

The process can be similarly continued to obtain $\tilde{J}_k(x, r_k)$ up to $k = 0$ by solving for each k the problem

$$\min_r \sum_{i=1}^{m} \big| \hat{J}_k(x^i) - \tilde{J}_k(x^i, r) \big|^2. \tag{4.7}$$

Given approximate cost-to-go functions $\tilde{J}_0(x, r_0), \ldots, \tilde{J}_{N-1}(x, r_{N-1})$, one may obtain a suboptimal policy by using at state-time pair (x, k) the control

$$\tilde{\mu}_k(x) = \arg \min_{u \in U(x)} \mathop{E}_{w} \big\{ g(x, u, w) + \tilde{J}_{k+1}\big(f(x, u, w), r_{k+1}\big) \big\}. \tag{4.8}$$

This control must be calculated on-line once the state x_k at time k becomes known.

Approximating Policies

An alternative to calculating on-line suboptimal controls based on the preceding equation is to approximate optimal policies directly. In particular, suppose that the control space is a Euclidean space, and that we obtain, for a finite number of states x^i, $i = 1, \ldots, m$, the minimizing controls

$$\hat{\mu}_k(x^i) = \arg \min_{u \in U(x^i)} \mathop{E}_{w} \big\{ g(x^i, u, w) + \tilde{J}_{k+1}\big(f(x^i, u, w), r_{k+1}\big) \big\}.$$

We can then approximate the optimal policy $\mu_k(x)$ by a function of some given form

$$\tilde{\mu}_k(x, s_k),$$

where s_k is a vector of parameters obtained by solving the problem

$$\min_s \sum_{i=1}^{m} \| \hat{\mu}_k(x^i) - \tilde{\mu}_k(x^i, s) \|^2. \tag{4.9}$$

In the case of deterministic optimal control problems, we can take advantage of the equivalence between open-loop and feedback control to carry out the approximation process more efficiently. In particular, for such problems we may select a representative finite subset of initial states, and generate an optimal open-loop trajectory starting from each of these states. (Gradient-based methods can often be used for this purpose.) Each of these trajectories yields a sequence of pairs $\big(x_k, J_k(x_k)\big)$ and a sequence of pairs $\big(x_k, \mu_k(x_k)\big)$, which can be used in the least-squares approximation procedures discussed above. In particular, we can use the exact values $J_k(x^i)$ and $\mu_k(x^i)$ obtained from the optimal open-loop trajectories in place of $\hat{J}_k(x^i)$ and $\hat{\mu}_k(x^i)$, respectively, in the least-squares problems of Eqs. (4.7) and (4.9).

Approximation Through Feature Extraction Mappings

Clearly, for the success of the approximation approach, it is very important to select a class of functions $\tilde{J}(x, r)$ that is suitable for the problem at hand. One particularly interesting type of cost approximation is provided by *feature extraction*, a process that maps the state x into some other vector y, called the *feature vector* associated with state x.

Feature vectors summarize, in a heuristic sense, important characteristics of the state, and they are very useful in incorporating the designer's prior knowledge or intuition about the problem and about the structure of the optimal controller. For example, in a queueing system involving several queues, a feature vector may involve for each queue a three-value indicator, that specifies whether the queue is "nearly empty," "moderately busy," or "nearly full." Also, feature extraction is closely related to the notion of *state aggregation*, which will be discussed in Vol. II, Section 1.3.

Feature vectors are normally associated with piecewise constant cost approximation. In particular, when a set of features is selected to represent states, the cost function is approximated by a function that is constant over the subset of states that are mapped into the same feature vector. The parameter vector r in the function $\tilde{J}(x, r)$ is the vector of the constant levels, that is, the costs associated with all the different feature vectors. In particular, in a system with m possible feature vectors, r is an m-dimensional vector $(r_1, \ldots, r_m)'$, and we have $\tilde{J}(x, r) = r_j$ for all states x that are mapped into the jth feature vector. More generally, one may consider cost approximations of the form $\hat{J}(y, r)$, where r is a parameter vector and the cost approximation depends on the state i through its feature vector y (see Fig. 6.4.1).

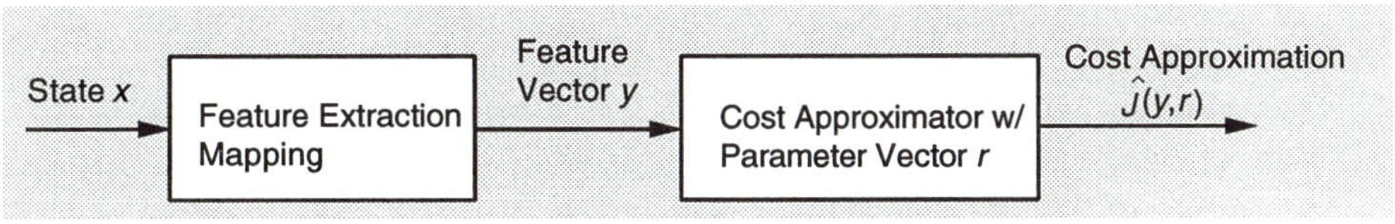

Figure 6.4.1 Using a feature extraction mapping to generate an input to a cost approximator.

A feature extraction mapping may also be useful in the context of *problems of imperfect state information*. In this context, a state estimator that provides some form of state reconstruction may be viewed as a feature extraction mapping. In fact, we may view a sufficient statistic as an "exact" feature extraction mapping, in the sense that it is equivalent to the full information vector for optimization purposes. This leads to the conjecture that good feature extraction mappings are provided by "approximate sufficient statistics," that is, mappings that in a heuristic sense are close to being a sufficient statistic for the given problem.

6.4.3 Other Approximations

We mention briefly two additional approaches for using approximations. In the first approach, the optimal cost-to-go functions $J_k(x_k)$ are approximated with functions $\tilde{J}_k(x_k, r_k)$, where $r = (r_0, r_1, \ldots, r_N)$ is a vector of unknown parameters, which is chosen to minimize some form of error in the DP equations; for example by solving the problem

$$\min_r \sum_{(x_k,k)\in\tilde{S}} \left| \tilde{J}_k(x_k, r_k) - \min_{u\in U(x_k)} \mathop{E}_{w_k} \Big\{ g(x_k, u, w_k) + \tilde{J}_{k+1}\big(f(x_k, u, w_k), r_{k+1}\big) \Big\} \right|^2 \tag{4.10}$$

where $\tilde{S}$ is a suitably chosen subset of "representative" state-time pairs. The above minimization can be attempted using some type of gradient method. Note that there is some difficulty in doing so because the cost function of Eq. (4.10) may be nondifferentiable for some values of r. However, there are adaptations of gradient methods that work with nondifferentiable cost functions, and for which we refer to the specialized literature. One possibility is to replace the nondifferentiable term

$$\min_{u\in U(x_k)} \mathop{E}_{w_k} \Big\{ g(x_k, u, w_k) + \tilde{J}_{k+1}\big(f(x_k, u, w_k), r_{k+1}\big) \Big\}$$

by a smooth approximation (see [Ber82b], Ch. 3). The approach of approximating cost-to-go functions by minimizing the error in the DP equations will also be discussed in more detail within an infinite horizon context (see Vol. II, Section 1.3).

The second approach is based on policy approximation. Thus in place of an overall optimal policy, we look for an optimal policy within some restricted class that is parameterized by some vector s of relatively low dimension. In particular, we consider policies of a given form

$$\pi_s = \big\{ \tilde{\mu}_0(x_0, s), \tilde{\mu}_1(x_1, s), \ldots, \tilde{\mu}_{N-1}(x_{N-1}, s) \big\},$$

where s is the unknown parameter vector, and $\tilde{\mu}_k$ are given functions with certain structure (e.g. linear in x_k). We then minimize over s the expected cost

$$\mathop{E}_{x_0} \big\{ J_{\pi_s}(x_0) \big\},$$

where the expectation is with respect to some suitable probability distribution of the initial state x_0. The advantage of this approach is that it applies even for complex problems where there is no explicit model for the system and the cost; instead the cost corresponding to a given policy may be calculated by simulation. One difficulty is that the gradient of the cost with respect to s may not be easily calculated. However, optimization methods that require only cost values (and not gradients) may be used. We refer to the specialized literature for description and analysis of such methods.

6.5 NOTES, SOURCES, AND EXERCISES

Excellent surveys of adaptive control are given in [Ast83] and [Kum85], which contain many other references. For textbook treatments, see [AsW90], [GoS84], [Her89], [KuV86], and [LjS83]. Self-tuning regulators received wide attention following the paper by Astrom and Wittenmark [AsW73].

Open-loop feedback control was suggested by Dreyfus [Dre65]. Its superiority over open-loop control was established by the author in the context of minimax control [Ber72]. A generalization of this result is given in [WhH80]. The POLFC was proposed in [Ber76].

For consistency analyses of various discretization and approximation procedures for discrete-time stochastic optimal control problems, see [Ber75], [Ber76a], [ChT89], [ChT91], [Fox71], [Whi78], [Whi79]. The discretization issues of continuous-time optimal control problems have been the subject of considerable research; see the monograph [KuD92], the survey [Kus90], and the sources cited there.

Whenever a suboptimal controller is used in practice it is desirable to know how close the resulting cost approximates the optimal. Tight bounds on the performance of the suboptimal controller are useful but are usually quite hard to obtain. For some interesting results in this direction, see [Wit69] and [Wit70].

E X E R C I S E S

6.1

Consider a problem with perfect state information involving the n-dimensional linear system of Section 4.1:

$$x_{k+1} = A_k x_k + B_k u_k + w_k, \qquad k = 0, 1, \ldots, N-1,$$

and a cost function of the form

$$\underset{\substack{w_k \\ k=0,1,\ldots,N-1}}{E} \left\{ g_N(c'x_N) + \sum_{k=0}^{N-1} g_k(u_k) \right\},$$

where $c \in R^n$ is a given vector. Show that the DP algorithm for this problem can be carried out over a one-dimensional state space.

6.2 [ThW66]

Consider the following two-dimensional, two-stage linear system with scalar control and disturbance

$$x_{k+1} = x_k + bu_k + dw_k, \qquad k = 0, 1,$$

where $b = (1,0)'$ and $d = (1/2, \sqrt{2}/2)'$. The initial state is $x_0 = 0$. The controls u_0 and u_1 are unconstrained. The disturbances w_0 and w_1 are independent random variables and each takes the values 1 and -1 with equal probability 1/2. Perfect state information prevails. The cost is

$$\underset{w_0, w_1}{E} \{ \|x_2\| \},$$

where $\|\cdot\|$ denotes the usual Euclidean norm. Show that the optimal open-loop cost and the optimal closed-loop cost are both $\sqrt{3}/2$, but the cost corresponding to the CEC is 1.

6.3

Consider a two-stage problem with perfect state information involving the scalar system

$$x_0 = 1, \qquad x_1 = x_0 + u_0 + w_0, \qquad x_2 = f(x_1, u_1).$$

The control constraints are $u_0,\ u_1 \in \{0, -1\}$. The random variable w_0 takes the values 1 and -1 with equal probability 1/2. The function f is defined by

$$f(1,0) = f(1,-1) = f(-1,0) = f(-1,-1) = 0.5,$$
$$f(2,0) = 0, \quad f(2,-1) = 2, \quad f(0,-1) = 0.6, \quad f(0,0) = 2.$$

The cost function is

$$\underset{w_0}{E}\{x_2\}.$$

(a) Show that one possible OLFC for this problem is

$$\overline{\mu}_0(x_0) = -1, \qquad \overline{\mu}_1(x_1) = \begin{cases} 0 & \text{if } x_1 = \pm 1, 2, \\ -1 & \text{if } x_1 = 0, \end{cases}$$

and the resulting cost is 0.5.

(b) Show that one possible CEC for this problem is

$$\tilde{\mu}_0(x_0) = 0, \qquad \tilde{\mu}_1(x_1) = \begin{cases} 0 & \text{if } x_1 = \pm 1, 2, \\ -1 & \text{if } x_1 = 0, \end{cases}$$

and the resulting cost is 0.3. Show also that this CEC is an optimal feedback controller.

6.4

Consider the system and cost function of Exercise 6.3 but with the difference that

$$f(0, -1) = 0.$$

(a) Show that the controller of part (a) of Exercise 6.3 is both an OLFC and a CEC, and that the corresponding cost is 0.5.

(b) Assume that the control constraint set for the first stage is $\{0\}$ rather than $\{0, -1\}$. Show that the controller of part (b) of Exercise 6.3 is both an OLFC and a CEC and that the corresponding cost is 0. *Note*: This problem illustrates a pathology that occurs generically in suboptimal control; that is, if the control constraint set is restricted, the performance of a suboptimal scheme may be improved. To see this, consider a problem and a suboptimal control scheme that is not optimal for the problem. Let $\pi^* = \{\mu_0^*, \ldots, \mu_{N-1}^*\}$ be an optimal policy. Restrict the control constraint set so that only the optimal control $\mu_k^*(x_k)$ is allowed at state x_k. Then the cost attained by the suboptimal control scheme will be improved.

6.5

Consider the ARMAX model

$$y_{k+1} + ay_k = bu_k + \epsilon_{k+1} + c\epsilon_k,$$

where the parameters a, b, and c are unknown. The controller hypothesizes a model of the form

$$y_{k+1} + ay_k = u_k + \epsilon_{k+1}$$

and uses at each k the minimum variance/certainty equivalent control

$$u_k^* = \hat{a}_k y_k,$$

where $\hat{a}_k$ is the least-squares estimate of a obtained as

$$\hat{a}_k = \arg\min_a \sum_{n=1}^{k} \left(y_n + ay_{n-1} - u_{n-1}^*\right)^2.$$

Write a computer program to test the hypothesis that the sequence $\{\hat{a}_k\}$ converges to the optimal value, which is $(c - a)/b$. Experiment with values $|a| < 1$ and $|a| > 1$.

6.6 (Semilinear Systems)

Consider the basic problem for semilinear systems (Exercise 1.13 in Chapter 1). Show that the CEC and the OLFC are optimal for this problem.

6.7

Consider the production control problem of Section 6.3.1 for the case where there is only one part type ($n = 1$), and assume that the cost per stage is a convex function g with $\lim_{|x| \to 1} g(x) = \infty$.

(a) Show that the cost-to-go function $J_k(x_k, \alpha_k)$ is convex as a function of x_k for each value of α_k.

(b) Show that for each k and α_k, there is a target value $\overline{x}_{k+1}$ such that for each x_k it is optimal to choose the control $u_k \in U_k(\alpha_k)$ that brings $x_{k+1} = x_k + u_k - d_k$ as close as possible to $\overline{x}_{k+1}$.

6.8

Provide a careful argument showing that searching a chess position with and without α-β pruning will give the same result.

6.9

In a version of the game of Nim, two players start with a stack of five pennies and take turns removing one, two, or three pennies from the stack. The player who removes the last penny loses. Construct the game tree and verify that the second player to move can win with optimal play.

6.10 (Continuous Space Shortest Path Problems)

Consider the two-dimensional system

$$\dot{x}_1(t) = u_1(t), \qquad \dot{x}_2(t) = u_2(t),$$

with the control constraint $\|u(t)\| = 1$. We want to find a state trajectory that starts at a given point $x(0)$, ends at another given point $x(T)$, and minimizes

$$\int_0^T r\big(x(t)\big)\,dt.$$

The function $r(\cdot)$ is nonnegative and continuous, and the final time T is subject to optimization. Suppose we discretize the plane with a mesh of size Δ that passes through $x(0)$ and $x(T)$, and we introduce a shortest path problem of going from $x(0)$ to $x(T)$ using moves of the following type: from each mesh point $\overline{x} = (\overline{x}_1, \overline{x}_2)$ we can go to each of the mesh points $(\overline{x}_1 + \Delta, \overline{x}_2)$, $(\overline{x}_1 - \Delta, \overline{x}_2)$, $(\overline{x}_1, \overline{x}_2 + \Delta)$, and $(\overline{x}_1, \overline{x}_2 - \Delta)$, at a cost $r(\overline{x})\Delta$. Show by example that this is a bad discretization of the original problem in the sense that the shortest distance need not approach the optimal cost of the original problem as $\Delta \to 0$.

7

Introduction to Infinite Horizon Problems

Contents

In this chapter, we provide an introduction to infinite horizon problems. These problems differ from those considered so far in two respects:

(a) The number of stages is infinite.

(b) The system is stationary, that is, the system equation, the cost per stage, and the random disturbance statistics do not change from one stage to the next.

The assumption of an infinite number of stages is never satisfied in practice, but constitutes a reasonable approximation for problems involving a finite but very large number of stages. The assumption of stationarity is often satisfied in practice, and in other cases it approximates well a situation where the system parameters vary slowly with time.

Infinite horizon problems are interesting because their analysis is elegant and insightful, and the implementation of optimal policies is often simple. For example, optimal policies are typically stationary; that is, the optimal rule for choosing controls does not change from one stage to the next.

On the other hand, infinite horizon problems generally require more sophisticated analysis than their finite horizon counterparts, because of the need to analyze limiting behavior as the horizon tends to infinity. This analysis is often nontrivial and at times reveals surprising possibilities. Our treatment will be limited to finite-state problems. A far more detailed development, together with applications from a variety of fields can be found in Vol. II of this work.

7.1 AN OVERVIEW

There are four principal classes of infinite horizon problems. In the first three classes, we try to minimize the *total cost over an infinite number of stages*, given by

$$J_\pi(x_0) = \lim_{N\to\infty} \mathop{E}_{\substack{w_k \\ k=0,1,\ldots}} \left\{ \sum_{k=0}^{N-1} \alpha^k g\big(x_k, \mu_k(x_k), w_k\big) \right\},$$

where $J_\pi(x_0)$ denotes the cost associated with an initial state x_0 and a policy $\pi = \{\mu_0, \mu_1, \ldots\}$, and α is a positive scalar with $0 < \alpha \leq 1$, called the *discount factor*. The meaning of $\alpha < 1$ is that future costs matter to us less than the same costs incurred at the present time. As an example, think of kth period dollars depreciated to initial period dollars by a factor of $(1+r)^{-k}$, where r is a rate of interest; here $\alpha = 1/(1+r)$. An important concern in total cost problems is that the limit in the definition of $J_\pi(x_0)$ be finite. In the first two of the following classes of problems, this is guaranteed through various assumptions on the problem structure and the discount factor. In the third class, the analysis is adjusted to deal with infinite cost for some of the policies. In the fourth class, this sum need not be finite for any policy, and for this reason, the cost is appropriately redefined.

(a) *Stochastic shortest path problems.* Here, $\alpha = 1$ but there is a special cost-free termination state; once the system reaches that state it remains there at no further cost. We will assume a problem structure such that termination is inevitable (this assumption will be relaxed somewhat in Chapter 2 of Vol. II). Thus the horizon is in effect finite, but its length is random and may be affected by the policy being used. These problems will be considered in the next section and their analysis will provide the foundation for the analysis of the other types of problems considered in this chapter.

(b) *Discounted problems with bounded cost per stage.* Here, $\alpha < 1$ and the absolute cost per stage $|g(x, u, w)|$ is bounded from above by some constant M; this makes the cost $J_\pi(x_0)$ well defined because it is the infinite sum of a sequence of numbers that are bounded in absolute value by the decreasing geometric progression $\{\alpha^k M\}$. We will consider these problems in Section 7.3.

(c) *Discounted and undiscounted problems with unbounded cost per stage.* Here the discount factor α may or may not be less than 1, and the cost per stage may be unbounded. These problems require a complicated analysis because the possibility of infinite cost for some of the policies is explicitly dealt with. We will not consider these problems here; see Chapter 3 of Vol. II.

(d) *Average cost per stage problems.* Minimization of the total cost $J_\pi(x_0)$ makes sense only if $J_\pi(x_0)$ is finite for at least some admissible policies π and some initial states x_0. Frequently, however, it turns out that $J_\pi(x_0) = \infty$ for every policy π and initial state x_0 (think of the case where $\alpha = 1$, and the cost for every state and control is positive). It turns out that in many such problems the *average cost per stage*, defined by

$$\lim_{N\to\infty} \frac{1}{N} \underset{\substack{w_k \\ k=0,1,\ldots}}{E} \left\{ \sum_{k=0}^{N-1} g\big(x_k, \mu_k(x_k), w_k\big) \right\},$$

is well defined and finite. We will consider some of these problems in Section 7.4.

A Preview of Infinite Horizon Results

There are several analytical and computational issues regarding infinite horizon problems. Many of these revolve around the relation between the optimal cost-to-go function J^* of the infinite horizon problem and the optimal cost-to-go functions of the corresponding N-stage problems. In particular, consider the case $\alpha = 1$ and let $J_N(x)$ denote the optimal cost of the problem involving N stages, initial state x, cost per stage $g(x, u, w)$, and zero terminal cost. The optimal N-stage cost is generated after N iterations of the DP algorithm

$$J_{k+1}(x) = \min_{u \in U(x)} \mathop{E}_{w} \left\{ g(x, u, w) + J_k\big(f(x, u, w)\big) \right\}, \qquad k = 0, 1, \ldots \quad (1.1)$$

starting from the initial condition $J_0(x) = 0$ for all x (note here that we have reversed the time indexing to suit our purposes). Since the infinite horizon cost of a given policy is, by definition, the limit of the corresponding N-stage costs as $N \to \infty$, it is natural to speculate that:

(1) The optimal infinite horizon cost is the limit of the corresponding N-stage optimal costs as $N \to \infty$; that is,

$$J^*(x) = \lim_{N \to \infty} J_N(x) \qquad (1.2)$$

for all states x. This relation is extremely valuable computationally and analytically, and, fortunately, it typically holds. In particular, it holds for the models of the next two sections [categories (a) and (b) above]. However, there are some unusual exceptions for problems in category (c) above, and this illustrates that infinite horizon problems should be approached with some care.

(2) The following limiting form of the DP algorithm should hold for all states x,

$$J^*(x) = \min_{u \in U(x)} \mathop{E}_{w} \left\{ g(x, u, w) + J^*\big(f(x, u, w)\big) \right\}, \qquad (1.3)$$

as suggested by Eqs. (1.1) and (1.2). This is not really an algorithm, but rather a system of equations (one equation per state), which has as solution the costs-to-go of all the states. It can also be viewed as a *functional equation* for the cost-to-go function J^*, and it is called *Bellman's equation*. Fortunately again, an appropriate form of this equation holds for every type of infinite horizon problem of interest.

(3) If $\mu(x)$ attains the minimum in the right-hand side of Bellman's equation for each x, then the policy $\{\mu, \mu, \ldots\}$ should be optimal. This is true for most infinite horizon problems of interest and in particular, for all the models discussed in the subsequent sections.

Most of the analysis of infinite horizon problems revolves around the above three issues and also around the issue of efficient computation of J^* and an optimal policy. In the next three sections we will provide a discussion of these issues for some of the simpler infinite horizon problems, all of which involve a finite state space.

Total Cost Problem Formulation

Throughout this chapter we assume a controlled discrete-time dynamic system whereby, at state i, the use of a control u specifies the transition probability $p_{ij}(u)$ to the next state j. Here the state i is an element of a finite state space, and the control u is constrained to take values in a given finite constraint set $U(i)$, which may depend on the current state i. As discussed in Section 1.1, the underlying system equation is

$$x_{k+1} = w_k,$$

where w_k is the disturbance. We will generally suppress w_k from the cost to simplify notation. Thus we will assume a kth stage cost $g(x_k, u_k)$ for using control u_k at state x_k. This amounts to averaging the cost per stage over all successor states in our calculations, which makes no essential difference in the subsequent analysis. Thus, if $\tilde{g}(i, u, j)$ is the cost of using u at state i and moving to state j, we use as cost per stage the expected cost $g(i, u)$ given by

$$g(i, u) = \sum_j p_{ij}(u)\tilde{g}(i, u, j).$$

The total expected cost associated with an initial state i and a policy $\pi = \{\mu_0, \mu_1, \ldots\}$ is

$$J_\pi(i) = \lim_{N\to\infty} E\left\{ \sum_{k=0}^{N-1} \alpha^k g\big(x_k, \mu_k(x_k)\big) \,\Big|\, x_0 = i \right\},$$

where α is a discount factor with $0 < \alpha \leq 1$. In the following two sections, we will impose assumptions that guarantee the existence of the above limit. The optimal cost from state i, that is, the minimum of $J_\pi(i)$ over all admissible π, is denoted by $J^*(i)$. A *stationary policy* is an admissible policy of the form $\pi = \{\mu, \mu, \ldots\}$, and its corresponding cost function is denoted by $J_\mu(i)$. For brevity, we refer to $\{\mu, \mu, \ldots\}$ as the stationary policy μ. We say that μ is optimal if $J_\mu(i) = J^*(i)$ for all states i.

7.2 STOCHASTIC SHORTEST PATH PROBLEMS

Here, we assume that there is no discounting ($\alpha = 1$), and to make the cost meaningful, we assume that *there is a special cost-free termination state t*. Once the system reaches that state, it remains there at no further cost, that is, $p_{tt}(u) = 1$ and $g(t, u) = 0$ for all $u \in U(t)$. We denote by $1, \ldots, n$ the states other than the termination state t.

We are interested in problems where reaching the termination state is inevitable, at least under an optimal policy. Thus, the essence of the problem is to reach the termination state with minimum expected cost. We call this problem the *stochastic shortest path problem*. The deterministic shortest path problem is obtained as the special case where for each state-control pair (i, u), the transition probability $p_{ij}(u)$ is equal to 1 for a unique state j that depends on (i, u). The reader may also verify that the finite horizon problem of Chapter 1 can be obtained as a special case by viewing as state the pair (x_k, k) (see also Section 3.6 of Vol. II).

Certain conditions are required to guarantee that, at least under an optimal policy, termination occurs with probability 1. We will make the following assumption that guarantees eventual termination under all policies:

Assumption 2.1: There exists an integer m such that regardless of the policy used and the initial state, there is positive probability that the termination state will be reached after no more that m stages; that is, for all admissible policies π we have

$$\rho_\pi = \max_{i=1,\ldots,n} P\{x_m \neq t \mid x_0 = i, \pi\} < 1. \tag{2.1}$$

We note, however, that the results to be presented are valid under more general circumstances.† Furthermore, it can be shown that if there exists an integer m with the property of Assumption 2.1, then there also exists an integer less or equal to n with this property (Exercise 7.12). Thus, we can always use $m = n$ in Assumption 2.1, if no smaller value of m is known. Let

$$\rho = \max_{\pi} \rho_\pi. \tag{2.2}$$

† Let us call a stationary policy π *proper* if the condition (2.1) is satisfied for some m, and call π *improper* otherwise. It can be shown that Assumption 2.1 is equivalent to the seemingly weaker assumption that all stationary policies are proper (see Vol. II, Exercise 2.3). However, the results of Prop. 2.1 can also be shown under the genuinely weaker assumption that there exists at least one proper policy, and furthermore, every improper policy results in infinite expected cost from at least one initial state (see [BeT89], [BeT91], or Vol. II, Chapter 2). These assumptions, when specialized to deterministic shortest path problems, are similar to the ones we used in Chapter 2. They imply that there is at least one path to the destination from every starting node and that all cycles have positive cost.

Since ρ_π depends only on the first m components of the policy π, it follows that there is only a finite number of distinct values of ρ_π so that

$$\rho < 1. \tag{2.3}$$

We therefore have for any π and any initial state i

$$\begin{aligned} P\{x_{2m} \neq t \mid x_0 = i, \pi\} &= P\{x_{2m} \neq t \mid x_m \neq t,\, x_0 = i, \pi\} \\ &\qquad \times P\{x_m \neq t \mid x_0 = i, \pi\} \\ &\leq \rho^2. \end{aligned}$$

More generally, for each admissible policy π, the probability of not reaching the termination state after km stages diminishes like ρ^k regardless of the initial state, that is,

$$P\{x_{km} \neq t \mid x_0 = i, \pi\} \leq \rho^k, \qquad i = 1, \ldots, n. \tag{2.4}$$

Thus the limit defining the associated total cost vector J_π exists and is finite, since the expected cost incurred in the m periods between km and $(k+1)m - 1$ is bounded in absolute value by

$$m\rho^k \max_{\substack{i=1,\ldots,n \\ u\in U(i)}} \big|g(i,u)\big|.$$

In particular, we have

$$\big|J_\pi(i)\big| \leq \sum_{k=0}^{\infty} m\rho^k \max_{\substack{i=1,\ldots,n \\ u\in U(i)}} \big|g(i,u)\big| = \frac{m}{1-\rho} \max_{\substack{i=1,\ldots,n \\ u\in U(i)}} \big|g(i,u)\big|. \tag{2.5}$$

The results of the following proposition are basic and are typical of many infinite horizon problems. The key idea of the proof is that the "tail" of the cost series,

$$\sum_{k=mK}^{\infty} E\big\{g\big(x_k, \mu_k(x_k)\big)\big\}$$

vanishes as K increases to ∞, since the probability that $x_{mK} \neq t$ decreases like ρ^K [cf. Eq. (2.4)].

Proposition 2.1 Under Assumption 2.1, the following hold for the stochastic shortest path problem:

(a) Given any initial conditions $J_0(1), \ldots, J_0(n)$, the sequence $J_k(i)$ generated by the DP iteration

$$J_{k+1}(i) = \min_{u\in U(i)} \left[g(i,u) + \sum_{j=1}^{n} p_{ij}(u) J_k(j) \right], \qquad i = 1, \ldots, n, \tag{2.6}$$

converges to the optimal cost $J^*(i)$ for each i.

(b) The optimal costs $J^*(1), \ldots, J^*(n)$ satisfy Bellman's equation,

$$J^*(i) = \min_{u \in U(i)} \left[g(i,u) + \sum_{j=1}^{n} p_{ij}(u) J^*(j) \right], \qquad i = 1, \ldots, n, \tag{2.7}$$

and in fact they are the unique solution of this equation.

(c) For any stationary policy μ, the costs $J_\mu(1), \ldots, J_\mu(n)$ are the unique solution of the equation

$$J_\mu(i) = g\big(i, \mu(i)\big) + \sum_{j=1}^{n} p_{ij}\big(\mu(i)\big) J_\mu(j), \qquad i = 1, \ldots, n.$$

Furthermore, given any initial conditions $J_0(1), \ldots, J_0(n)$, the sequence $J_k(i)$ generated by the DP iteration

$$J_{k+1}(i) = g\big(i, \mu(i)\big) + \sum_{j=1}^{n} p_{ij}\big(\mu(i)\big) J_k(j), \qquad i = 1, \ldots, n,$$

converges to the cost $J_\mu(i)$ for each i.

(d) A stationary policy μ is optimal if and only if for every state i, $\mu(i)$ attains the minimum in Bellman's equation (2.7).

Proof: (a) For every positive integer K, initial state x_0, and policy $\pi = \{\mu_0, \mu_1, \ldots\}$, we break down the cost $J_\pi(x_0)$ into the portions incurred over the first mK stages and over the remaining stages

$$\begin{aligned} J_\pi(x_0) &= \lim_{N \to \infty} E\left\{ \sum_{k=0}^{N-1} g\big(x_k, \mu_k(x_k)\big) \right\} \\ &= E\left\{ \sum_{k=0}^{mK-1} g\big(x_k, \mu_k(x_k)\big) \right\} \\ &\quad + \lim_{N \to \infty} E\left\{ \sum_{k=mK}^{N-1} g\big(x_k, \mu_k(x_k)\big) \right\}. \end{aligned}$$

Let M denote the following upper bound on the cost of an m-stage cycle, assuming termination does not occur during the cycle,

$$M = m \max_{\substack{i=1,\ldots,n \\ u \in U(i)}} \big| g(i,u) \big|.$$

The expected cost during the Kth m-stage cycle [stages Km to $(K+1)m-1$] is upper bounded by $M\rho^K$ [cf. Eqs. (2.4) and (2.5)], so that

$$\left| \lim_{N\to\infty} E\left\{ \sum_{k=mK}^{N-1} g\big(x_k, \mu_k(x_k)\big) \right\} \right| \leq M \sum_{k=K}^{\infty} \rho^k = \frac{\rho^K M}{1-\rho}.$$

Also, denoting $J_0(t) = 0$, let us view J_0 as a terminal cost function and bound its expected value under π after mK stages. We have

$$\begin{aligned} \big|E\{J_0(x_{mK})\}\big| &= \left| \sum_{i=1}^{n} P(x_{mK} = i \mid x_0, \pi) J_0(i) \right| \\ &\leq \left(\sum_{i=1}^{n} P(x_{mK} = i \mid x_0, \pi) \right) \max_{i=1,\ldots,n} \big|J_0(i)\big| \\ &\leq \rho^K \max_{i=1,\ldots,n} \big|J_0(i)\big|, \end{aligned}$$

since the probability that $x_{mK} \neq t$ is less or equal to ρ^K for any policy. Combining the preceding relations, we obtain

$$\begin{aligned} -\rho^K & \max_{i=1,\ldots,n} \big|J_0(i)\big| + J_\pi(x_0) \\ &\leq E\left\{ J_0(x_{mK}) + \sum_{k=0}^{mK-1} g\big(x_k, \mu_k(x_k)\big) \right\} + \frac{\rho^K M}{1-\rho} \\ &\leq \rho^K \max_{i=1,\ldots,n} \big|J_0(i)\big| + J_\pi(x_0). \end{aligned} \tag{2.8}$$

Note that the expected value in the middle term of the above inequalities is the mK-stage cost of policy π starting from state x_0, with a terminal cost $J_0(x_{mK})$; the minimum of this cost over all π is equal to the value $J_{mK}(x_0)$, which is generated by the DP recursion (2.6) after mK iterations. Thus, by taking the minimum over π in Eq. (2.8), we obtain for all x_0 and K,

$$\begin{aligned} -\rho^K \max_{i=1,\ldots,n} \big|J_0(i)\big| + J^*(x_0) &\leq J_{mK}(x_0) + \frac{\rho^K M}{1-\rho} \\ &\leq \rho^K \max_{i=1,\ldots,n} \big|J_0(i)\big| + J^*(x_0), \end{aligned} \tag{2.9}$$

and by taking the limit as $K \to \infty$, we obtain $\lim_{K\to\infty} J_{mK}(x_0) = J^*(x_0)$ for all x_0. Since

$$\big|J_{mK+q}(x_0) - J_{mK}(x_0)\big| \leq \rho^K M, \qquad q = 1, \ldots, m,$$

we see that $\lim_{K\to\infty} J_{mK+q}(x_0)$ is the same for all $q = 1, \ldots, m$, so that we have $\lim_{k\to\infty} J_k(x_0) = J^*(x_0)$.

(b) By taking the limit as $k \to \infty$ in the DP iteration (2.6) and using the result of part (a), we see that $J^*(1), \ldots, J^*(n)$ satisfy Bellman's equation. To show uniqueness, observe that if $J(1), \ldots, J(n)$ satisfy Bellman's equation, then the DP iteration (2.6) starting from $J(1), \ldots, J(n)$ just replicates $J(1), \ldots, J(n)$. It follows from the convergence result of part (a) that $J(i) = J^*(i)$ for all i.

(c) Given the stationary policy μ, we can consider a modified stochastic shortest path problem, which is the same as the original except that the control constraint set contains only one element for each state i, the control $\mu(i)$; that is, the control constraint set is $\tilde{U}(i) = \{\mu(i)\}$ instead of $U(i)$. From part (b) we then obtain that $J_\mu(1), \ldots, J_\mu(n)$ solve uniquely Bellman's equation for this modified problem, that is,

$$J_\mu(i) = g\big(i, \mu(i)\big) + \sum_{j=1}^{n} p_{ij}\big(\mu(i)\big) J_\mu(j), \qquad i = 1, \ldots, n,$$

and from part (a) it follows that the corresponding DP iteration converges to $J_\mu(i)$.

(d) We have that $\mu(i)$ attains the minimum in Eq. (2.7) if and only if we have

$$\begin{aligned} J^*(i) &= \min_{u \in U(i)} \left[g(i,u) + \sum_{j=1}^{n} p_{ij}(u) J^*(j) \right] \\ &= g\big(i, \mu(i)\big) + \sum_{j=1}^{n} p_{ij}\big(\mu(i)\big) J^*(j), \qquad i = 1, \ldots, n. \end{aligned}$$

Part (c) and the above equation imply that $J_\mu(i) = J^*(i)$ for all i. Conversely, if $J_\mu(i) = J^*(i)$ for all i, parts (b) and (c) imply the above equation. **Q.E.D.**

Example 2.1 (Minimizing the Expected Time to Termination)

The case where

$$g(i,u) = 1, \qquad i = 1, \ldots, n, \quad u \in U(i),$$

corresponds to a problem where the objective is to terminate as fast as possible on the average, while the corresponding optimal cost $J^*(i)$ is the minimum expected time to termination starting from state i. Under our assumptions, the costs $J^*(i)$ uniquely solve Bellman's equation, which has the form

$$J^*(i) = \min_{u \in U(i)} \left[1 + \sum_{j=1}^{n} p_{ij}(u) J^*(j) \right], \qquad i = 1, \ldots, n.$$

In the special case where there is only one control at each state, $J^*(i)$ represents the mean first passage time from i to t (see Appendix D). These times, denoted m_i, are the unique solution of the equations

$$m_i = 1 + \sum_{j=1}^{n} p_{ij} m_j, \qquad i = 1, \ldots, n.$$

Example 2.2

A spider and a fly move along a straight line at times $k = 0, 1, \ldots$ The initial positions of the fly and the spider are integer. At each time period, the fly moves one unit to the left with probability p, one unit to the right with probability p, and stays where it is with probability $1-2p$. The spider, knows the position of the fly at the beginning of each period, and will always move one unit towards the fly if its distance from the fly is more that one unit. If the spider is one unit away from the fly, it will either move one unit towards the fly or stay where it is. If the spider and the fly land in the same position at the end of a period, then the spider captures the fly and the process terminates. The spider's objective is to capture the fly in minimum expected time.

We view as state the distance between spider and fly. Then the problem can be formulated as a stochastic shortest path problem with states $0, 1, \ldots, n$, where n is the initial distance. State 0 is the termination state where the spider captures the fly. Let us denote $p_{1j}(M)$ and $p_{1j}(\overline{M})$ the transition probabilities from state 1 to state j if the spider moves and does not move, respectively, and let us denote by p_{ij} the transition probabilities from a state $i \geq 2$. We have

$$p_{ii} = p, \quad p_{i(i-1)} = 1 - 2p, \quad p_{i(i-2)} = p, \qquad i \geq 2,$$

$$p_{11}(M) = 2p, \quad p_{10}(M) = 1 - 2p,$$

$$p_{12}(\overline{M}) = p, \quad p_{11}(\overline{M}) = 1 - 2p, \quad p_{10}(\overline{M}) = p,$$

with all other transition probabilities being 0.

For states $i \geq 2$, Bellman's equation is written as

$$J^*(i) = 1 + pJ^*(i) + (1 - 2p)J^*(i - 1) + pJ^*(i - 2), \quad i \geq 2, \tag{2.10}$$

where $J^*(0) = 0$ by definition. The only state where the spider has a choice is when it is one unit away from the fly, and for that state Bellman's equation is given by

$$J^*(1) = 1 + \min\big[2pJ^*(1),\, pJ^*(2) + (1 - 2p)J^*(1)\big], \tag{2.11}$$

where the first and the second expression within the bracket above are associated with the spider moving and not moving, respectively. By writing Eq. (2.10) for $i = 2$, we obtain

$$J^*(2) = 1 + pJ^*(2) + (1 - 2p)J^*(1),$$

from which

$$J^*(2) = \frac{1}{1-p} + \frac{(1-2p)J^*(1)}{1-p}. \tag{2.12}$$

Substituting this expression in Eq. (2.11), we obtain

$$J^*(1) = 1 + \min\left[2pJ^*(1),\ \frac{p}{1-p} + \frac{p(1-2p)J^*(1)}{1-p} + (1-2p)J^*(1)\right],$$

or equivalently,

$$J^*(1) = 1 + \min\left[2pJ^*(1),\ \frac{p}{1-p} + \frac{(1-2p)J^*(1)}{1-p}\right].$$

To solve the above equation, we consider the two cases where the first expression within the bracket is larger and is smaller than the second expression. Thus we solve for $J^*(1)$ in the two cases where

$$J^*(1) = 1 + 2pJ^*(1), \tag{2.13a}$$

$$2pJ^*(1) \le \frac{p}{1-p} + \frac{(1-2p)J^*(1)}{1-p}, \tag{2.13b}$$

and

$$J^*(1) = 1 + \frac{p}{1-p} + \frac{(1-2p)J^*(1)}{1-p}, \tag{2.14a}$$

$$2pJ^*(1) \ge \frac{p}{1-p} + \frac{(1-2p)J^*(1)}{1-p}. \tag{2.14b}$$

The solution of Eq. (2.13a) is seen to be $J^*(1) = 1/(1-2p)$, and by substitution in Eq. (2.13b), we find that this solution is valid when

$$\frac{2p}{1-2p} \le \frac{p}{1-p} + \frac{1}{1-p},$$

or equivalently (after some calculation), $p \le 1/3$. Thus for $p \le 1/3$, it is optimal for the spider to move when it is one unit away from the fly.

Similarly, the solution of Eq. (2.14a) is seen to be $J^*(1) = 1/p$, and by substitution in Eq. (2.13b), we find that this solution is valid when

$$2 \ge \frac{p}{1-p} + \frac{1-2p}{p(1-p)},$$

or equivalently (after some calculation), $p \ge 1/3$. Thus, for $p \ge 1/3$ it is optimal for the spider not to move when it is one unit away from the fly.

The minimal expected number of steps for capture when the spider is one unit away from the fly was calculated earlier to be

$$J^*(1) = \begin{cases} 1/(1-2p) & \text{if } p \le 1/3, \\ 1/p & \text{if } p \ge 1/3. \end{cases}$$

Given the value of $J^*(1)$, we can calculate from Eq. (2.12) the minimal expected number of steps for capture when two units away, $J^*(2)$, and we can then obtain the remaining values $J^*(i)$, $i = 3, \ldots, n$, from Eq. (2.10).

Value Iteration and Error Bounds

The DP iteration

$$J_{k+1}(i) = \min_{u \in U(i)} \left[g(i,u) + \sum_{j=1}^{n} p_{ij}(u) J_k(j) \right], \qquad i = 1, \ldots, n, \tag{2.15}$$

is called *value iteration* and is a principal method for calculating the optimal cost function J^*. Generally, value iteration requires an infinite number of iterations, although there are important special cases where it terminates finitely (see Vol. II, Section 2.2). Note that from Eq. (2.9) we obtain that the error

$$\big| J_{mK}(i) - J^*(i) \big|$$

is bounded by a constant multiple of ρ^K.

The value iteration algorithm can sometimes be strengthened with the use of some error bounds. In particular, it can be shown (see Exercise 7.13) that for all k and j, we have

$$J_{k+1}(j) + \big(N_*(j) - 1\big)\, \underline{c}_k \le J^*(j) \le J_{\mu^k}(j) \le J_{k+1}(j) + \big(N_k(j) - 1\big)\, \overline{c}_k, \tag{2.16}$$

where μ^k is such that $\mu^k(i)$ attains the minimum in the kth value iteration (2.15) for all i, and

$N_*(j)$: The average number of stages to reach t starting from j and using some optimal stationary policy,

$N_k(j)$: The average number of stages to reach t starting from j and using the stationary policy μ^k,

$$\underline{c}_k = \min_{i=1,\ldots,n} \big[J_{k+1}(i) - J_k(i) \big], \qquad \overline{c}_k = \max_{i=1,\ldots,n} \big[J_{k+1}(i) - J_k(i) \big].$$

Unfortunately, the values $N_*(j)$ and $N_k(j)$ are easily computed or approximated only in the presence of special problem structure (see for example the next section). Despite this fact, the bounds (2.16) often provide a useful guideline for stopping the value iteration algorithm while being assured that J_k approximates J^* with sufficient accuracy.

Policy Iteration

There is an alternative to value iteration, which always terminates finitely. This algorithm is called *policy iteration* and operates as follows: we start with a stationary policy μ^0, and we generate a sequence of new policies $\mu^1, \mu^2, \ldots$ Given the policy μ^k, we perform a *policy evaluation step*, that computes $J_{\mu^k}(i)$, $i = 1, \ldots, n$, as the solution of the (linear) system of equations

$$J(i) = g\big(i, \mu^k(i)\big) + \sum_{j=1}^{n} p_{ij}\big(\mu^k(i)\big) J(j), \qquad i = 1, \ldots, n, \tag{2.17}$$

in the n unknowns $J(1), \ldots, J(n)$ [cf. Prop. 2.1(c)]. We then perform a *policy improvement step*, which computes a new policy μ^{k+1} as

$$\mu^{k+1}(i) = \arg\min_{u \in U(i)} \left[g(i,u) + \sum_{j=1}^{n} p_{ij}(u) J_{\mu^k}(j) \right], \qquad i = 1, \ldots, n. \tag{2.18}$$

The process is repeated with μ^{k+1} used in place of μ^k, unless we have $J_{\mu^{k+1}}(i) = J_{\mu^k}(i)$ for all i, in which case the algorithm terminates with the policy μ^k. The following proposition establishes the validity of policy iteration.

Proposition 2.2 Under Assumption 2.1, the policy iteration algorithm for the stochastic shortest path problem generates an improving sequence of policies [that is, $J_{\mu^{k+1}}(i) \leq J_{\mu^k}(i)$ for all i and k] and terminates with an optimal policy.

Proof: For any k, consider the sequence generated by the recursion

$$J_{N+1}(i) = g\big(i, \mu^{k+1}(i)\big) + \sum_{j=1}^{n} p_{ij}\big(\mu^{k+1}(i)\big) J_N(j), \qquad i = 1, \ldots, n,$$

where $N = 0, 1, \ldots$, and

$$J_0(i) = J_{\mu^k}(i), \qquad i = 1, \ldots, n.$$

From Eqs. (2.17) and (2.18), we have

$$\begin{aligned} J_0(i) &= g\big(i, \mu^k(i)\big) + \sum_{j=1}^{n} p_{ij}\big(\mu^k(i)\big) J_0(j) \\ &\geq g\big(i, \mu^{k+1}(i)\big) + \sum_{j=1}^{n} p_{ij}\big(\mu^{k+1}(i)\big) J_0(j) \\ &= J_1(i), \end{aligned}$$

for all i. By using the above inequality we obtain (compare with the monotonicity property of DP, Exercise 1.23 in Chapter 1)

$$\begin{aligned} J_1(i) &= g\big(i, \mu^{k+1}(i)\big) + \sum_{j=1}^{n} p_{ij}\big(\mu^{k+1}(i)\big) J_0(j) \\ &\geq g\big(i, \mu^{k+1}(i)\big) + \sum_{j=1}^{n} p_{ij}\big(\mu^{k+1}(i)\big) J_1(j) \\ &= J_2(i), \end{aligned}$$

for all i, and by continuing similarly we have

$$J_0(i) \geq J_1(i) \geq \cdots \geq J_N(i) \geq J_{N+1}(i) \geq \cdots, \qquad i = 1, \ldots, n. \tag{2.19}$$

Since by Prop. 2.1(c), $J_N(i) \to J_{\mu^{k+1}}(i)$, we obtain $J_0(i) \geq J_{\mu^{k+1}}(i)$ or

$$J_{\mu^k}(i) \geq J_{\mu^{k+1}}(i), \qquad i = 1, \ldots, n, \quad k = 0, 1, \ldots.$$

Thus the sequence of generated policies is improving, and since the number of stationary policies is finite, we must after a finite number of iterations, say $k+1$, obtain $J_{\mu^k}(i) = J_{\mu^{k+1}}(i)$ for all i. Then we will have equality holding throughout in Eq. (2.19), which means that

$$J_{\mu^k}(i) = \min_{u \in U(i)} \left[g(i,u) + \sum_{j=1}^{n} p_{ij}(u) J_{\mu^k}(j) \right], \qquad i = 1, \ldots, n.$$

Thus the costs $J_{\mu^k}(1), \ldots, J_{\mu^k}(n)$ solve Bellman's equation, and by Prop. 2.1(b), it follows that $J_{\mu^k}(i) = J^*(i)$ and that μ^k is optimal. **Q.E.D.**

The linear system of equations (2.17) of the policy evaluation step can be solved by standard methods such as Gaussian elimination, but when the number of states is large, this is cumbersome and time-consuming. A typically more efficient alternative is to approximate the policy evaluation step with a few value iterations aimed at solving the corresponding system (2.17). One can show that the policy iteration method that uses such approximate policy evaluation yields in the limit the optimal costs and an optimal stationary policy, even if we evaluate each policy using any positive number of value iterations (see Vol. II, Section 1.3).

Linear Programming

Suppose that we use value iteration to generate a sequence of vectors $J_k = \big(J_k(1), \ldots, J_k(n)\big)$ starting with an initial condition vector $J_0 = \big(J_0(1), \ldots, J_0(n)\big)$ such that

$$J_0(i) \leq \min_{u \in U(i)} \left[g(i,u) + \sum_{j=1}^{n} p_{ij}(u) J_0(j) \right], \qquad i = 1, \ldots, n.$$

Then we will have $J_k(i) \leq J_{k+1}(i)$ for all k and i (the monotonicity property of DP; Exercise 1.23 in Chapter 1). It follows from Prop. 2.1(a) that we will also have $J_0(i) \leq J^*(i)$ for all i. Thus J^* is the "largest" J that satisfies the constraint

$$J(i) \leq g(i,u) + \sum_{j=1}^{n} p_{ij}(u) J(j), \qquad \text{for all } i = 1, \ldots, n \text{ and } u \in U(i).$$

In particular, $J^*(1), \ldots, J^*(n)$ solve the linear program of maximizing $\sum_{i=1}^{n} J(i)$ subject to the above constraint (see Fig. 7.2.1). Unfortunately, for large n the dimension of this program can be very large and its solution can be impractical, particularly in the absence of special structure.

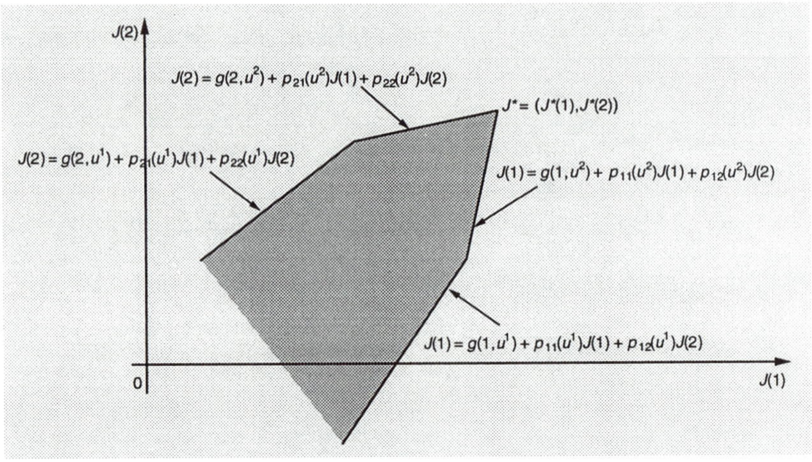

Figure 7.2.1 Linear program associated with a two-state stochastic shortest path problem. The constraint set is shaded, and the objective to maximize is $J(1) + J(2)$.

7.3 DISCOUNTED PROBLEMS

We now consider a discounted problem, where there is a discount factor $\alpha < 1$. We will show that this problem can be converted to a stochastic shortest path problem for which the analysis of the preceding section holds. To see this, let $i = 1, \ldots, n$ be the states, and consider an associated stochastic shortest path problem involving the states $1, \ldots, n$ plus an extra termination state t, with state transitions and costs obtained as follows: From a state $i \neq t$, when control u is applied, a cost $g(i, u)$ is incurred, and the next state is j with probability $\alpha p_{ij}(u)$ and t with probability $1-\alpha$; see Fig. 7.3.1. Note that Assumption 2.1 of the preceding section is satisfied for the associated stochastic shortest path problem.

Suppose now that we use the same policy in the discounted problem and in the associated stochastic shortest path problem. Then, as long as termination has not occurred, the state evolution in the two problems is governed by the same transition probabilities. Furthermore, the expected cost of the kth stage of the associated shortest path problem is $g\big(x_k, \mu_k(x_k)\big)$ multiplied by the probability that state t has not yet been reached, which is α^k. This is also the expected cost of the kth stage for the discounted problem. We conclude that the cost of any policy starting from a given state, is the same for the original discounted problem and for the associated stochastic shortest path problem. We can thus apply the results of the preceding section to the latter problem and obtain the following:

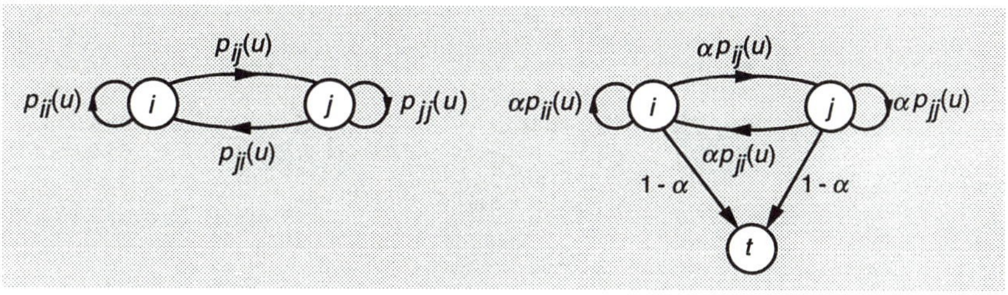

Figure 7.3.1 Transition probabilities for an α-discounted problem and its associated stochastic shortest path problem. In the latter problem, the probability that the state is not t after k stages is α^k. The expected cost per stage at each state $i = 1, \ldots, n$ is $g(i, u)$ for both problems, but it must be multiplied by α^k because of discounting (in the discounted case) or because it is incurred with probability α^k when termination has not yet been reached (in the stochastic shortest path case).

Proposition 3.1 The following hold for the discounted cost problem:

(a) The value iteration algorithm

$$J_{k+1}(i) = \min_{u \in U(i)} \left[g(i, u) + \alpha \sum_{j=1}^{n} p_{ij}(u) J_k(j) \right], \qquad i = 1, \ldots, n, \tag{3.1}$$

converges to the optimal costs $J^*(i)$, $i = 1, \ldots, n$, starting from arbitrary initial conditions $J_0(1), \ldots, J_0(n)$.

(b) The optimal costs $J^*(1), \ldots, J^*(n)$ of the discounted problem satisfy Bellman's equation,

$$J^*(i) = \min_{u \in U(i)} \left[g(i, u) + \alpha \sum_{j=1}^{n} p_{ij}(u) J^*(j) \right], \qquad i = 1, \ldots, n, \tag{3.2}$$

and in fact they are the unique solution of this equation.

(c) For any stationary policy μ, the costs $J_\mu(1), \ldots, J_\mu(n)$ are the unique solution of the equation

$$J_\mu(i) = g\big(i, \mu(i)\big) + \alpha \sum_{j=1}^{n} p_{ij}\big(\mu(i)\big) J_\mu(j), \qquad i = 1, \ldots, n.$$

Furthermore, given any initial conditions $J_0(1), \ldots, J_0(n)$, the sequence $J_k(i)$ generated by the DP iteration

$$J_{k+1}(i) = g\big(i, \mu(i)\big) + \alpha \sum_{j=1}^{n} p_{ij}\big(\mu(i)\big) J_k(j), \qquad i = 1, \ldots, n,$$

converges to the cost $J_\mu(i)$ for each i.

(d) A stationary policy μ is optimal if and only if for every state i, $\mu(i)$ attains the minimum in Bellman's equation (3.2).

(e) The policy iteration algorithm given by

$$\mu^{k+1}(i) = \arg \min_{u \in U(i)} \left[g(i, u) + \alpha \sum_{j=1}^{n} p_{ij}(u) J_{\mu^k}(j) \right], \quad i = 1, \ldots, n,$$

generates an improving sequence of policies and terminates with an optimal policy.

Proof: Parts (a)-(d) and part (e) are proved by applying parts (a)-(d) of Prop. 2.1, and Prop. 2.2, respectively, to the associated stochastic shortest path problem described above. **Q.E.D.**

Bellman's equation (3.2) has a familiar DP interpretation. At state i, the optimal cost $J^*(i)$ is the minimum over all controls of the sum of the expected current stage cost and the expected optimal cost of all future stages. The former cost is $g(i, u)$. The latter cost is $J^*(j)$, but since this cost starts accumulating after one stage, it is discounted by multiplication with α.

As in the case of stochastic shortest path problems [see Eq. (2.9) and the discussion following the proof of Prop. 2.1], we can show that the error

$$\big| J_k(i) - J^*(i) \big|$$

is bounded by a constant times α^k. Furthermore, the error bounds (2.16) become

$$J_{k+1}(j) + \frac{\alpha}{1-\alpha} \underline{c}_k \leq J^*(j) \leq J_{\mu^k}(j) \leq J_{k+1}(j) + \frac{\alpha}{1-\alpha} \bar{c}_k, \tag{3.3}$$

where μ^k is such that $\mu^k(i)$ attains the minimum in the kth value iteration (3.1) for all i, and

$$\underline{c}_k = \min_{i=1,\ldots,n} \big[J_{k+1}(i) - J_k(i) \big], \qquad \bar{c}_k = \max_{i=1,\ldots,n} \big[J_{k+1}(i) - J_k(i) \big],$$

since for the associated stochastic shortest path problem it can be shown that for every policy and starting state, the expected number of stages to reach the termination state t is $1/(1-\alpha)$, so that the terms $N_*(j) - 1$ and $N_k(j) - 1$ appearing in Eq. (2.16) are equal to $\alpha/(1-\alpha)$. We note also that there are a number of additional enhancements to the value iteration algorithm for the discounted problem (see Section 1.3 of Vol. II). There are also discounted cost variants of the approximate policy iteration and linear programming approaches discussed for stochastic shortest path problems.

Example 3.1 (Asset Selling)

Consider an infinite horizon version of the asset selling example of Section 4.4, assuming the set of possible offers is finite. Here, if accepted, the amount x_k offered in period k, will be invested at a rate of interest r. By depreciating the sale amount to period 0 dollars, we view $(1+r)^{-k}x_k$ as the reward for selling the asset in period k at a price x_k, where $r > 0$ is the rate of interest. Then we have a total discounted reward problem with discount factor $\alpha = 1/(1+r)$. The analysis of the present section is applicable, and the optimal value function J^* is the unique solution of Bellman's equation

$$J^*(x) = \max \left[x, \frac{E\{J^*(w)\}}{1+r} \right],$$

[cf. Eq. (4.4) in Section 4.4]. The optimal reward function is characterized by the critical number

$$\bar{\alpha} = \frac{E\{J^*(w)\}}{1+r},$$

which can be calculated as in Section 4.4. An optimal policy is to sell if and only if the current offer x_k is greater than or equal to $\bar{\alpha}$.

Example 3.2

A manufacturer at each time period receives an order for her product with probability p and receives no order with probability $1-p$. At any period she has a choice of processing all the unfilled orders in a batch, or process no order at all. The cost per unfilled order at each time period is $c > 0$, and the setup cost to process the unfilled orders is $K > 0$. The manufacturer wants to find a processing policy that minimizes the total expected cost, assuming the discount factor is $\alpha < 1$ and the maximum number of orders that can remain unfilled is n.

Here the state is the number of unfilled orders at the beginning of each period, and Bellman's equation takes the form

$$J^*(i) = \min\left[K + \alpha(1-p)J^*(0) + \alpha p J^*(1),\ ci + \alpha(1-p)J^*(i) + \alpha p J^*(i+1)\right], \tag{3.4}$$

for the states $i = 0, 1, \ldots, n-1$, and takes the form

$$J^*(n) = K + \alpha(1-p)J^*(0) + \alpha p J^*(1) \tag{3.5}$$

for state n. The first expression within brackets in Eq. (3.4) corresponds to processing the i unfilled orders, while the second expression corresponds to leaving the orders unfilled for one more period. When the maximum n of unfilled orders is reached, the orders must necessarily be processed, as indicated by Eq. (3.5).

To solve the problem, we observe that the optimal cost $J^*(i)$ is monotonically nondecreasing in i. This is intuitively clear, and can be rigorously proved by using the value iteration method. In particular, we can show by using the (finite horizon) DP algorithm that the k-stage optimal cost functions $J_k(i)$ are monotonically nondecreasing in i for all k (Exercise 7.7), and then argue that the optimal infinite horizon cost function $J^*(i)$ is also monotonically nondecreasing in i, since $J^*(i) = \lim_{k\to\infty} J_k(i)$ by Prop. 3.1(a). Given that $J^*(i)$ is monotonically nondecreasing in i, from Eq. (3.4) we have that if processing a batch of m orders is optimal, that is,

$$K + \alpha(1-p)J^*(0) + \alpha p J^*(1) \leq cm + \alpha(1-p)J^*(m) + \alpha p J^*(m+1),$$

then processing a batch of $m+1$ orders is also optimal. Therefore a *threshold policy*, that is, a policy that processes the orders if their number exceeds some threshold integer m^*, is optimal.

We leave it as Exercise 7.8 for the reader to verify that if we start the policy iteration algorithm with a threshold policy, every subsequently generated policy will be a threshold policy. Since there are $n+1$ distinct threshold policies, and the sequence of generated policies is improving, it follows that the policy iteration algorithm will yield an optimal policy after at most n iterations.

7.4 AVERAGE COST PER STAGE PROBLEMS

The methodology of the last two sections applies mainly to problems where the optimal total expected cost is finite either because of discounting or because of a cost-free termination state that the system eventually enters. In many situations, however, discounting is inappropriate and there is no natural cost-free termination state. In such situations it is often meaningful to optimize the average cost per stage starting from a state i, which is defined by

$$J_\pi(i) = \lim_{N\to\infty} \frac{1}{N} E\left\{ \sum_{k=0}^{N-1} g\big(x_k, \mu_k(x_k)\big) \,\Big|\, x_0 = i \right\}.$$

Let us first argue heuristically that *for most problems of interest the average cost per stage of a policy and the optimal average cost per stage are independent of the initial state.*

To this end we note that the average cost per stage of a policy primarily expresses cost incurred in the long term. Costs incurred in the early stages do not matter since their contribution to the average cost per stage is reduced to zero as $N \to \infty$; that is,

$$\lim_{N\to\infty} \frac{1}{N} E\left\{ \sum_{k=0}^{K} g\big(x_k, \mu_k(x_k)\big) \right\} = 0, \tag{4.1}$$

for any fixed K. Consider now a stationary policy μ and two states i and j such that the system will, under μ, eventually reach j with probability 1 starting from i. Then intuitively, it is clear that the average costs per stage starting from i and from j cannot be different, since the costs incurred in the process of reaching j from i do not contribute essentially to the average cost per stage. More precisely, let $K_{ij}(\mu)$ be the first passage time from i to j under μ, that is, the first index k for which $x_k = j$ starting from $x_0 = i$ under μ (see Appendix D). Then the average cost per stage corresponding to initial condition $x_0 = i$ can be expressed as

$$\begin{aligned} J_\mu(i) = {} & \lim_{N\to\infty} \frac{1}{N} E\left\{ \sum_{k=0}^{K_{ij}(\mu)-1} g\big(x_k, \mu(x_k)\big) \right\} \\ & + \lim_{N\to\infty} \frac{1}{N} E\left\{ \sum_{k=K_{ij}(\mu)}^{N} g\big(x_k, \mu(x_k)\big) \right\}. \end{aligned}$$

If $E\big\{K_{ij}(\mu)\big\} < \infty$ (which is equivalent to assuming that the system eventually reaches j starting from i with probability 1; see Appendix D), then it can be seen that the first limit is zero [cf. Eq. (4.1)], while the second limit is equal to $J_\mu(j)$. Therefore,

$$J_\mu(i) = J_\mu(j), \qquad \text{for all } i, j \text{ with } E\big\{K_{ij}(\mu)\big\} < \infty.$$

The preceding argument suggests that the optimal cost $J^*(i)$ should also be independent of the initial state i under normal circumstances. To see this, assume that for any two states i and j, there exists a stationary policy μ (dependent on i and j) such that j can be reached from i with probability 1 under μ. Then it is impossible that

$$J^*(j) < J^*(i),$$

since when starting from i we can adopt the policy μ up to the time when j is first reached and then switch to a policy that is optimal when starting from j, thereby achieving an average cost starting from i that is equal to $J^*(j)$. Indeed, it can be shown that

$$J^*(i) = J^*(j), \qquad \text{for all } i, j = 1, \ldots, n,$$

under the preceding assumption (see Vol. II, Section 4.2).

The Associated Stochastic Shortest Path Problem

The results to be developed in this section can be proved under a variety of different assumptions (see Chapter 4 in Vol. II). Here, we will make the following assumption, which will allow us to use the stochastic shortest path analysis of Section 7.2.

> **Assumption 4.1:** One of the states, by convention state n, is such that for some integer $m > 0$, and for all initial states and all policies, n is visited with positive probability at least once within the first m stages.

Assumption 4.1 can be shown to be equivalent to the assumption that the special state n is recurrent in the Markov chain corresponding to each stationary policy (see Appendix D).

Under Assumption 4.1 we will make an important connection of the average cost problem with an associated stochastic shortest path problem. To motivate this connection, consider a sequence of generated states, and divide it into cycles marked by successive visits to the state n. The first cycle includes the transitions from the initial state to the first visit to state n, and the kth cycle, $k = 2, 3, \ldots$, includes the transitions from the $(k-1)$th to the kth visit to state n. Each of the cycles can be viewed as a state trajectory of a corresponding stochastic shortest path problem with the termination state being essentially n. More precisely, this problem is obtained by leaving unchanged all transition probabilities $p_{ij}(u)$ for $j \neq n$, by setting all transition probabilities $p_{in}(u)$ to 0, and by introducing an artificial termination state t to which we move from each state i with probability $p_{in}(u)$; see Fig. 7.4.1. Note that Assumption 4.1 is equivalent to the Assumption 2.1 of Section 7.2 under which the results of Section 7.2 on stochastic shortest path problems were shown.

We have not yet fixed the cost structure of the stochastic shortest path problem. We will next argue that if we fix the expected stage cost incurred at state i to be

$$g(i, u) - \lambda^*,$$

where λ^* is the optimal average cost per stage starting from the special state n, then the associated stochastic shortest path problem becomes essentially equivalent to the original average cost per stage problem. Furthermore, Bellman's equation for the associated stochastic shortest path problem can be viewed as Bellman's equation for the original average cost per stage problem.

For a heuristic argument of why this is so, note that under all stationary policies there will be an infinite number of cycles marked by successive visits to state n. From this, it can be conjectured (and it can also be shown)

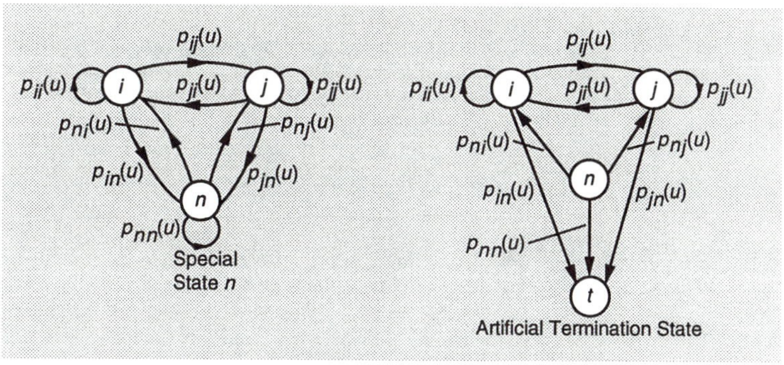

Figure 7.4.1 Transition probabilities for an average cost problem and its associated stochastic shortest path problem. The latter problem is obtained by introducing, in addition to $1, \ldots, n$, an artificial termination state t to which we move from each state i with probability $p_{in}(u)$, by setting all transition probabilities $p_{in}(u)$ to 0, and by leaving unchanged all other transition probabilities.

that the average cost problem amounts to finding a stationary policy μ that minimizes the average cycle cost

$$\lambda_\mu = \frac{C_{nn}(\mu)}{N_{nn}(\mu)}, \tag{4.2}$$

where for a fixed μ,

$C_{nn}(\mu)$: expected cost starting from n up to the first return to n,

$N_{nn}(\mu)$: expected number of stages to return to n starting from n.

The optimal average cost per stage λ^* starting from n, satisfies $\lambda_\mu \geq \lambda^*$ for all μ, so from Eq. (4.2) we obtain

$$C_{nn}(\mu) - N_{nn}(\mu)\lambda^* \geq 0, \tag{4.3}$$

with equality holding if μ is optimal. Thus, to attain such an optimal μ, we must minimize over μ the expression $C_{nn}(\mu) - N_{nn}(\mu)\lambda^*$, which is the expected cost of μ starting from n in the associated stochastic shortest path problem with state costs

$$g(i,u) - \lambda^*, \qquad i = 1, \ldots, n. \tag{4.4}$$

Let us denote by $h^*(i)$ the optimal cost of this stochastic shortest path problem when starting at the nontermination states $i = 1, \ldots, n$. Then by Prop. 2.1(b), $h^*(1), \ldots, h^*(n)$ solve uniquely the corresponding Bellman's equation, which has the form

$$h^*(i) = \min_{u \in U(i)} \left[g(i,u) - \lambda^* + \sum_{j=1}^{n-1} p_{ij}(u) h^*(j) \right], \qquad i = 1, \ldots, n, \tag{4.5}$$

since in the stochastic shortest path problem, the transition probability from i to $j \neq n$ is $p_{ij}(u)$ and the transition probability from i to n is zero under all u. If μ^* is an optimal stationary policy for the average cost problem, then this policy must satisfy, in view of Eq. (4.2),

$$C_{nn}(\mu^*) - N_{nn}(\mu^*)\lambda^* = 0,$$

and from Eq. (4.3), this policy must also be optimal for the associated stochastic shortest path problem. Thus, we must have

$$h^*(n) = C_{nn}(\mu^*) - N_{nn}(\mu^*)\lambda^* = 0. \tag{4.6}$$

By using this equation, we can now write Bellman's equation (4.5) as

$$\lambda^* + h^*(i) = \min_{u \in U(i)} \left[g(i,u) + \sum_{j=1}^{n} p_{ij}(u) h^*(j) \right], \qquad i = 1, \ldots, n. \tag{4.7}$$

Equation (4.7), which is really Bellman's equation for the associated stochastic shortest path problem, will be viewed as Bellman's equation for the average cost per stage problem. The preceding argument indicates that this equation has a unique solution as long as we impose the constraint $h^*(n) = 0$. Furthermore, by minimization of its right-hand side we should obtain an optimal stationary policy. We will now prove these facts formally.

Bellman's Equation

The following proposition provides the main results regarding Bellman's equation.

Proposition 4.1 Under Assumption 4.1 the following hold for the average cost per stage problem:

(a) The optimal average cost λ^* is the same for all initial states and together with some vector $h^* = \{h^*(1), \ldots, h^*(n)\}$ satisfies Bellman's equation

$$\lambda^* + h^*(i) = \min_{u \in U(i)} \left[g(i,u) + \sum_{j=1}^{n} p_{ij}(u) h^*(j) \right], \qquad i = 1, \ldots, n. \tag{4.8}$$

Furthermore, if $\mu(i)$ attains the minimum in the above equation for all i, the stationary policy μ is optimal. In addition, out of all vectors h^* satisfying this equation, there is a unique vector for which $h^*(n) = 0$.

(b) If a scalar λ and a vector $h = \{h(1), \ldots, h(n)\}$ satisfy Bellman's equation, then λ is the average optimal cost per stage for each initial state.

(c) Given a stationary policy μ with corresponding average cost per stage λ_μ, there is a unique vector $h_\mu = \{h_\mu(1), \ldots, h_\mu(n)\}$ such that $h_\mu(n) = 0$ and

$$\lambda_\mu + h_\mu(i) = g\big(i, \mu(i)\big) + \sum_{j=1}^{n} p_{ij}\big(\mu(i)\big) h_\mu(j), \qquad i = 1, \ldots, n. \tag{4.9}$$

Proof: (a) Let us denote

$$\tilde{\lambda} = \min_{\mu} \frac{C_{nn}(\mu)}{N_{nn}(\mu)}, \tag{4.10}$$

where $C_{nn}(\mu)$ and $N_{nn}(\mu)$ have been defined earlier, and the minimum is taken over the finite set of all stationary policies. Note that $C_{nn}(\mu)$ and $N_{nn}(\mu)$ are finite in view of Assumption 4.1 and the results of Section 7.2. Then we have

$$C_{nn}(\mu) - N_{nn}(\mu)\tilde{\lambda} \geq 0, \tag{4.11}$$

with equality holding for all μ that attain the minimum in Eq. (4.10). Consider the associated stochastic shortest path problem when the expected stage cost incurred at state i is

$$g(i, u) - \tilde{\lambda}. \tag{4.12}$$

Then by Prop. 2.1(b), the costs $h^*(1), \ldots, h^*(n)$ solve uniquely the corresponding Bellman's equation

$$h^*(i) = \min_{u \in U(i)} \left[g(i, u) - \tilde{\lambda} + \sum_{j=1}^{n-1} p_{ij}(u) h^*(j) \right], \tag{4.13}$$

since the transition probability from i to n is zero in the associated stochastic shortest path problem. An optimal stationary policy must minimize the cost $C_{nn}(\mu) - N_{nn}(\mu)\tilde{\lambda}$ and reduce it to zero [in view of Eq. (4.10)], so we see that

$$h^*(n) = 0. \tag{4.14}$$

Thus, Eq. (4.13) is written as

$$\tilde{\lambda} + h^*(i) = \min_{u \in U(i)} \left[g(i, u) + \sum_{j=1}^{n} p_{ij}(u) h^*(j) \right], \qquad i = 1, \ldots, n. \tag{4.15}$$

We will show that this relation implies that $\tilde{\lambda} = \lambda^*$.

Indeed, let $\pi = \{\mu_0, \mu_1, \ldots\}$ be any admissible policy, let N be a positive integer, and for all $k = 0, \ldots, N-1$, define $J_k(i)$ using the following recursion

$$J_0(i) = h^*(i), \qquad i = 1, \ldots, n,$$

$$J_{k+1}(i) = g\big(i, \mu_{N-k-1}(i)\big) + \sum_{j=1}^{n} p_{ij}\big(\mu_{N-k-1}(i)\big) J_k(j), \quad i = 1, \ldots, n. \tag{4.16}$$

Note that $J_N(i)$ is the N-stage cost of π when the starting state is i and the terminal cost function is h^*. From Eq. (4.15), we have

$$\tilde{\lambda} + h^*(i) \le g\big(i, \mu_{N-1}(i)\big) + \sum_{j=1}^{n} p_{ij}\big(\mu_{N-1}(i)\big) h^*(j), \quad i = 1, \ldots, n,$$

or equivalently, using Eq. (4.16) for $k = 0$ and the definition of J_0,

$$\tilde{\lambda} + J_0(i) \le J_1(i), \qquad i = 1, \ldots, n.$$

Using this relation, we have

$$\begin{aligned} g(i, \mu_{N-2}(i)) + \tilde{\lambda} + \sum_{j=1}^{n} p_{ij}\big(\mu_{N-2}(i)\big) J_0(j) \\ \le g\big(i, \mu_{N-2}(i)\big) + \sum_{j=1}^{n} p_{ij}\big(\mu_{N-2}(i)\big) J_1(j), \quad i = 1, \ldots, n. \end{aligned}$$

By Eq. (4.15) and the definition $J_0(j) = h^*(j)$, the left-hand side of the above inequality is no less than $2\tilde{\lambda} + h^*(i)$, while by Eq. (4.16), the right-hand side is equal to $J_2(i)$. We thus obtain

$$2\tilde{\lambda} + h^*(i) \le J_2(i), \qquad i = 1, \ldots, n.$$

By repeating this argument several times, we obtain

$$k\tilde{\lambda} + h^*(i) \le J_k(i), \qquad k = 0, \ldots, N, \ i = 1, \ldots, n,$$

and in particular, for $k = N$,

$$\tilde{\lambda} + \frac{1}{N} \le \frac{1}{N} J_N(i), \qquad i = 1, \ldots, n. \tag{4.17}$$

Furthermore, equality holds in the above relation if $\mu_k(i)$ attains the minimum in Eq. (4.15) for all i and k.

Let us now take the limit as $N \to \infty$ in Eq. (4.17). The left-hand side tends to $\tilde{\lambda}$. We claim that the right-hand side tends to $J_\pi(i)$, the average cost per stage of π starting at state i. The reason is that from the definition (4.16), $J_N(i)$ is the N-stage cost of π starting at i, when the terminal cost function is h^*; when we take the limit of $(1/N)J_N(i)$, the dependence on the terminal cost function h^* disappears. Thus, by taking the limit as $N \to \infty$ in Eq. (4.17), we obtain

$$\tilde{\lambda} \leq J_\pi(i), \qquad i = 1, \ldots, n,$$

for all admissible π, with equality if π is a stationary policy μ such that $\mu(i)$ attains the minimum in Eq. (4.15) for all i and k. It follows that

$$\tilde{\lambda} = \min_\pi J_\pi(i) = \lambda^*, \qquad i = 1, \ldots, n,$$

and from Eq. (4.15), we obtain the desired Eq. (4.8).

Finally, Eqs. (4.14) and (4.15) are equivalent to Bellman's equation (4.13) for the associated stochastic shortest path problem. Since the solution to the latter equation is unique, the same is true for the solution of Eqs. (4.14) and (4.15).

(b) The proof of this part is obtained by using the argument of the proof of part (a) following Eq. (4.15).

(c) The proof of this part is obtained by specializing part (a) to the case where the constraint set at each state i is $\tilde{U}(i) = \{\mu(i)\}$. **Q.E.D.**

An examination of the preceding proof shows that the unique vector h^* in Bellman's equation (4.8) for which $h^*(n) = 0$ is the optimal cost vector for the associated stochastic shortest path problem when the expected stage cost at state i is

$$g(i, u) - \lambda^*,$$

[cf. Eq. (4.13)]. Consequently, $h^*(i)$ has the interpretation of a *relative or differential cost*; it is the minimum of the difference between the expected cost to reach n from i for the first time and the cost that would be incurred if the cost per stage was the average λ^*. We note that the relation between optimal policies of the stochastic shortest path and the average cost problems is clarified in Exercise 7.15.

We finally mention that Prop. 4.1 can be shown under considerably weaker conditions (see Section 4.2 of Vol. II). In particular, Prop. 4.1 can be shown assuming that all stationary policies have a single recurrent class, even if their corresponding recurrent classes do not have state n in common. The proof, however, requires a more sophisticated use of the connection with an associated stochastic shortest path problem. Proposition 4.1 can also be shown assuming that for every pair of states i and j, there exists a stationary policy under which there is positive probability of reaching j starting from i. In this case, however, an associated stochastic shortest path problem cannot be defined and the corresponding connection with the average cost per stage problem cannot be made. The analysis of Chapter 4 of Vol. II relies on another connection that exists between the average cost per stage problem and the discounted cost problem, but to establish this connection and to fully explore its ramifications, a much more sophisticated analysis is required.

Example 4.1

Consider the average cost version of the manufacturer's problem of Example 3.2. Here, state 0 plays the role of the special state n in Assumption 4.1. Bellman's equation takes the form

$$\lambda^* + h^*(i) = \min\big[K + (1-p)h^*(0) + ph^*(1),\ ci + (1-p)h^*(i) + ph^*(i+1)\big], \tag{4.18}$$

for the states $i = 0, 1, \ldots, n-1$, and takes the form

$$\lambda^* + h^*(n) = K + (1-p)h^*(0) + ph^*(1)$$

for state n. The first expression within brackets in Eq. (4.18) corresponds to processing the i unfilled orders, while the second expression corresponds to leaving the orders unfilled for one more period. The optimal policy is to process i unfilled orders if

$$K + (1-p)h^*(0) + ph^*(1) \le ci + (1-p)h^*(i) + ph^*(i+1).$$

If we view $h^*(i)$, $i = 1, \ldots, n$, as differential costs associated with an optimal policy, it is intuitively clear that $h^*(i)$ is monotonically nondecreasing with i (this can also be proved with an analysis based on the theory presented in Vol. II, Section 4.2). As in Example 3.2, the monotonicity property of $h^*(i)$ implies that a threshold policy is optimal.

Value Iteration

The most natural version of the value iteration method for the average cost problem is simply to select arbitrarily a terminal cost function, say J_0, and to generate successively the corresponding optimal k-stage costs $J_k(i)$, $k = 1, 2, \ldots$ This can be done by executing the DP algorithm starting with J_0, that is, by using the recursion

$$J_{k+1}(i) = \min_{u \in U(i)} \left[g(i,u) + \sum_{j=1}^{n} p_{ij}(u) J_k(j) \right], \qquad i = 1, \ldots, n. \tag{4.19}$$

It is natural to expect that the ratios $J_k(i)/k$ should converge to the optimal average cost per stage λ^* as $k \to \infty$, that is,

$$\lim_{k \to \infty} \frac{J_k(i)}{k} = \lambda^*.$$

To show this, let us define the recursion

$$J_{k+1}^*(i) = \min_{u \in U(i)} \left[g(i,u) + \sum_{j=1}^{n} p_{ij}(u) J_k^*(j) \right], \qquad i = 1, \ldots, n, \tag{4.20}$$

with the initial condition

$$J_0^*(i) = h^*(i), \qquad i = 1, \ldots, n, \tag{4.21}$$

where h^* is a differential cost vector satisfying Bellman's equation

$$\lambda^* + h^*(i) = \min_{u \in U(i)} \left[g(i,u) + \sum_{j=1}^{n} p_{ij}(u) h^*(j) \right], \qquad i = 1, \ldots, n. \tag{4.22}$$

Using this equation, it can be shown by induction that for all k we have

$$J_k^*(i) = k\lambda^* + h^*(i), \qquad i = 1, \ldots, n. \tag{4.23}$$

On the other hand, it can be seen that for all k,

$$\big| J_k(i) - J_k^*(i) \big| \leq \max_{j=1,\ldots,n} \big| J_0(j) - h^*(j) \big|, \qquad i = 1, \ldots, n. \tag{4.24}$$

The reason is that $J_k(i)$ and $J_k^*(i)$ are optimal costs for two k-stage problems that differ only in the corresponding terminal cost functions, which are J_0 and h^*, respectively. From the preceding two equations, we see that for all k,

$$\big| J_k(i) - k\lambda^* \big| \leq \max_{j=1,\ldots,n} \big| J_0(j) - h^*(j) \big| + \max_{j=1,\ldots,n} \big| h^*(j) \big|, \qquad i = 1, \ldots, n. \tag{4.25}$$

so that $J_k(i)/k$ converges to λ^* at the rate of a constant divided by k. Note that the above proof shows that $J_k(i)/k$ converges to λ^* under any conditions that guarantee that Bellman's equation (4.22) holds for some vector h^*.

The value iteration method just described is simple and straightforward, but has two drawbacks. First, since typically some of the components of J_k diverge to ∞ or $-\infty$, direct calculation of $\lim_{k \to \infty} J_k(i)/k$ is numerically cumbersome. Second, this method will not provide us with a corresponding differential cost vector h^*. We can bypass both difficulties by subtracting the same constant from all components of the vector J_k, so that the difference, call it h_k, remains bounded. In particular, we can consider the algorithm

$$h_k(i) = J_k(i) - J_k(s), \qquad i = 1, \ldots, n. \tag{4.26}$$

where s is some fixed state. By using Eq. (4.19), we then obtain

$$\begin{aligned} h_{k+1}(i) &= J_{k+1}(i) - J_{k+1}(s) \\ &= \min_{u\in U(i)} \left[g(i,u) + \sum_{j=1}^{n} p_{ij}(u) J_k(j) \right] \\ &\quad - \min_{u\in U(s)} \left[g(s,u) + \sum_{j=1}^{n} p_{sj}(u) J_k(j) \right], \end{aligned}$$

from which in view of the relation $h_k(j) = J_k(j) - J_k(s)$, we have

$$\begin{aligned} h_{k+1}(i) = {}& \min_{u\in U(i)} \left[g(i,u) + \sum_{j=1}^{n} p_{ij}(u) h_k(j) \right] \\ & - \min_{u\in U(s)} \left[g(s,u) + \sum_{j=1}^{n} p_{sj}(u) h_k(j) \right], \qquad i = 1, \ldots, n. \end{aligned} \tag{4.27}$$

The above algorithm, known as *relative value iteration*, is mathematically equivalent to the value iteration method (4.19) that generates $J_k(i)$. The iterates generated by the two methods merely differ by a constant [cf. Eq. (4.26)], and the minimization problems involved in the corresponding iterations of the two methods are mathematically equivalent. However, under Assumption 4.1, it can be shown that the iterates $h_k(i)$ generated by the relative value iteration method are bounded, while this is typically not true for the value iteration method.

It can be seen that if the relative value iteration (4.27) converges to some vector h, then we have

$$\lambda + h(i) = \min_{u\in U(i)} \left[g(i,u) + \sum_{j=1}^{n} p_{ij}(u) h(j) \right],$$

where

$$\lambda = \min_{u\in U(s)} \left[g(s,u) + \sum_{j=1}^{n} p_{sj}(u) h(j) \right].$$

By Prop. 4.1(b), this implies that λ is the optimal average cost per stage for all initial states, and h is an associated differential cost vector. Unfortunately, the convergence of the relative value iteration is not guaranteed under Assumption 4.1 (see Exercise 7.14 for a counterexample). A stronger assumption is required. It turns out, however, that there is a simple variant of the relative value iteration for which convergence is guaranteed under Assumption 4.1. This variant is given by

$$
\begin{aligned}
h_{k+1}(i) = (1-\tau)h_k(i) + \min_{u\in U(i)}\left[g(i,u) + \tau\sum_{j=1}^{n} p_{ij}(u)h_k(j)\right] \\
- \min_{u\in U(s)}\left[g(s,u) + \tau\sum_{j=1}^{n} p_{sj}(u)h_k(j)\right], \quad i = 1,\ldots,n,
\end{aligned}
\tag{4.28}
$$

where τ is a scalar such that $0 < \tau < 1$. Note that for $\tau = 1$, we obtain the relative value iteration (4.27). The convergence proof of this algorithm is is somewhat complicated. It can be found in Section 4.3 of Vol. II.

Policy Iteration

It is possible to use a policy iteration algorithm for the average cost problem. This algorithm operates similar to the policy iteration algorithms of the preceding sections: given a stationary policy, we obtain an improved policy by means of a minimization process until no further improvement is possible. In particular, at the typical step of the algorithm, we have a stationary policy μ^k. We then perform a *policy evaluation* step; that is, we obtain corresponding average and differential costs λ^k and $h^k(i)$ satisfying

$$
\lambda^k + h^k(i) = g\big(i,\mu^k(i)\big) + \sum_{j=1}^{n} p_{ij}\big(\mu^k(i)\big)h^k(j), \qquad i = 1,\ldots,n, \tag{4.29}
$$

$$
h^k(n) = 0. \tag{4.30}
$$

We subsequently perform a *policy improvement* step; that is, we find a stationary policy μ^{k+1}, where for all i, $\mu^{k+1}(i)$ is such that

$$
\begin{aligned}
g\big(i,\mu^{k+1}(i)\big) + \sum_{j=1}^{n} p_{ij}\big(\mu^{k+1}(i)\big)h^k(j) \\
= \min_{u\in U(i)}\left[g(i,u) + \sum_{j=1}^{n} p_{ij}(u)h^k(j)\right].
\end{aligned}
\tag{4.31}
$$

If $\lambda^{k+1} = \lambda^k$ and $h^{k+1}(i) = h^k(i)$ for all i, the algorithm terminates; otherwise, the process is repeated with μ^{k+1} replacing μ^k.

To prove that the policy iteration algorithm terminates, it is sufficient that each iteration makes some irreversible progress towards optimality, since there are finitely many stationary policies. The type of irreversible progress that we can demonstrate is described in the following proposition, which also shows that an optimal policy is obtained upon termination.

Proposition 4.2 Under Assumption 4.1, in the policy iteration algorithm, for each k we either have

$$\lambda^{k+1} < \lambda^k \tag{4.32}$$

or else we have

$$\lambda^{k+1} = \lambda^k, \qquad h^{k+1}(i) \leq h^k(i), \quad i = 1, \ldots, n. \tag{4.33}$$

Furthermore, the algorithm terminates and the policies μ^k and μ^{k+1} obtained upon termination are optimal.

Proof: To simplify notation, denote $\mu^k = \mu$, $\mu^{k+1} = \overline{\mu}$, $\lambda^k = \lambda$, $\lambda^{k+1} = \overline{\lambda}$, $h^k(i) = h(i)$, $h^{k+1}(i) = \overline{h}(i)$. Define for $N = 1, 2, \ldots$

$$h_N(i) = g(i, \overline{\mu}(i)) + \sum_{j=1}^n p_{ij}\big(\overline{\mu}(i)\big) h_{N-1}(j), \qquad i = 1, \ldots, n, \tag{4.34}$$

where

$$h_0(i) = h(i), \qquad i = 1, \ldots, n. \tag{4.35}$$

Note that $h_N(i)$ is the N-stage cost of policy $\overline{\mu}$ starting from i when the terminal cost function is h. Thus we have

$$\overline{\lambda} = J_{\overline{\mu}}(i) = \lim_{N\to\infty} \frac{1}{N} h_N(i), \qquad i = 1, \ldots, n, \tag{4.36}$$

since the contribution of the terminal cost to $(1/N)h_N(i)$ vanishes when $N \to \infty$. By the definition of $\overline{\mu}$ and Prop. 4.1(c), we have for all i

$$\begin{aligned} h_1(i) &= g\big(i, \overline{\mu}(i)\big) + \sum_{j=1}^n p_{ij}\big(\overline{\mu}(i)\big) h_0(j) \\ &\leq g\big(i, \mu(i)\big) + \sum_{j=1}^n p_{ij}\big(\mu(i)\big) h_0(j)) \\ &= \lambda + h_0(i). \end{aligned} \tag{4.37}$$

From the above equation, we also obtain

$$\begin{aligned} h_2(i) &= g\big(i, \overline{\mu}(i)\big) + \sum_{j=1}^n p_{ij}\big(\overline{\mu}(i)\big) h_1(j) \\ &\leq g\big(i, \overline{\mu}(i)\big) + \sum_{j=1}^n p_{ij}\big(\overline{\mu}(i)\big)\big(\lambda + h_0(j)\big) \end{aligned}$$

$$
\begin{aligned}
&= \lambda + g\big(i, \overline{\mu}(i)\big) + \sum_{j=1}^{n} p_{ij}\big(\overline{\mu}(i)\big) h_0(j) \\
&\leq \lambda + g\big(i, \mu(i)\big) + \sum_{j=1}^{n} p_{ij}\big(\mu(i)\big) h_0(j) \\
&= 2\lambda + h_0(i),
\end{aligned} \tag{4.38}
$$

and by proceeding similarly, we see that for all i and N we have

$$h_N(i) \leq N\lambda + h_0(i).$$

Thus,

$$\frac{1}{N} h_N(i) \leq \lambda + \frac{1}{N} h_0(i),$$

and by taking the limit as $N \to \infty$ and using Eq. (4.36), we obtain $\overline{\lambda} \leq \lambda$.

If $\overline{\lambda} = \lambda$, then it is seen that the iteration that produces μ^{k+1} is a policy improvement step for the associated stochastic shortest path problem with cost per stage

$$g(i, u) - \lambda.$$

Furthermore, $h(i)$ and $\overline{h}(i)$ are the optimal costs starting from i and corresponding to μ and $\overline{\mu}$, respectively, in this associated stochastic shortest path problem. Thus, by Prop. 2.2, we must have $\overline{h}(i) \leq h(i)$ for all i.

Let us now show that when the algorithm terminates with $\overline{\lambda} = \lambda$ and $\overline{h}(i) = h(i)$ for all i, the policies $\overline{\mu}$ and μ are optimal. Indeed, upon termination we have for all i

$$
\begin{aligned}
\lambda + h(i) &= \overline{\lambda} + \overline{h}(i) \\
&= g\big(i, \overline{\mu}(i)\big) + \sum_{j=1}^{n} p_{ij}\big(\overline{\mu}(i)\big) \overline{h}(j) \\
&= g\big(i, \overline{\mu}(i)\big) + \sum_{j=1}^{n} p_{ij}\big(\overline{\mu}(i)\big) h(j) \\
&= \min_{u \in U(i)} \left[g(i, u) + \sum_{j=1}^{n} p_{ij}(u) h(j) \right].
\end{aligned}
$$

Thus λ and h satisfy Bellman's equation, and by Prop. 4.1(b), λ must be equal to the optimal average cost. Furthermore, $\overline{\mu}(i)$ attains the minimum in the right-hand side of Bellman's equation, so by Prop. 4.1(a), $\overline{\mu}$ is optimal. Since we also have for all i

$$\lambda + h(i) = g\big(i, \mu(i)\big) + \sum_{j=1}^{n} p_{ij}\big(\mu(i)\big) h(j),$$

the same is true for μ. **Q.E.D.**

We note that policy iteration can be shown to terminate under less restrictive conditions than Assumption 4.1 (see Vol. II, Section 4.3).

7.5 NOTES, SOURCES, AND EXERCISES

This chapter is an introduction to infinite horizon problems. There is an extensive theory for these problems with interesting mathematical and computational content. Volume II provides a comprehensive treatment and gives many references to the literature. The line of development given here is original in that it uses the stochastic shortest path problem as the starting point for the analysis of the other problems. Also, the error bounds (2.16) are new.

EXERCISES

7.1

A tennis player has a Fast serve and a Slow serve, denoted F and S, respectively. The probability of F (or S) landing in bounds is p_F (or p_S, respectively). The probability of winning the point assuming the serve landed in bounds is q_F (or q_S, respectively). We assume that $p_F < p_S$ and $q_F > q_S$. The problem is to find the serve to be used at each possible scoring situation during a single game in order to maximize the probability of winning that game.

(a) Formulate this as a stochastic shortest path problem, argue that Assumption 2.1 of Section 7.2 holds, and write Bellman's equation.

(b) Computer assignment: Assume that $q_F = 0.6$, $q_S = 0.4$, and $p_S = 0.95$. Use value iteration to calculate and plot (in increments of 0.05) the probability of the server winning a game with optimal serve selection as a function of p_F.

7.2

A quarterback can choose between running and passing the ball on any given play. The number of yards gained by running is integer and is Poisson distributed with parameter λ_r. A pass is incomplete with probability p, is intercepted with probability q, and is completed with probability $1-p-q$. When completed, a pass gains an integer number of yards that is Poisson distributed with parameter λ_p. We assume that the probability of scoring a touchdown on a single play starting i yards from the goal is equal to the probability of gaining a number of yards greater than or equal to i. We assume also that yardage cannot be lost on any play and that there are no penalties. The ball is turned over to the other team on a fourth down or when an interception occurs.

(a) Formulate the problem as a stochastic shortest path problem, argue that Assumption 2.1 of Section 7.2 holds, and write Bellman's equation.

(b) Computer assignment: Use value iteration to compute the quarterback's play-selection policy that maximizes the probability of scoring a touchdown on any single drive for $\lambda_r = 3$, $\lambda_p = 10$, $p = 0.4$, and $q = 0.05$.

7.3

A computer manufacturer can be in one of two states. In state 1 his product sells well, while in state 2 his product sells poorly. While in state 1 he can advertise his product in which the one-stage reward is 4 units, and the transition probabilities are $p_{11} = 0.8$ and $p_{12} = 0.2$. If in state 1, he does not advertise, the reward is 6 units and the transition probabilities are $p_{11} = p_{12} = 0.5$. While in state 2, he can do research to improve his product, in which case the one-stage reward is -5 units, and the transition probabilities are $p_{21} = 0.7$ and $p_{22} = 0.3$. If in state 2 he does not do research, the reward is -3, and the transition probabilities are $p_{21} = 0.4$ and $p_{22} = 0.6$. Consider the infinite horizon, discounted version of this problem.

(a) Show that when the discount factor α is sufficiently small, the computer manufacturer should follow the "shortsighted" policy of not advertising (not doing research) while in state 1 (state 2). By contrast, when α is sufficiently close to unity, he should follow the "farsighted" policy of advertising (doing research) while in state 1 (state 2).

(b) For $\alpha = 0.9$ calculate the optimal policy using policy iteration.

(c) For $\alpha = 0.99$, use a computer to solve the problem by value iteration, with and without the error bounds (3.3).

7.4

An energetic salesman works every day of the week. He can work in only one of two towns A and B on each day. For each day he works in town A (or B) his expected reward is r_A (or r_B, respectively). The cost for changing towns is c. Assume that $c > r_A > r_B$ and that there is a discount factor $\alpha < 1$.

(a) Show that for α sufficiently small, the optimal policy is to stay in the town he starts in, and that for α sufficiently close to 1, the optimal policy is to move to town A (if not starting there) and stay in A for all subsequent times.

(b) Solve the problem for $c = 3$, $r_A = 2$, $r_B = 1$, and $\alpha = 0.9$ using policy iteration.

(c) Use a computer to solve the problem of part (b) by value iteration, with and without the error bounds (3.3).

7.5

A person has an umbrella that she takes from home to office and vice versa. There is a probability p of rain at the time she leaves home or office independently of earlier weather. If the umbrella is in the place where she is and it rains, she takes the umbrella to go to the other place (this involves no cost). If there is no umbrella and it rains, there is a cost W for getting wet. If the umbrella is in the place where she is but it does not rain, she may take the umbrella to go to the other place (this involves an inconvenience cost V) or she may leave the umbrella behind (this involves no cost). Costs are discounted at a factor $\alpha < 1$.

(a) Formulate this as an infinite horizon total cost discounted problem. *Hint*: Try to use as few states as possible.

(b) Characterize the optimal policy as best as you can.

7.6

For the tennis player's problem (Exercise 7.1), show that it is optimal (regardless of score) to use F on both serves if $(p_F q_F)/(p_S q_S) > 1$, to use S on both serves if $(p_F q_F)/(p_S q_S) < 1 + p_F - p_S$, and to use F on the first serve and S on the second otherwise.

7.7

Consider the value iteration method for the Example 3.2:

$$J_{k+1}(i) = \min\big[K + \alpha(1-p)J_k(0) + \alpha p J_k(1),\\ ci + \alpha(1-p)J_k(i) + \alpha p J_k(i+1)\big], \qquad i = 0, 1, \ldots, n-1,$$

$$J_{k+1}(n) = K + \alpha(1-p)J_k(0) + \alpha p J_k(1),$$

where $J_0(i) = 0$ for all i. Show by induction that $J_k(i)$ is monotonically nondecreasing in i.

7.8

Consider the policy iteration algorithm for the problem of Example 3.2.

(a) Show that if we start the algorithm with a threshold policy, every subsequently generated policy will be a threshold policy.

(b) Carry out the algorithm for the case $c = 1$, $K = 5$, $n = 10$, $p = 0.5$, $\alpha = 0.9$, and an initial policy that always processes the unfilled orders.

7.9

Solve the average cost version ($\alpha = 1$) of the computer manufacturer's problem (Exercise 7.3).

7.10

An unemployed worker receives a job offer at each time period, which she may accept or reject. The offered salary takes one of n possible values $w^1, \ldots, w^n$ with given probabilities, independently of preceding offers. If she accepts the offer, she must keep the job for the rest of her life at the same salary level. If she rejects the offer, she receives unemployment compensation c for the current period and is eligible to accept future offers. Assume that income is discounted by a factor $\alpha < 1$.

(a) Show that there is a threshold $\overline{w}$ such that it is optimal to accept an offer if and only if its salary is larger than $\overline{w}$, and characterize $\overline{w}$.

(b) Consider the variant of the problem where there is a given probability p_i that the worker will be fired from her job at any one period if her salary is w^i. Show that the result of part (a) holds in the case where p_i is the same for all i. Analyze the case where p_i depends on i.

7.11

Do part (b) of Exercise 7.10 for the case where income is not discounted and the worker maximizes her average income per period.

7.12

Show that one can always take $m = n$ in Assumption 2.1. *Hint*: For any π and i, let $S_k(i)$ be the set of states that are reachable with positive probability from i under π in k stages or less. Show that under Assumption 2.1, we cannot have $S_k(i) = S_{k+1}(i)$ while $t \neq S_k(i)$.

7.13

Show the error bounds (2.16). *Hint*: Complete the details of the following argument. Let $\mu^k(i)$ attain the minimum in the value iteration (2.15) for all i. Then, in vector form, we have

$$J_{k+1} = g_k + P_k J_k,$$

where J_k and g_k are the vectors with components $J_k(i)$, $i = 1, \ldots, n$, and $g_k\big(i, \mu^k(i)\big)$, $i = 1, \ldots, n$, respectively, and P_k is the matrix with elements $p_{ij}\big(\mu^k(i)\big)$. Also from Bellman's equation, we have

$$J^* \le g_k + P_k J^*,$$

where the vector inequality above is meant to hold separately for each component. Let $e = (1, \dots, 1)'$ be the unit vector. Using the above two relations, we have

$$J^* - J_k \le J^* - J_{k+1} + \overline{c}_k e \le P_k(J^* - J_k) + \overline{c}_k e. \tag{5.1}$$

Multiplying this relation with P_k and adding $\overline{c}_k e$, we obtain

$$P_k(J^* - J_k) + \overline{c}_k e \le P_k^2(J^* - J_k) + \overline{c}_k(I + P_k)e.$$

Similarly continuing, we have for all $r \ge 1$

$$J^* - J_{k+1} + \overline{c}_k e \le P_k^r(J^* - J_k) + \overline{c}_k(I + P_k + \cdots + P_k^{r-1})e.$$

For $s = 1, 2, \dots$, the ith component of the vector $P_k^s e$ is equal to the probability $P\{x_s \ne t \mid x_0 = i, \, \mu^k\}$ that t has not been reached after s stages starting from i and using the stationary policy μ^k. Thus, Assumption 2.1 implies that $\lim_{r \to \infty} P_k^r = 0$, while we have

$$\lim_{r \to \infty} (I + P_k + \cdots + P_k^{r-1})e = N_k,$$

where N_k is the vector $\big(N_k(1), \dots, N_k(n)\big)'$. Combining the above two relations, we obtain

$$J^* \le J_{k+1} + \overline{c}_k(N_k - e),$$

proving the desired upper bound.

The lower bound is proved similarly, by using in place of μ^k, an optimal stationary policy μ^*. In particular, in place of Eq. (5.1), we can show that

$$J_k - J^* \le J_{k+1} - J^* - \underline{c}_k e \le P^*(J_k - J^*) - \underline{c}_k e,$$

where P^* is the matrix with elements $p_{ij}\big(\mu^*(i)\big)$. We similarly obtain for all $r \ge 1$

$$J_{k+1} - J^* - \underline{c}_k e \le (P^*)^r(J_k - J^*) - \underline{c}_k\big(I + P^* + \cdots + (P^*)^{r-1}\big)e,$$

from which $J_{k+1} + \underline{c}_k(N^* - e) \le J^*$, where N^* is the vector $\big(N^*(1), \dots, N^*(n)\big)'$.

7.14

Apply the relative value iteration algorithm (4.27) for the case where there are two states and only one control per state. The transition probabilities are $p_{11} = \epsilon$, $p_{12} = 1 - \epsilon$, $p_{21} = 1 - \epsilon$, and $p_{22} = \epsilon$, where $0 \le \epsilon < 1$. Show that if $0 < \epsilon$ the algorithm converges, but if $\epsilon = 0$, the algorithm may not converge. Show also that the variation (4.28) converges when $\epsilon = 0$.

7.15

Consider the average cost problem and its associated stochastic shortest path problem when the expected cost incurred at state i is $g(i, u) - \lambda^*$.

(a) Use Prop. 2.1(d) and Prop. 4.1(a) to show that if a stationary policy is optimal for the latter problem it is also optimal for the former.

(b) Show by example that the reverse of part (a) need not be true.

APPENDIX A: Mathematical Review

The purpose of this appendix is to provide a list of mathematical definitions, notations, and results that are used frequently in the text. For detailed expositions, the reader may consult references [HoK61], [Roc70], [Roy68], [Rud64], and [Str76].

.1 SETS

If x is a member of the set S, we write $x \in S$. We write $x \notin S$ if x is not a member of S. A set S may be specified by listing its elements within braces. For example, by writing $S = \{x_1, x_2, \ldots, x_n\}$ we mean that the set S consists of the elements $x_1, x_2, \ldots, x_n$. A set S may also be specified in the generic form

$$S = \{x \mid x \text{ satisfies } P\}$$

as the set of elements satisfying property P. For example,

$$S = \{x \mid x: \text{ real}, 0 \leq x \leq 1\}$$

denotes the set of all real numbers x satisfying $0 \leq x \leq 1$.

The *union* of two sets S and T is denoted by $S \cup T$ and the *intersection* of S and T is denoted by $S \cap T$. The union and the intersection of a sequence of sets $S_1, S_2, \ldots, S_k, \ldots$ are denoted by $\cup_{k=1}^{\infty} S_k$ and $\cap_{k=1}^{\infty} S_k$, respectively. If S is a subset of T (i.e., if every element of S is also an element of T), we write $S \subset T$ or $T \supset S$.

Finite and Countable Sets

A set S is said to be *finite* if it consists of a finite number of elements. It is said to be *countable* if there exists a one-to-one function from S into the set of nonnegative integers. Thus, according to our definition, a finite set is also countable but not conversely. A countable set S that is not finite may be represented by listing its elements $x_0, x_1, x_2, \ldots$ (i.e., $S = \{x_0, x_1, x_2, \ldots\}$). A countable union of countable sets is countable, that is, if $A = \{a_0, a_1, \ldots\}$ is a countable set and $S_{a_0}, S_{a_1}, \ldots$ are each countable sets, then $\cup_{k=0}^{\infty} S_{a_k}$ is also a countable set.

Sets of Real Numbers

If a and b are real numbers or $+\infty$, $-\infty$, we denote by $[a, b]$ the set of numbers x satisfying $a \leq x \leq b$ (including the possibility $x = +\infty$ or $x = -\infty$). A rounded, instead of square, bracket denotes strict inequality in the definition. Thus $(a, b]$, $[a, b)$, and (a, b) denote the set of all x satisfying $a < x \leq b$, $a \leq x < b$, and $a < x < b$, respectively.

If S is a set of real numbers that is bounded above, then there is a smallest real number y such that $x \leq y$ for all $x \in S$. This number is called the *least upper bound* or *supremum* of S and is denoted by $\sup\{x \mid x \in S\}$ or $\max\{x \mid x \in S\}$. (This is somewhat inconsistent with normal mathematical usage, where the use of max in place of sup indicates that the supremum is attained by some element of S.) Similarly, the greatest real number z such that $z \leq x$ for all $x \in S$ is called the *greatest lower bound* or *infimum* of S and is denoted by $\inf\{x \mid x \in S\}$ or $\min\{x \mid x \in S\}$. If S is unbounded above, we write $\sup\{x \mid x \in S\} = +\infty$, and if it is unbounded below, we write $\inf\{x \mid x \in S\} = -\infty$. If S is the empty set, then by convention we write $\inf\{x \mid x \in S\} = +\infty$ and $\sup\{x \mid x \in S\} = -\infty$.

A.2 EUCLIDEAN SPACE

The set of all n-tuples $x = (x_1, \ldots, x_n)$ of real numbers constitutes the *n-dimensional Euclidean space*, denoted by $\Re^n$. The elements of $\Re^n$ are referred to as n-dimensional vectors or simply vectors when confusion cannot arise. The one-dimensional Euclidean space $\Re^1$ consists of all the real numbers and is denoted by $\Re$. Vectors in $\Re^n$ can be added by adding their corresponding components. They can be multiplied by a scalar by multiplication of each component by a scalar. The *inner product* of two vectors $x = (x_1, \ldots, x_n)$ and $y = (y_1, \ldots, y_n)$ is denoted by $x'y$ and is equal to $\sum_{i=1}^n x_i y_i$. The *norm* of a vector $x = (x_1, \ldots, x_n) \in \Re^n$ is denoted by $\|x\|$ and is equal to $(x'x)^{1/2} = \left(\sum_{i=1}^n x_i^2\right)^{1/2}$.

A set of vectors $a_1, a_2, \ldots, a_k$ is said to be *linearly dependent* if there exist scalars $\lambda_1, \lambda_2, \ldots, \lambda_k$, not all zero, such that $\sum_{i=1}^{k} \lambda_i a_i = 0$. If no such set of scalars exists, the vectors are said to be *linearly independent*.

.3 MATRICES

An $m \times n$ *matrix* is a rectangular array of numbers, referred to as elements, which are arranged in m rows and n columns. If $m = n$ the matrix is said to be *square*. The element in the ith row and jth column of a matrix A is denoted by a subscript ij, such as a_{ij}, in which case we write $A = [a_{ij}]$. The $n \times n$ *identity matrix*, denoted by I, is the matrix with elements $a_{ij} = 0$ for $i \neq j$ and $a_{ii} = 1$, for $i = 1, \ldots, n$. The *sum* of two $m \times n$ matrices A and B is written as $A + B$ and is the matrix whose elements are the sum of the corresponding elements in A and B. The *product of a matrix A and a scalar* λ, written as λA or $A\lambda$, is obtained by multiplying each element of A by λ. The *product AB* of an $m \times n$ matrix A and an $n \times p$ matrix B is the $m \times p$ matrix C with elements $c_{ij} = \sum_{k=1}^{n} a_{ik} b_{kj}$. If b is an n-dimensional column vector and A is an $m \times n$ matrix, then Ab is an m-dimensional column vector.

The *transpose* of an $m \times n$ matrix A is the $n \times m$ matrix A' with elements $a'_{ij} = a_{ji}$. The elements of a given row (or column) of A constitute a vector called a row vector (or column vector, respectively) of A. A square matrix A is *symmetric* if $A' = A$. An $n \times n$ matrix A is called *nonsingular* or *invertible* if there is an $n \times n$ matrix called the *inverse* of A and denoted by A^{-1}, such that $A^{-1}A = I = AA^{-1}$, where I is the $n \times n$ identity matrix. An $n \times n$ matrix is nonsingular if and only if its n row vectors are linearly independent or, equivalently, if its n column vectors are linearly independent. Thus, an $n \times n$ matrix A is nonsingular if and only if the relation $Av = 0$, where $v \in \Re^n$, implies that $v = 0$.

Rank of a Matrix

The *rank* of a matrix A is equal to the maximum number of linearly independent row vectors of A. It is also equal to the maximum number of linearly independent column vectors. Thus, the rank of an $m \times n$ matrix is at most equal to the minimum of the dimensions m and n. An $m \times n$ matrix is said to be of *full rank* if its rank is maximal, that is, if its rank is equal to the minimum of m and n. A square matrix is of full rank if and only if it is nonsingular.

Eigenvalues

Given a square $n \times n$ matrix A, the determinant of the matrix $\gamma I - A$, where I is the $n \times n$ identity matrix and γ is a scalar, is an nth degree polynomial. The n roots of this polynomial are called the *eigenvalues* of A. Thus, γ is an eigenvalue of A if and only if the matrix $\gamma I - A$ is singular, or equivalently, if and only if there exists a nonzero vector v such that $Av = \gamma v$. Such a vector v is called an *eigenvector* corresponding to γ. The eigenvalues and eigenvectors of A can be complex even if A is real. A matrix A is singular if and only if it has an eigenvalue that is equal to zero. If A is nonsingular, then the eigenvalues of A^{-1} are the reciprocals of the eigenvalues of A. The eigenvalues of A and A' coincide.

If $\gamma_1, \ldots, \gamma_n$ are the eigenvalues of A, then the eigenvalues of $cI + A$, where c is a scalar and I is the identity matrix, are $c + \gamma_1, \ldots, c + \gamma_n$. The eigenvalues of A^k, where k is any positive integer, are equal to $\gamma_1^k, \ldots, \gamma_n^k$. From this it follows that $\lim_{k \to 0} A^k = 0$ if and only if all the eigenvalues of A lie strictly within the unit circle of the complex plane. Furthermore, if the latter condition holds, the iteration $x_{k+1} = Ax_k + b$, where b is a given vector, converges to $\overline{x} = (I - A)^{-1} b$, which is the unique solution of the equation $x = Ax + b$.

If all the eigenvalues of A are distinct, then their number is exactly n, and there exists a set of corresponding linearly independent eigenvectors. In this case, if $\gamma_1, \ldots, \gamma_n$ are the eigenvalues and $v_1, \ldots, v_n$ are such eigenvectors, every vector $x \in \Re^n$ can be decomposed as

$$x = \sum_{i=1}^{n} \xi_i v_i,$$

where ξ_i are some unique (possibly complex) numbers. Furthermore, we have for all positive integers k,

$$A^k x = \sum_{i=1}^{n} \gamma_i^k \xi_i v_i.$$

If A is a transition probability matrix, that is, all the elements of A are nonnegative and the sum of the elements of each of its rows is equal to one, then all the eigenvalues of A lie within the unit circle of the complex plane. Furthermore, 1 is an eigenvalue of A and the unit vector $(1, 1, \ldots, 1)$ is a corresponding eigenvector.

Positive Definite and Semidefinite Symmetric Matrices

A square symmetric $n \times n$ matrix A is said to be *positive semidefinite* if $x'Ax \geq 0$ for all $x \in \Re^n$. It is said to be *positive definite* if $x'Ax > 0$ for all nonzero $x \in \Re^n$. The matrix A is said to be *negative semidefinite (definite)* if $-A$ is *positive semidefinite (definite)*. In this book, the notions of positive definiteness and semidefiniteness will be used only in connection with symmetric matrices.

A positive definite symmetric matrix is invertible and its inverse is also positive definite symmetric. Also, an invertible positive semidefinite symmetric matrix is positive definite. Analogous results hold for negative definite and semidefinite symmetric matrices. If A and B are $n \times n$ positive semidefinite (definite) symmetric matrices, then the matrix $\lambda A + \mu B$ is also positive semidefinite (definite) symmetric for all $\lambda > 0$ and $\mu > 0$. If A is an $n \times n$ positive semidefinite symmetric matrix and C is an $m \times n$ matrix, then the matrix CAC' is positive semidefinite symmetric. If A is positive definite symmetric, and C has rank m (equivalently, $m \leq n$ and C has full rank), then CAC' is positive definite symmetric.

An $n \times n$ positive definite symmetric matrix A can be written as CC' where C is a square invertible matrix. If A is positive semidefinite symmetric and its rank is m, then it can be written as CC', where C is an $n \times m$ matrix of full rank.

A symmetric $n \times n$ matrix A has real eigenvalues and a set of n real linearly independent eigenvectors, which are orthogonal (the inner product of any pair is 0). If A is positive semidefinite (definite) symmetric, its eigenvalues are nonnegative (respectively, positive).

Partitioned Matrices

It is often convenient to partition a matrix into submatrices. For example, the matrix

$$A = \begin{pmatrix} a_{11} & a_{12} & a_{13} & a_{14} \\ a_{21} & a_{22} & a_{23} & a_{24} \\ a_{31} & a_{32} & a_{33} & a_{34} \end{pmatrix}$$

may be partitioned into

$$A = \begin{pmatrix} A_{11} & A_{12} \\ A_{21} & A_{22} \end{pmatrix},$$

where

$$A_{11} = (\, a_{11} \quad a_{12} \,), \qquad A_{12} = (\, a_{13} \quad a_{14} \,),$$

$$A_{21} = \begin{pmatrix} a_{21} & a_{22} \\ a_{31} & a_{32} \end{pmatrix}, \qquad A_{22} = \begin{pmatrix} a_{23} & a_{24} \\ a_{33} & a_{34} \end{pmatrix}.$$

We separate the components of a partitioned matrix by a space, as in $(B\ C)$, or by a comma, as in (B, C). The transpose of the partitioned matrix A is

$$A' = \begin{pmatrix} A'_{11} & A'_{21} \\ A'_{12} & A'_{22} \end{pmatrix}.$$

Partitioned matrices may be multiplied just as nonpartitioned matrices, provided the dimensions involved in the partitions are compatible. Thus if

$$A = \begin{pmatrix} A_{11} & A_{12} \\ A_{21} & A_{22} \end{pmatrix}, \qquad B = \begin{pmatrix} B_{11} & B_{12} \\ B_{21} & B_{22} \end{pmatrix},$$

then

$$AB = \begin{pmatrix} A_{11}B_{11} + A_{12}B_{21} & A_{11}B_{12} + A_{12}B_{22} \\ A_{21}B_{11} + A_{22}B_{21} & A_{21}B_{12} + A_{22}B_{22} \end{pmatrix},$$

provided the dimensions of the submatrices are such that the preceding products $A_{ij}B_{jk}$, $i, j, k = 1, 2$ can be formed.

Matrix Inversion Formulas

Let A and B be square invertible matrices, and let C be a matrix of appropriate dimension. Then, if all the following inverses exist, we have

$$(A + CBC')^{-1} = A^{-1} - A^{-1}C(B^{-1} + C'A^{-1}C)^{-1}C'A^{-1}.$$

The equation can be verified by multiplying the right-hand side by $A + CBC'$ and showing that the product is the identity matrix.

Consider a partitioned matrix M of the form

$$M = \begin{pmatrix} A & B \\ C & D \end{pmatrix}.$$

Then we have

$$M^{-1} = \begin{pmatrix} Q & -QBD^{-1} \\ -D^{-1}CQ & D^{-1} + D^{-1}CQBD^{-1} \end{pmatrix},$$

where

$$Q = (A - BD^{-1}C)^{-1},$$

provided all the inverses exist. The proof is obtained by multiplying M with the expression given for M^{-1} and verifying that the product yields the identity matrix.

A.4 ANALYSIS

Convergence of Sequences

A sequence of vectors $x_0, x_1, \ldots, x_k, \ldots$ in $\Re^n$, denoted by $\{x_k\}$, is said to converge to a *limit* x if $\|x_k - x\| \to 0$ as $k \to \infty$ (i.e., if, given any $\epsilon > 0$, there is an integer N such that for all $k \geq N$ we have $\|x_k - x\| < \epsilon$). If $\{x_k\}$ converges to x, we write $x_k \to x$ or $\lim_{k\to\infty} x_k = x$. We have $Ax_k + By_k \to Ax + By$ if $x_k \to x$, $y_k \to y$, and A, B are matrices of appropriate dimension.

A vector x is said to be a *limit point* of a sequence $\{x_k\}$ if there is a subsequence of $\{x_k\}$ that converges to x, that is, if there is an infinite subset $\mathcal{K}$ of the nonnegative integers such that for any $\epsilon > 0$, there is an integer N such that for all $k \in \mathcal{K}$ with $k \geq N$ we have $\|x_k - x\| < \epsilon$.

A sequence of real numbers $\{r_k\}$, which is monotonically nondecreasing (nonincreasing), that is, satisfies $r_k \leq r_{k+1}$ for all k, must either converge to a real number or be unbounded above (below). In the latter case we write $\lim_{k\to\infty} r_k = \infty$ $(-\infty)$. Given any bounded sequence of real numbers $\{r_k\}$, we may consider the sequence $\{s_k\}$, where $s_k = \sup\{r_i \mid i \geq k\}$. Since this sequence is monotonically nonincreasing and bounded, it must have a limit. This limit is called the *limit superior* of $\{r_k\}$ and is denoted by $\limsup_{k\to\infty} r_k$. The *limit inferior* of $\{r_k\}$ is similarly defined and is denoted by $\liminf_{k\to\infty} r_k$. If $\{r_k\}$ is unbounded above, we write $\limsup_{k\to\infty} r_k = \infty$, and if it is unbounded below, we write $\liminf_{k\to\infty} r_k = -\infty$. We also use this notation if $r_k \in [-\infty, \infty]$ for all k.

Open, Closed, and Compact Sets

A subset S of $\Re^n$ is said to be *open* if for every vector $x \in S$ one can find an $\epsilon > 0$ such that $\big\{z \mid \|z - x\| < \epsilon\big\} \subset S$. A set S is *closed* if and only if every convergent sequence $\{x_k\}$ with elements in S converges to a point that also belongs to S. A set S is said to be *compact* if and only if it is both closed and bounded (i.e., it is closed and for some $M > 0$ we have $\|x\| \leq M$ for all $x \in S$). A set S is compact if and only if every sequence $\{x_k\}$ with elements in S has at least one limit point that belongs to S. Another important fact is that if $S_0, S_1, \ldots, S_k, \ldots$ is a sequence of nonempty compact sets in $\Re^n$ such that $S_k \supset S_{k+1}$ for all k, then the intersection $\cap_{k=0}^{\infty} S_k$ is a nonempty and compact set.

Continuous Functions

A function f mapping a set S_1 into a set S_2 is denoted by $f : S_1 \to S_2$. A function $f : \Re^n \to \Re^m$ is said to be *continuous* if for all x, $f(x_k) \to f(x)$ whenever $x_k \to x$. Equivalently, f is continuous if, given $x \in \Re^n$ and $\epsilon > 0$, there is a $\delta > 0$ such that whenever $\|y - x\| < \delta$, we have $\|f(y) - f(x)\| < \epsilon$. The function

$$(a_1 f_1 + a_2 f_2)(\cdot) = a_1 f_1(\cdot) + a_2 f_2(\cdot)$$

is continuous for any two scalars a_1, a_2 and any two continuous functions $f_1, f_2 : \Re^n \to \Re^m$. If S_1, S_2, S_3 are any sets and $f_1 : S_1 \to S_2$, $f_2 : S_2 \to S_3$ are functions, the function $f_2 \cdot f_1 : S_1 \to S_3$ defined by $(f_2 \cdot f_1)(x) = f_2\big(f_1(x)\big)$ is called the *composition* of f_1 and f_2. If $f_1 : \Re^n \to \Re^m$ and $f_2 : \Re^m \to \Re^p$ are continuous, then $f_2 \cdot f_1$ is also continuous.

Derivatives

Let $f : \Re^n \mapsto \Re$ be some function. For a fixed $x \in \Re^n$, the first partial derivative of f at the point x with respect to the ith coordinate is defined by

$$\frac{\partial f(x)}{\partial x_i} = \lim_{\alpha \to 0} \frac{f(x + \alpha e_i) - f(x)}{\alpha},$$

where e_i is the ith unit vector, and we assume that the above limit exists. If the partial derivatives with respect to all coordinates exist, f is called differentiable at x and its *gradient* at x is defined to be the column vector

$$\nabla f(x) = \begin{pmatrix} \frac{\partial f(x)}{\partial x_1} \\ \vdots \\ \frac{\partial f(x)}{\partial x_n} \end{pmatrix}.$$

The function f is called differentiable if it is differentiable at every $x \in \Re^n$. If $\nabla f(x)$ exists for every x and is a continuous function of x, f is said to be *continuously differentiable*. Such a function admits, for every fixed x, the first order expansion

$$f(x + y) = f(x) + y' \nabla f(x) + o(\|y\|),$$

where $o(\|y\|)$ is a function of y with the property $\lim_{\|y\| \to 0} o(\|y\|)/\|y\| = 0$.

A vector-valued function $f : \Re^n \mapsto \Re^m$ is called differentiable (respectively, continuously differentiable) if each component f_i of f is differentiable (respectively, continuously differentiable). The *gradient matrix* of f, denoted by $\nabla f(x)$, is the $n \times m$ matrix whose ith column is the gradient $\nabla f_i(x)$ of f_i. Thus,

$$\nabla f(x) = \Big[\nabla f_1(x) \cdots \nabla f_m(x)\Big].$$

The transpose of ∇f is the *Jacobian* of f; it is the matrix whose ijth entry is equal to the partial derivative $\partial f_i / \partial x_j$.

If the gradient $\nabla f(x)$ is itself a differentiable function, then f is said to be twice differentiable. We denote by $\nabla^2 f(x)$ the Hessian matrix of f at x, that is, the matrix

$$\nabla^2 f(x) = \left[\frac{\partial^2 f(x)}{\partial x^i \partial x^j}\right]$$

the elements of which are the second partial derivatives of f at x.

Let $f : \Re^k \mapsto \Re^m$ and $g : \Re^m \mapsto \Re^n$ be continuously differentiable functions, and let $h(x) = g\big(f(x)\big)$. The *chain rule* for differentiation states that

$$\nabla h(x) = \nabla f(x) \nabla g\big(f(x)\big), \qquad \text{for all } x \in \Re^k.$$

For example, if A and B are given matrices, then if $h(x) = Ax$, we have $\nabla h(x) = A'$ and if $h(x) = ABx$, we have $\nabla h(x) = B'A'$.

A.5 CONVEX SETS AND FUNCTIONS

A subset C of $\Re^n$ is said to be *convex* if for every $x, y \in C$ and every scalar α with $0 \leq \alpha \leq 1$, we have $\alpha x + (1-\alpha)y \in C$. In words, C is convex if the line segment connecting any two points in C belongs to C. A function $f : C \to \Re$ defined over a convex subset C of $\Re^n$ is said to be *convex* if for every $x, y \in C$ and every scalar α with $0 \leq \alpha \leq 1$ we have

$$f\big(\alpha x + (1-\alpha)y\big) \leq \alpha f(x) + (1-\alpha)f(y).$$

The function f is said to be *concave* if $(-f)$ is convex, or equivalently if for every $x, y \in C$ and every scalar α with $0 \leq \alpha \leq 1$ we have

$$f\big(\alpha x + (1-\alpha)y\big) \geq \alpha f(x) + (1-\alpha)f(y).$$

If $f : C \to \Re$ is convex, then the sets $\Gamma_\lambda = \big\{x \mid x \in C,\ f(x) \leq \lambda\big\}$ are also convex for every scalar λ. An important property is that a real-valued convex function on $\Re^n$ is continuous.

If $f_1, f_2, \ldots, f_m$ are convex functions over a convex subset C of $\Re^n$ and $\alpha_1, \alpha_2, \ldots, \alpha_m$ are nonnegative scalars, then the function $\alpha_1 f_1 + \cdots + \alpha_m f_m$ is also convex over C. If $f : \Re^m \to \Re$ is convex, A is an $m \times n$ matrix, and b is a vector in $\Re^m$, the function $g : \Re^n \to \Re$ defined by $g(x) = f(Ax + b)$ is also convex. If $f : \Re^n \to \Re$ is convex, then the function $g(x) = E_w\big\{f(x+w)\big\}$, where w is a random vector in $\Re^n$, is a convex function provided the expected value is finite for every $x \in \Re^n$.

For functions $f : \Re^n \to \Re$ that are differentiable, there are alternative characterizations of convexity. Thus, the function f is convex if and only if

$$f(y) \geq f(x) + \nabla f(x)'(y-x), \qquad \text{for all } x, y \in \Re^n.$$

If f is twice continuously differentiable, then f is convex if and only if $\nabla^2 f(x)$ is a positive semidefinite symmetric matrix for every $x \in \Re^n$.

APPENDIX B:
On Optimization Theory

The purpose of this appendix is to provide a few definitions and results of deterministic optimization. For detailed expositions, see references such as [Lue84] and [Ber95].

B.1 OPTIMAL SOLUTIONS

Given a set S, a real-valued function $f : S \to \Re$, and a subset $X \subset S$, the optimization problem

$$\begin{aligned} &\text{minimize} \quad f(x) \\ &\text{subject to} \quad x \in X, \end{aligned} \tag{B.1}$$

is to find an element $x^* \in X$ (called a *minimizing element* or an *optimal solution*) such that

$$f(x^*) \le f(x), \qquad \text{for all } x \in X.$$

For any minimizing element x^*, we write

$$x^* = \arg\min_{x \in X} f(x).$$

Note that a minimizing element need not exist. For example, the scalar functions $f(x) = x$ and $f(x) = e^x$ have no minimizing elements over the set of real numbers. The first function decreases without bound to $-\infty$ as x tends toward $-\infty$, while the second decreases toward 0 as x tends toward $-\infty$ but always takes positive values. Given the range of values that $f(x)$ takes as x ranges over X, that is, the set of real numbers

$$\{f(x) \mid x \in X\},$$

there are two possibilities:

1. The set $\{f(x) \mid x \in X\}$ is unbounded below (i.e., contains arbitrarily small real numbers) in which case we write

$$\min\{f(x) \mid x \in X\} = -\infty \quad \text{or} \quad \min_{x \in X} f(x) = -\infty.$$

2. The set $\{f(x) \mid x \in X\}$ is bounded below; that is, there exists a scalar M such that $M \leq f(x)$ for all $x \in X$. The greatest lower bound of $\{f(x) \mid x \in X\}$ is also denoted by

$$\min\{f(x) \mid x \in X\} \quad \text{or} \quad \min_{x \in X} f(x).$$

In either case we call $\min_{x \in X} f(x)$ the *optimal value* of problem (B.1).

A maximization problem of the form

$$\begin{aligned} &\text{maximize} \;\; f(x) \\ &\text{subject to} \;\; x \in X \end{aligned}$$

may be converted to the minimization problem

$$\begin{aligned} &\text{minimize} \;\; -f(x) \\ &\text{subject to} \;\; x \in X, \end{aligned}$$

in the sense that both problems have the same optimal solutions, and the optimal value of one is equal to minus the optimal value of the other. The optimal value for the maximization problem is denoted by $\max_{x \in X} f(x)$.

Existence of at least one optimal solution in problem (B.1) is guaranteed if $f : \Re^n \to \Re$ is a continuous function and X is a compact subset of $\Re^n$. This is the *Weierstrass theorem*. By a related result, existence of an optimal solution is guaranteed if $f : \Re^n \to \Re$ is a continuous function, X is closed, and $f(x) \to \infty$ if $\|x\| \to \infty$.

B.2 OPTIMALITY CONDITIONS

Optimality conditions are available when f is a differentiable function on $\Re^n$ and X is a convex subset of $\Re^n$ (possibly $X = \Re^n$). In particular, if x^* is an optimal solution of problem (B.1), $f : \Re^n \to \Re$ is a continuously differentiable function on $\Re^n$, and X is convex, we have

$$\nabla f(x^*)'(x - x^*) \geq 0, \qquad \text{for all } x \in X, \tag{B.2}$$

where $\nabla f(x^*)$ denotes the gradient of f at x^*. When $X = \Re^n$ (i.e., the minimization is unconstrained), the necessary condition (B.2) is equivalent to

$$\nabla f(x^*) = 0. \tag{B.3}$$

When f is twice continuously differentiable and $X = \Re^n$, an additional necessary condition is that the *Hessian matrix* $\nabla^2 f(x^*)$ *be positive semidefinite at* x^*. An important fact is that *if* $f : \Re^n \to \Re$ *is a convex function and* X *is convex, then Eq. (B.2) is both a necessary and a sufficient condition for optimality of* x^*.

B.3 MINIMIZATION OF QUADRATIC FORMS

Let $f : \Re^n \to \Re$ be a quadratic form

$$f(x) = \tfrac{1}{2}x'Qx + b'x,$$

where Q is a symmetric $n \times n$ matrix and $b \in \Re^n$. Its gradient is given by

$$\nabla f(x) = Qx + b.$$

The function f is convex if and only if Q is positive semidefinite. If Q is positive definite, then f is convex and Q is invertible, so by Eq. (B.3), a vector x^* minimizes f if and only if

$$\nabla f(x^*) = Qx^* + b = 0,$$

or equivalently

$$x^* = -Q^{-1}b.$$

APPENDIX C:
On Probability Theory

This appendix lists selectively some of the basic probabilistic notions that we will be using. Its main purpose is to familiarize the reader with some of our terminology. It is not meant to be exhaustive, and the reader should consult references such as [Ash70], [Fel68], [Pap65], and [Ros85] for detailed treatments. For fairly accessible treatments of measure theoretic probability, see [AdG86] and [Ash72].

C.1 PROBABILITY SPACES

A *probability space* consists of

(a) A set Ω.

(b) A collection $\mathcal{F}$ of subsets of Ω, called *events*, which includes Ω and has the following properties:

   (1) If A is an event, then the complement $\overline{A} = \{\omega \in \Omega \mid \omega \notin A\}$ is also an event. (The complement of Ω is the empty set and is considered to be an event.)

   (2) If A_1 and A_2 are events, then $A_1 \cap A_2$ and $A_1 \cup A_2$ are also events.

(3) If $A_1, A_2, \ldots, A_k, \ldots$ are events, then $\cup_{k=1}^{\infty} A_k$ and $\cap_{k=1}^{\infty} A_k$ are also events.

(c) A function $P(\cdot)$ assigning to each event A a real number $P(A)$, called the *probability of the event* A, and satisfying:

(1) $P(A) \geq 0$ for every event A.

(2) $P(\Omega) = 1$.

(3) $P(A_1 \cup A_2) = P(A_1) + P(A_2)$ for every pair of disjoint events A_1, A_2.

(4) $P(\cup_{k=1}^{\infty} A_k) = \sum_{k=1}^{\infty} P(A_k)$ for every sequence of mutually disjoint events $A_1, A_2, \ldots, A_k, \ldots$

The function P is referred to as a *probability measure.*

Convention for Finite and Countable Probability Spaces

The case of a probability space where the set Ω is a countable (possibly finite) set is encountered frequently in this book. When we specify that Ω is finite or countable, we implicitly assume that the associated collection of events is the collection of *all* subsets of Ω (including Ω and the empty set). Then, if Ω is a finite set, $\Omega = \{\omega_1, \omega_2, \ldots, \omega_n\}$, the probability space is specified by the probabilities $p_1, p_2, \ldots, p_n$, where p_i denotes the probability of the event consisting of ω_i. Similarly, if $\Omega = \{\omega_1, \omega_2, \ldots, \omega_k, \ldots\}$, the probability space is specified by the corresponding probabilities $p_1, p_2, \ldots, p_k, \ldots$ In either case we refer to $(p_1, p_2, \ldots, p_n)$ or $(p_1, p_2, \ldots, p_k, \ldots)$ as a *probability distribution over* Ω.

C.2 RANDOM VARIABLES

A *random variable* on a probability space$(\Omega, \mathcal{F}, P)$ is a function $x : \Omega \to \Re$ such that for every scalar λ the set

$$\{\omega \in \Omega \mid x(\omega) \leq \lambda\}$$

is an event (i.e., belongs to the collection $\mathcal{F}$). An n-dimensional *random vector* $x = (x_1, x_2, \ldots, x_n)$ is an n-tuple of random variables $x_1, x_2, \ldots, x_n$, each defined on the same probability space.

We define the *distribution function* $F : \Re \to \Re$ of a random variable x by

$$F(z) = P\big(\{\omega \in \Omega \mid x(\omega) \leq z\}\big);$$

that is, $F(z)$ is the probability that the random variable takes a value less than or equal to z. We define the distribution function $F : \Re^n \to \Re$ of a random vector $x = (x_1, x_2, \ldots, x_n)$ by

$$F(z_1, z_2, \ldots, z_n) = P\big(\{\omega \in \Omega \mid x_1(\omega) \le z_1, x_2(\omega) \le z_2, \ldots, x_n(\omega) \le z_n\}\big).$$

Given the distribution function of a random vector $x = (x_1, \ldots, x_n)$, the (marginal) distribution function of each random variable x_i is obtained from

$$F_i(z_i) = \lim_{z_j \to \infty,\, j \neq i} F(z_1, z_2, \ldots, z_n).$$

The random variables $x_1, \ldots, x_n$ are said to be *independent* if

$$F(z_1, z_2, \ldots, z_n) = F_1(z_1) F_2(z_2) \cdots F_n(z_n),$$

for all scalars $z_1, \ldots, z_n$.

The *expected value of a random variable* x with distribution function F is defined by

$$E\{x\} = \int_{-\infty}^{\infty} z dF(z)$$

provided the integral is well defined. The *expected value of a random vector* $x = (x_1, \ldots, x_n)$ is the vector

$$E\{x\} = \big(E\{x_1\}, E\{x_2\}, \ldots, E\{x_n\}\big).$$

The *covariance matrix* of a random vector $x = (x_1, \ldots, x_n)$ with expected value $E\{x\} = (\overline{x}_1, \ldots, \overline{x}_n)$ is defined to be the $n \times n$ positive semidefinite symmetric matrix

$$\begin{pmatrix} E\{(x_1 - \overline{x}_1)^2\} & \cdots & E\{(x_1 - \overline{x}_1)(x_n - \overline{x}_n)\} \\ \vdots & \vdots & \vdots \\ E\{(x_n - \overline{x}_n)(x_1 - \overline{x}_1)\} & \cdots & E\{(x_n - \overline{x}_n)^2\} \end{pmatrix},$$

provided the expectations are well defined.

Two random vectors x and y are said to be *uncorrelated* if

$$E\{\big(x - E\{x\}\big)\big(y - E\{y\}\big)'\} = 0,$$

where $\big(x - E\{x\}\big)$ is viewed as a column vector and $\big(y - E\{y\}\big)'$ is viewed as a row vector.

The random vector $x = (x_1, \ldots, x_n)$ is said to be characterized by a *probability density function* $f : \Re^n \to \Re$ if

$$F(z_1, z_2, \ldots, z_n) = \int_{-\infty}^{z_1} \int_{-\infty}^{z_2} \cdots \int_{-\infty}^{z_n} f(y_1, \ldots, y_n) dy_1 \ldots dy_n,$$

for every $z_1, \ldots, z_n$.

C.3 CONDITIONAL PROBABILITY

We shall restrict ourselves to the case where the underlying probability space Ω is a countable (possibly finite) set and the set of events is the set of all subsets of Ω.

Given two events A and B, we define the *conditional probability of B given A* by

$$P(B \mid A) = \begin{cases} \frac{P(A \cap B)}{P(A)} & \text{if } P(A) > 0, \\ 0 & \text{if } P(A) = 0. \end{cases}$$

We also use the notation $P\{B \mid A\}$ in place of $P(B \mid A)$. If $B_1, B_2, \ldots$ are a countable (possibly finite) collection of mutually exclusive and exhaustive events (i.e., the sets B_i are disjoint and their union is Ω) and A is an event, then we have

$$P(A) = \sum_i P(A \cap B_i).$$

From the two preceding relations it is seen that

$$P(A) = \sum_i P(B_i) P(A \mid B_i).$$

We thus obtain for every k,

$$P(B_k \mid A) = \frac{P(A \cap B_k)}{P(A)} = \frac{P(B_k) P(A \mid B_k)}{\sum_i P(B_i) P(A \mid B_i)},$$

assuming that $P(A) > 0$. This relation is referred to as *Bayes' rule.*

Consider now two random vectors x and y taking values in $\Re^n$ and $\Re^m$, respectively [i.e., $x(\omega) \in \Re^n$, $y(\omega) \in \Re^m$ for all $\omega \in \Omega$]. Given two subsets X and Y of $\Re^n$ and $\Re^m$, respectively, we denote

$$P(X \mid Y) = P\big(\{\omega \mid x(\omega) \in X\} \mid \{\omega \mid y(\omega) \in Y\}\big).$$

For a fixed vector $v \in \Re^n$, we define the *conditional distribution function* of x given v by

$$F(z \mid v) = P\big(\{\omega \mid x(\omega) \leq z\} \mid \{\omega \mid y(\omega) = v\}\big),$$

and the *conditional expectation* of x given v by

$$E\{x \mid v\} = \int_{\Re^n} z dF(z \mid v),$$

assuming that the integral is well defined. Note that $E\{x \mid v\}$ is a function mapping v into $\Re^n$.

Finally, let us derive Bayes' rule for random vectors. If $\omega_1, \omega_2, \ldots$ are the elements of Ω, denote

$$z_i = x(\omega_i), \qquad v_i = y(\omega_i), \qquad i = 1, 2, \ldots$$

Also, for any vectors $z \in \Re^n$, $v \in \Re^m$, let us denote

$$P(z) = P\big(\{\omega \mid x(\omega) = z\}\big), \qquad P(v) = P\big(\{\omega \mid y(\omega) = v\}\big).$$

We have $P(z) = 0$ if $z \neq z_i$, $i = 1, 2, \ldots$, and $P(v) = 0$ if $v \neq v_i$, $i = 1, 2, \ldots$ Denote also

$$P(z \mid v) = P\big(\{\omega \mid x(\omega) = z\} \mid \{\omega \mid y(\omega) = v\}\big),$$

$$P(v \mid z) = P\big(\{\omega \mid y(\omega) = v\} \mid \{\omega \mid x(\omega) = z\}\big).$$

Then, for all $k = 1, 2, \ldots$, Bayes' rule yields

$$P(z_k \mid v) = \begin{cases} \frac{P(z_k)P(v|z_k)}{\sum_i P(z_i)P(v|z_i)} & \text{if } P(v) > 0, \\ 0 & \text{if } P(v) = 0. \end{cases}$$

APPENDIX D: On Finite-State Markov Chains

This appendix provides some of the basic probabilistic notions related to stationary Markov chains with a finite number of states. For detailed presentations, see [Ash70], [Chu60], [KeS60], and [Ros85].

D.1 STATIONARY MARKOV CHAINS

A square $n \times n$ matrix $[p_{ij}]$ is said to be a *stochastic* matrix if all its elements are nonnegative, that is, $p_{ij} \geq 0$, $i, j = 1, \ldots, n$, and the sum of the elements of each of its rows is equal to 1, that is, $\sum_{j=1}^{n} p_{ij} = 1$ for all $i = 1, \ldots, n$.

Suppose we are given a stochastic $n \times n$ matrix P together with a finite set of states $S = \{1, \ldots, n\}$. The pair (S, P) will be referred to as a *stationary finite-state Markov chain.* We associate with (S, P) a process whereby an initial state $x_0 \in S$ is chosen in accordance with some initial probability distribution

$$r_0 = (r_0^1, r_0^2, \ldots, r_0^n).$$

Subsequently, a transition is made from state x_0 to a new state $x_1 \in S$ in accordance with a probability distribution specified by P as follows. The probability that the new state will be j is equal to p_{ij} whenever the initial state is i, that is,

$$P(x_1 = j \mid x_0 = i) = p_{ij}, \qquad i, j = 1, \ldots, n.$$

Similarly, subsequent transitions produce states $x_2, x_3, \ldots$ in accordance with

$$P(x_{k+1} = j \mid x_k = i) = p_{ij}, \qquad i, j = 1, \ldots, n. \tag{D.1}$$

The probability that after the kth transition the state x_k will be j, given that the initial state x_0 is i, is denoted by

$$p_{ij}^k = P(x_k = j \mid x_0 = i), \qquad i, j = 1, \ldots, n. \tag{D.2}$$

A straightforward calculation shows that these probabilities are equal to the elements of the matrix P^k (P raised to the kth power), in the sense that p_{ij}^k is the element in the ith row and jth column of P^k:

$$P^k = [p_{ij}^k]. \tag{D.3}$$

Given the initial probability distribution p_0 of the state x_0 (viewed as a row vector in $\Re^n$), the probability distribution of the state x_k after k transitions

$$r_k = (r_k^1, r_k^2, \ldots, r_k^n)$$

(viewed again as a row vector) is given by

$$r_k = r_0 P^k, \qquad k = 1, 2, \ldots \tag{D.4}$$

This relation follows from Eqs. (D.2) and (D.3) once we write

$$r_k^j = \sum_{i=1}^n P(x_k = j \mid x_0 = i) r_0^i = \sum_{i=1}^n p_{ij}^k r_0^i.$$

D.2 CLASSIFICATION OF STATES

Given a stationary finite-state Markov chain (S, P), we say that two states i and j *communicate* if there exist two positive integers k_1 and k_2 such that $p_{ij}^{k_1} > 0$ and $p_{ji}^{k_2} > 0$. In words, states i and j communicate if one can be reached from the other with positive probability.

Let $\tilde{S} \subset S$ be a subset of states such that:

1. All states in $\tilde{S}$ communicate.

2. If $i \in \tilde{S}$ and $j \notin \tilde{S}$, then $p_{ij}^k = 0$ for all k.

Then we say that $\tilde{S}$ forms a *recurrent class* of states.

If S forms by itself a recurrent class (i.e., all states communicate with each other), then we say that the Markov chain is *irreducible*. It is possible that there exist several recurrent classes. It is also possible to prove that at least one recurrent class must exist. A state that belongs to some recurrent class is called *recurrent*; otherwise it is called *transient*. We have

$$\lim_{k\to\infty} p_{ii}^k = 0 \quad \text{if and only if} \quad i \text{ is transient.}$$

In other words, if the process starts at a transient state, the probability of returning to the same state after k transitions diminishes to zero as k tends to infinity.

The definitions imply that if the process starts within a recurrent class, it stays within that class. If it starts at a transient state, it eventually (with probability one) enters a recurrent class after a number of transitions and subsequently remains there.

D.3 LIMITING PROBABILITIES

An important property of any stochastic matrix P is that the matrix P^* defined by

$$P^* = \lim_{N\to\infty} \frac{1}{N} \sum_{k=0}^{N-1} P^k \tag{D.5}$$

exists [in the sense that the sequences of the elements of $(1/N)\sum_{k=0}^{N-1} P^k$ converge to the corresponding elements of P^*]. The elements p_{ij}^* of P^* satisfy

$$p_{ij}^* \geq 0, \qquad \sum_{j=1}^{n} p_{ij}^* = 1, \qquad i,j = 1,\ldots,n.$$

Thus, P^* is a stochastic matrix. A proof of this is given in Prop. A.1 of Appendix A in Vol. II.

Note that the (i,j)th element of the matrix P^k is the probability that the state will be j after k transitions starting from state i. With this in mind, it can be seen from the definition (D.5) that p_{ij}^* can be interpreted as the long term fraction of time that the state is j given that the initial state is i. This suggests that for any two states i and i' in the same recurrent class we have $p_{ij}^* = p_{i'j}^*$, and this can indeed be proved. In particular, if a Markov chain is irreducible, the matrix P^* has identical rows. Also, if j is a transient state, we have

$$p_{ij}^* = 0, \qquad \text{for all } i = 1,\ldots,n,$$

so the columns of the matrix P^* corresponding to transient states consist of zeroes.

D.4 FIRST PASSAGE TIMES

Let us denote by q_{ij}^k the probability that the state will be j for the first time after exactly $k \geq 1$ transitions given that the initial state is i, that is,

$$q_{ij}^k = P(x_k = j,\, x_m \neq j,\, 1 \leq m < k \mid x_0 = i).$$

Denote also, for fixed i and j,

$$K_{ij} = \min\{k \geq 1 \mid x_k = j,\, x_0 = i\}.$$

Then K_{ij}, called the *first passage time from* i *to* j, may be viewed as a random variable. We have, for every $k = 1, 2, \ldots,$

$$P(K_{ij} = k) = q_{ij}^k,$$

and we write

$$P(K_{ij} = \infty) = P(x_k \neq j, k = 1, 2, \ldots \mid x_0 = i) = 1 - \sum_{k=1}^{\infty} q_{ij}^k.$$

Note that it is possible that $\sum_{k=1}^{\infty} q_{ij}^k < 1$. This will occur, for example, if j cannot be reached from i, in which case $q_{ij}^k = 0$ for all $k = 1, 2, \ldots$ The *mean first passage time* from i to j is the expected value of K_{ij}:

$$E\{K_{ij}\} = \begin{cases} \sum_{k=1}^{\infty} k q_{ij}^k & \text{if } \sum_{k=1}^{k} q_{ij}^k = 1, \\ \infty & \text{if } \sum_{k=1}^{\infty} q_{ij}^k < 1. \end{cases}$$

It may be proved that if i and j belong to the same recurrent class then

$$E\{K_{ij}\} < \infty.$$

In fact if there is only one recurrent class and t is a state of that class, the mean first passage times $E\{K_{it}\}$ are the unique solution of the following linear system of equations

$$E\{K_{it}\} = 1 + \sum_{j=1,\, j \neq t}^{n} p_{ij} E\{K_{jt}\}, \qquad i = 1, \ldots, n,\, i \neq t;$$

see Example 2.1 in Section 7.2. If i and j belong to two different recurrent classes, then $E\{K_{ij}\} = E\{K_{ji}\} = \infty$. If i belongs to a recurrent class and j is transient, we have $E\{K_{ij}\} = \infty$.

APPENDIX E: Kalman Filtering

In this appendix we present the basic principles of least-squares estimation and their application in estimating the state of a linear discrete-time dynamic system using measurements that are linear in the state variables.

Fundamentally, the problem is the following. There are two random vectors x and y, which are related through their joint probability distribution so that the value of one provides information about the value of the other. We get to know the value of y, and we want to estimate the value of x so that the average squared error between x and its estimate is minimized. A related problem is to find the best estimate of x within the class of all estimates that are *linear* in the measured vector y. We will specialize these problems to a case where there is an underlying linear dynamic system. In particular, we will estimate the state of the system using measurements that are obtained sequentially in time. By exploiting the special structure of the problem, the computation of the state estimate can be organized conveniently in a recursive algorithm – the Kalman filter.

E.1 LEAST-SQUARES ESTIMATION

Consider two jointly distributed random vectors x and y taking values in $\Re^n$ and $\Re^m$, respectively. We view y as a measurement that provides some information about x. Thus, while prior to knowing y our estimate of

x may have been the expected value $E\{x\}$, once the value of y is known, we want to form an updated estimate $x(y)$ of the value x. This updated estimate depends, of course, on the value of y, so we are interested in a rule that gives us the estimate for each possible value of y, i.e., we are interested in a function $x(\cdot)$, where $x(y)$ is the estimate of x given y. Such a function $x(\cdot) : \Re^m \to \Re^m$ is called an *estimator*. We are seeking an estimator that is optimal in some sense and the criterion we shall employ is based on minimization of

$$\mathop{E}_{x,y}\big\{\|x - x(y)\|^2\big\}. \tag{E.1}$$

Here, $\|\cdot\|$ denotes the usual norm in $\Re^n$ ($\|z\|^2 = z'z$ for $z \in \Re^n$). Furthermore, throughout the appendix, we assume that all encountered expected values are finite.

An estimator that minimizes the expected squared error above over all $x(\cdot) : \Re^n \to \Re^m$ is called a *least-squares estimator* and is denoted by $x^*(\cdot)$. Since

$$\mathop{E}_{x,y}\big\{\|x - x(y)\|^2\big\} = \mathop{E}_{y}\big\{\mathop{E}_{x}\big\{\|x - x(y)\|^2 \mid y\big\}\big\},$$

it is clear that $x^*(\cdot)$ is a least-squares estimator if $x^*(y)$ minimizes the conditional expectation in the right-hand side above for every $y \in \Re^m$, that is,

$$\mathop{E}_{x}\big\{\|x - x^*(y)\|^2 \mid y\big\} = \min_{z \in \Re^m} \mathop{E}_{x}\big\{\|x - z\|^2 \mid y\big\}, \qquad \text{for all } y \in \Re^m. \tag{E.2}$$

By carrying out this minimization, we obtain the following proposition.

Proposition E.1: The least-squares estimator $x^*(\cdot)$ is given by

$$x^*(y) = \mathop{E}_{x}\{x \mid y\}, \qquad \text{for all } y \in \Re^m. \tag{E.3}$$

Proof: We have for every fixed $z \in \Re^n$

$$\mathop{E}_{x}\big\{\|x - z\|^2 \mid y\big\} = \mathop{E}_{x}\big\{\|x\|^2 \mid y\big\} - 2z' \mathop{E}_{x}\{x \mid y\} + \|z\|^2.$$

By setting to zero the derivative with respect to z, we see that the above expression is minimized by $z = E_x\{x \mid y\}$, and the result follows. **Q.E.D.**

E.2 LINEAR LEAST-SQUARES ESTIMATION

The least-squares estimator $E_x\{x \mid y\}$ may be a complicated nonlinear function of y. As a result its practical calculation may be difficult. This motivates finding optimal estimators within the restricted class of *linear* estimators, i.e., estimators of the form

$$x(y) = Ay + b, \tag{E.4}$$

where A is an $n \times m$ matrix and b is an n-dimensional vector. An estimator

$$\hat{x}(y) = \hat{A}y + \hat{b}$$

where $\hat{A}$ and $\hat{b}$ minimize

$$\mathop{E}_{x,y}\left\{\|x - Ay - b\|^2\right\}$$

over all $n \times m$ matrices A and vectors $b \in \Re^n$ is called a *linear least-squares estimator*.

In the special case where x and y are jointly Gaussian random vectors it turns out that the conditional expectation $E_x\{x \mid y\}$ is a linear function of y (plus a constant vector), and as a result, a linear least-squares estimator is also a least-squares estimator. This is shown in the next proposition.

Proposition E.2: If x, y are jointly Gaussian random vectors, then the least-squares estimate $E_x\{x \mid y\}$ of x given y is linear in y.

Proof: Consider the random vector $z \in \Re^{n+m}$

$$z = \begin{pmatrix} x \\ y \end{pmatrix}$$

and assume that z is Gaussian with mean

$$\overline{z} = E\{z\} = \begin{pmatrix} E\{x\} \\ E\{y\} \end{pmatrix} = \begin{pmatrix} \overline{x} \\ \overline{y} \end{pmatrix} \tag{E.5}$$

and covariance matrix

$$\begin{aligned} \Sigma = E\big\{(z - \overline{z})(z - \overline{z})'\big\} &= \begin{pmatrix} E\big\{(x - \overline{x})(x - \overline{x})'\big\} & E\big\{(x - \overline{x})(y - \overline{y})'\big\} \\ E\big\{(y - \overline{y})(x - \overline{x})'\big\} & E\big\{(y - \overline{y})(y - \overline{y})'\big\} \end{pmatrix} \\ &= \begin{pmatrix} \Sigma_{xx} & \Sigma_{xy} \\ \Sigma_{yx} & \Sigma_{yy} \end{pmatrix}. \end{aligned} \tag{E.6}$$

To simplify our proof we assume that Σ is a positive definite symmetric matrix so that it possesses an inverse; the result, however, holds without this assumption. Since z is Gaussian, its probability density function is of the form

$$p(z) = p(x, y) = ce^{-\frac{1}{2}(z-\overline{z})'\Sigma^{-1}(z-\overline{z})},$$

where

$$c = (2\pi)^{-(n+m)/2}(\det \Sigma)^{-1/2}$$

and $\det \Sigma$ denotes the determinant of Σ. Similarly the probability density functions of x and y are of the form

$$p(x) = c_1 e^{-\frac{1}{2}(x-\overline{x})'\Sigma_{xx}^{-1}(x-\overline{x})},$$

$$p(y) = c_2 e^{-\frac{1}{2}(y-\overline{y})'\Sigma_{yy}^{-1}(y-\overline{y})},$$

where c_1 and c_2 are appropriate constants. By Bayes' rule the conditional probability density function of x given y is

$$p(x \mid y) = \frac{p(x,y)}{p(y)} = \frac{c}{c_2} e^{-\frac{1}{2}\left((z-\overline{z})'\Sigma^{-1}(z-\overline{z})-(y-\overline{y})'\Sigma_{yy}^{-1}(y-\overline{y})\right)}. \tag{E.7}$$

It can now be seen that there exist a positive definite symmetric $n \times n$ matrix D, an $n \times m$ matrix A, a vector $b \in \Re^n$, and a scalar s such that

$$(z-\overline{z})'\Sigma^{-1}(z-\overline{z}) - (y-\overline{y})'\Sigma_{yy}^{-1}(y-\overline{y}) = (x - Ay - b)'D^{-1}(x - Ay - b) + s. \tag{E.8}$$

This is because by substitution of the expressions for $\overline{z}$ and Σ of Eqs. (E.5) and (E.6), the left-hand side of Eq. (E.8) becomes a quadratic form in x and y, which can be put in the form indicated in the right-hand side of Eq. (E.8). In fact, by computing the inverse of Σ using the partitioned matrix inversion formula (Appendix A) it can be verified that A, b, D, and s in Eq. (E.8) have the form

$$A = \Sigma_{xy}\Sigma_{yy}^{-1}, \qquad b = \overline{x} - \Sigma_{xy}\Sigma_{yy}^{-1}\overline{y}, \qquad D = \Sigma_{xx} - \Sigma_{xy}\Sigma_{yy}^{-1}\Sigma_{yx}, \qquad s = 0.$$

Now it follows from Eqs. (E.8) and (E.7) that the conditional expectation $E_x\{x \mid y\}$ is of the form $Ay + b$, where A is some $n \times m$ matrix and $b \in \Re^n$. **Q.E.D.**

We now turn to the characterization of the linear least-squares estimator.

Proposition E.3: Let x, y be random vectors taking values in $\Re^n$ and $\Re^m$, respectively, with given joint probability distribution. The expected values and covariance matrices of x, y are denoted by

$$E\{x\} = \overline{x} \qquad E\{y\} = \overline{y}, \tag{E.9}$$

$$E\{(x-\overline{x})(x-\overline{x})'\} = \Sigma_{xx}, \qquad E\{(y-\overline{y})(y-\overline{y})'\} = \Sigma_{yy}, \tag{E.10}$$

$$E\{(x-\overline{x})(y-\overline{y})'\} = \Sigma_{xy}, \qquad E\{(y-\overline{y})(x-\overline{x})'\} = \Sigma_{xy}', \tag{E.11}$$

and we assume that Σ_{yy} is invertible. Then the linear least-squares estimator of x given y is

$$\hat{x}(y) = \overline{x} + \Sigma_{xy}\Sigma_{yy}^{-1}(y-\overline{y}). \tag{E.12}$$

The corresponding error covariance matrix is given by

$$\underset{x,y}{E}\left\{\big(x-\hat{x}(y)\big)\big(x-\hat{x}(y)\big)'\right\} = \Sigma_{xx} - \Sigma_{xy}\Sigma_{yy}^{-1}\Sigma_{yx}. \tag{E.13}$$

Proof: The linear least-squares estimator is defined as

$$\hat{x}(y) = \hat{A}y + \hat{b},$$

where $\hat{A}, \hat{b}$ minimize the function $f(A,b) = E_{x,y}\{\|x - Ay - b\|^2\}$ over A and b. Taking the derivatives of $f(A,b)$ with respect to A and b and setting them to zero, we obtain the two conditions

$$0 = \left.\frac{\partial f}{\partial A}\right|_{\hat{A},\hat{b}} = 2\,\underset{x,y}{E}\left\{(\hat{b} + \hat{A}y - x)y'\right\}, \tag{E.14}$$

$$0 = \left.\frac{\partial f}{\partial b}\right|_{\hat{A},\hat{b}} = 2\,\underset{x,y}{E}\{\hat{b} + \hat{A}y - x\}. \tag{E.15}$$

The second condition yields

$$\hat{b} = \overline{x} - \hat{A}\overline{y}, \tag{E.16}$$

and by substitution in the first, we obtain

$$\underset{x,y}{E}\left\{y\big(\hat{A}(y-\overline{y}) - (x-\overline{x})\big)'\right\} = 0. \tag{E.17}$$

We have

$$\underset{x,y}{E}\left\{\hat{A}(y-\overline{y}) - (x-\overline{x})\right\}' = 0,$$

so that

$$\overline{y} \underset{x,y}{E}\{\hat{A}(y-\overline{y}) - (x-\overline{x})\}' = 0. \tag{E.18}$$

By subtracting Eq. (E.18) from Eq. (E.17), we obtain

$$\underset{x,y}{E}\{(y-\overline{y})\big(\hat{A}(y-\overline{y}) - (x-\overline{x})\big)'\} = 0.$$

Equivalently,

$$\Sigma_{yy}\hat{A}' - \Sigma_{yx} = 0,$$

from which

$$\hat{A} = \Sigma_{yx}'\Sigma_{yy}^{-1} = \Sigma_{xy}\Sigma_{yy}^{-1}. \tag{E.19}$$

Using the expressions (E.16) and (E.19) for $\hat{b}$ and $\hat{A}$, respectively, we obtain

$$\hat{x}(y) = \hat{A}y + \hat{b} = \overline{x} + \Sigma_{xy}\Sigma_{yy}^{-1}(y-\overline{y}),$$

which was to be proved. The desired Eq. (E.13) for the error covariance follows upon substitution of the expression for $\hat{x}(y)$ obtained above. **Q.E.D.**

We list some of the properties of the least-squares estimator as corollaries.

Corollary E.3.1: The linear least-squares estimator is unbiased, i.e.,

$$\underset{y}{E}\{\hat{x}(y)\} = \overline{x}.$$

Proof: This follows from Eq. (E.12). **Q.E.D.**

Corollary E.3.2: The estimation error $x - \hat{x}(y)$ is uncorrelated with both y and $\hat{x}(y)$, i.e.,

$$\underset{x,y}{E}\{y\big(x-\hat{x}(y)\big)'\} = 0, \qquad \underset{x,y}{E}\{\hat{x}(y)\big(x-\hat{x}(y)\big)'\} = 0.$$

Proof: The first equality is Eq. (E.14). The second equality can be written as $E_{x,y}\{(\hat{A}y+\hat{b})\big(x-\hat{x}(y)\big)'\} = 0$ and follows from the first equality and Cor. E.3.1. **Q.E.D.**

Corollary E.3.2 is known as the *orthogonal projection principle*. It states a property that characterizes the linear least-squares estimate and forms the basis for an alternative treatment of least-squares estimation as a problem of projection in a Hilbert space of random variables (see [Lue69]).

Corollary E.3.3: Consider in addition to x and y, the random vector z defined by

$$z = Cx,$$

where C is a given $p \times m$ matrix. Then the linear least-squares estimate of z given y is

$$\hat{z}(y) = C\hat{x}(y),$$

and the corresponding error covariance matrix is given by

$$\mathop{E}_{z,y}\big\{\big(z - \hat{z}(y)\big)\big(z - \hat{z}(y)\big)'\big\} = C \mathop{E}_{x,y}\big\{\big(x - \hat{x}(y)\big)\big(x - \hat{x}(y)\big)'\big\}C'.$$

Proof: We have $E\{z\} = \overline{z} = C\overline{x}$ and

$$\Sigma_{zz} = \mathop{E}_{z}\big\{(z - \overline{z})(z - \overline{z})'\big\} = C\Sigma_{xx}C',$$

$$\Sigma_{zy} = \mathop{E}_{z,y}\big\{(z - \overline{z})(y - \overline{y})'\big\} = C\Sigma_{xy},$$

$$\Sigma_{yz} = \Sigma'_{zy} = \Sigma_{yx}C'.$$

By Prop. E.3 we have

$$\hat{z}(y) = \overline{z} + \Sigma_{zy}\Sigma_{yy}^{-1}(y - \overline{y}) = C\overline{x} + C\Sigma_{xy}\Sigma_{yy}^{-1}(y - \overline{y}) = C\hat{x}(y),$$

$$\begin{aligned}\mathop{E}_{x,y}\big\{\big(z - \hat{z}(y)\big)\big(z - \hat{z}(y)\big)'\big\} &= \Sigma_{zz} - \Sigma_{zy}\Sigma_{yy}^{-1}\Sigma_{yz} \\ &= C\big(\Sigma_{xx} - \Sigma_{xy}\Sigma_{yy}^{-1}\Sigma_{yx}\big)C' \\ &= C \mathop{E}_{x,y}\big\{\big(x - \hat{x}(y)\big)\big(x - \hat{x}(y)\big)'\big\}C'.\end{aligned}$$

Q.E.D.

Corollary E.3.4: Consider in addition to x and y, an additional random vector z of the form

$$z = Cy + u, \tag{E.20}$$

where C is a given $p \times m$ matrix of rank p and u is a given vector in $\Re^p$. Then the linear least-squares estimate $\hat{x}(z)$ of x given z is

$$\hat{x}(z) = \overline{x} + \Sigma_{xy}C'(C\Sigma_{yy}C')^{-1}(z - C\overline{y} - u), \tag{E.21}$$

and the corresponding error covariance matrix is

$$\underset{x,z}{E}\left\{\big(x - \hat{x}(z)\big)\big(x - \hat{x}(z)\big)'\right\} = \Sigma_{xx} - \Sigma_{xy}C'(C\Sigma_{yy}C')^{-1}C\Sigma_{yx}. \tag{E.22}$$

Proof: We have

$$\overline{z} = E\{z\} = C\overline{y} + u, \tag{E.23a}$$

$$\Sigma_{zz} = E\big\{(z - \overline{z})(z - \overline{z})'\big\} = C\Sigma_{yy}C', \tag{E.23b}$$

$$\Sigma_{zx} = E\big\{(z - \overline{z})(x - \overline{x})'\big\} = C\Sigma_{yx}, \tag{E.23c}$$

$$\Sigma_{xz} = E\big\{(x - \overline{x})(z - \overline{z})'\big\} = \Sigma_{xy}C'. \tag{E.23d}$$

From Prop. E.3 we have

$$\hat{x}(z) = \overline{x} + \Sigma_{xz}\Sigma_{zz}^{-1}(z - \overline{z}), \tag{E.24a}$$

$$\underset{x,z}{E}\left\{\big(x - \hat{x}(z)\big)\big(x - \hat{x}(z)\big)'\right\} = \Sigma_{xx} - \Sigma_{xz}\Sigma_{zz}^{-1}\Sigma_{zx}, \tag{E.24b}$$

where $\Sigma_{zz} = C\Sigma_{yy}C'$ has an inverse, since Σ_{yy} is invertible and C has rank p. By substituting the relations (E.23) into Eqs. (E.24a) and (E.24b) the result follows. **Q.E.D.**

Frequently we want to estimate a vector of parameters $x \in \Re^n$ given a measurement vector $z \in \Re^m$ of the form $z = Cx + v$, where C is a given $m \times n$ matrix, and $v \in \Re^m$ is a random measurement error vector. The following corollary gives the linear least-squares estimate $\hat{x}(z)$ and its error covariance.

Corollary E.3.5: Let

$$z = Cx + v,$$

where C is a given $m \times n$ matrix, and the random vectors $x \in \Re^n$ and $v \in \Re^m$ are uncorrelated. Denote

$$E\{x\} = \overline{x}, \qquad E\big\{(x - \overline{x})(x - \overline{x})'\big\} = \Sigma_{xx},$$

$$E\{v\} = \overline{v}, \qquad E\big\{(v - \overline{v})(v - \overline{v})'\big\} = \Sigma_{vv},$$

and assume further that Σ_{vv} is a positive definite matrix. Then

$$\hat{x}(z) = \overline{x} + \Sigma_{xx}C'(C\Sigma_{xx}C' + \Sigma_{vv})^{-1}(z - C\,\overline{x} - \overline{v}),$$

$$\underset{x,v}{E}\left\{\big(x - \overline{x}(z)\big)\big(x - \overline{x}(z)\big)'\right\} = \Sigma_{xx} - \Sigma_{xx}C'(C\Sigma_{xx}C' + \Sigma_{vv})^{-1}C\Sigma_{xx}.$$

Proof: Define

$$y = (\,x' \quad v'\,)', \qquad \overline{y} = (\,\overline{x}' \quad \overline{v}'\,)', \qquad \tilde{C} = (\,C \quad I\,).$$

Then we have $z = \tilde{C}y$, and by Cor. E.3.3,

$$\hat{x}(z) = (\,I \quad 0\,)\,\hat{y}(z),$$

$$E\left\{\big(x - \hat{x}(z)\big)\big(x - \hat{x}(z)\big)'\right\} = (\,I \quad 0\,)\,E\left\{\big(y - \hat{y}(z)\big)\big(y - \hat{y}(z)\big)'\right\}\begin{pmatrix} I \\ 0 \end{pmatrix},$$

where $\hat{y}(z)$ is the linear least-squares estimate of y given z. By applying Cor. E.3.4 with $u = 0$ and $x = y$ we obtain

$$\hat{y}(z) = \overline{y} + \Sigma_{yy}\tilde{C}'(\tilde{C}\Sigma_{yy}\tilde{C}')^{-1}(z - \tilde{C}\,\overline{y}),$$

$$E\left\{\big(y - \hat{y}(z)\big)\big(y - \hat{y}(z)\big)'\right\} = \Sigma_{yy} - \Sigma_{yy}\tilde{C}'(\tilde{C}\Sigma_{yy}\tilde{C}')^{-1}\tilde{C}\Sigma_{yy}.$$

By using the equations

$$\Sigma_{yy} = \begin{pmatrix} \Sigma_{xx} & 0 \\ 0 & \Sigma_{vv} \end{pmatrix}, \qquad \tilde{C} = (\,C \quad I\,),$$

and by carrying out the straightforward calculation the result follows. **Q.E.D.**

The next two corollaries deal with least-squares estimates involving multiple measurement vectors that are obtained sequentially. In particular, the corollaries show how to modify an existing least-squares estimate $\hat{x}(y)$ to obtain $\hat{x}(y,z)$ once an additional vector z becomes known. This is a central operation in Kalman filtering.

Corollary E.3.6: Consider in addition to x and y, an additional random vector z taking values in $\Re^p$, which is uncorrelated with y. Then the linear least-squares estimate $\hat{x}(y,z)$ of x given y *and* z [i.e., given the composite vector (y,z)] has the form

$$\hat{x}(y,z) = \hat{x}(y) + \hat{x}(z) - \overline{x}, \tag{E.25}$$

where $\hat{x}(y)$ and $\hat{x}(z)$ are the linear least-squares estimates of x given y and given z, respectively. Furthermore,

$$\underset{x,y,z}{E}\left\{\big(x-\hat{x}(y,z)\big)\big(x-\hat{x}(y,z)\big)'\right\} = \Sigma_{xx} - \Sigma_{xy}\Sigma_{yy}^{-1}\Sigma_{yx} - \Sigma_{xz}\Sigma_{zz}^{-1}\Sigma_{zx}, \tag{E.26}$$

where

$$\Sigma_{xz} = \underset{x,z}{E}\left\{(x-\overline{x})(z-\overline{z})'\right\}, \qquad \Sigma_{zx} = \underset{x,z}{E}\left\{(z-\overline{z})(x-\overline{x})'\right\},$$

$$\Sigma_{zz} = \underset{z}{E}\left\{(z-\overline{z})(z-\overline{z})'\right\}, \qquad \overline{z} = \underset{z}{E}\{z\},$$

and it is assumed that Σ_{zz} is invertible.

Proof: Let

$$w = \begin{pmatrix} y \\ z \end{pmatrix}, \qquad \overline{w} = \begin{pmatrix} \overline{y} \\ \overline{z} \end{pmatrix}.$$

By Eq. (E.12) we have

$$\hat{x}(w) = \overline{x} + \Sigma_{xw}\Sigma_{ww}^{-1}(w-\overline{w}). \tag{E.27}$$

Furthermore

$$\Sigma_{xw} = [\Sigma_{xy},\, \Sigma_{xz}],$$

and since y and z are uncorrelated, we have

$$\Sigma_{ww} = \begin{pmatrix} \Sigma_{yy} & 0 \\ 0 & \Sigma_{zz} \end{pmatrix}.$$

Substituting the above expressions in Eq. (E.27), we obtain

$$\hat{x}(w) = \overline{x} + \Sigma_{xy}\Sigma_{yy}^{-1}(y-\overline{y}) + \Sigma_{xz}\Sigma_{zz}^{-1}(z-\overline{z}) = \hat{x}(y) + \hat{x}(z) - \overline{x},$$

and Eq. (E.25) is proved. The proof of Eq. (E.26) is similar by using the relations above and the covariance Eq. (E.13). **Q.E.D.**

Corollary E.3.7: Let z be as in the preceding corollary and assume that y and z are not necessarily uncorrelated, that is, we may have

$$\Sigma_{yz} = \Sigma_{zy}' = \underset{y,z}{E}\left\{(y-\overline{y})(z-\overline{z})'\right\} \neq 0.$$

Then

$$\hat{x}(y,z) = \hat{x}(y) + \hat{x}\big(z - \hat{z}(y)\big) - \overline{x}, \tag{E.28}$$

where $\hat{x}\big(z - \hat{z}(y)\big)$ denotes the linear least-squares estimate of x given the random vector $z - \hat{z}(y)$ and $\hat{z}(y)$ is the linear least-squares estimate of z given y. Furthermore,

$$\mathop{E}_{x,y,z}\left\{\big(x - \hat{x}(y,z)\big)\big(x - \hat{x}(y,z)\big)'\right\} = \Sigma_{xx} - \Sigma_{xy}\Sigma_{yy}^{-1}\Sigma_{yx} - \hat{\Sigma}_{xz}\hat{\Sigma}_{zz}^{-1}\hat{\Sigma}_{zx}, \tag{E.29}$$

where

$$\hat{\Sigma}_{xz} = \mathop{E}_{x,y,z}\left\{(x - \overline{x})\big(z - \hat{z}(y)\big)'\right\},$$

$$\hat{\Sigma}_{zz} = \mathop{E}_{y,z}\left\{\big(z - \hat{z}(y)\big)\big(z - \hat{z}(y)\big)'\right\},$$

$$\hat{\Sigma}_{zx} = \mathop{E}_{x,y,z}\left\{\big(z - \hat{z}(y)\big)(x - \overline{x})'\right\}.$$

Proof: It can be seen that, since $\hat{z}(y)$ is a linear function of y, the linear least-squares estimate of x given y and z is the same as the linear least-squares estimate of x given y and $z - \hat{z}(y)$. By Cor. E.3.2 the random vectors y and $z - \hat{z}(y)$ are uncorrelated. Given this observation the result follows by application of the preceding corollary. **Q.E.D.**

E.3 STATE ESTIMATION – THE KALMAN FILTER

Consider now a linear dynamic system of the type considered in Section 5.2 but without a control vector ($u_k \equiv 0$)

$$x_{k+1} = A_k x_k + w_k, \qquad k = 0, 1, \ldots, N-1, \tag{E.30}$$

where $x_k \in \Re^n$ and $w_k \in \Re^n$ denote the state and random disturbance vectors, respectively, and the matrices A_k are known. Consider also the measurement equation

$$z_k = C_k x_k + v_k, \qquad k = 0, 1, \ldots, N-1, \tag{E.31}$$

where $z_k \in \Re^s$ and $v_k \in \Re^s$ are the observation and observation noise vectors, respectively.

We assume that $x_0, w_0, \ldots, w_{N-1}, v_0, \ldots, v_{N-1}$ are independent random vectors with given probability distributions and that

$$E\{w_k\} = E\{v_k\} = 0, \qquad k = 0, 1, \ldots, N-1. \tag{E.32}$$

We use the notation

$$S = E\{(x_0 - E\{x_0\})(x_0 - E\{x_0\})'\}, \quad M_k = E\{w_k w_k'\}, \quad N_k = E\{v_k v_k'\}, \tag{E.33}$$

and we assume that N_k is positive definite for all k.

A Nonrecursive Least-Squares Estimate

We first give a straightforward but somewhat tedious method to derive the linear least-squares estimate of x_{k+1} or x_k given the values of $z_0, z_1, \ldots, z_k$. Let us denote

$$Z_k = (z_0', z_1', \ldots, z_k')', \qquad r_{k-1} = (x_0', w_0', w_1', \ldots, w_{k-1}')'.$$

In this method, we first find the linear least-squares estimate of r_{k-1} given Z_k, and we then obtain the linear least-squares estimate of x_k given Z_k after expressing x_k as a linear function of r_{k-1}.

For each i with $0 \le i \le k$ we have, by using the system equation,

$$x_{i+1} = L_i r_i,$$

where L_i is the $n \times \big(n(i+1)\big)$ matrix

$$L_i = (\, A_i \cdots A_0, \quad A_i \cdots A_1, \quad \ldots \quad, A_i, \quad I\,).$$

As a result we may write

$$Z_k = \Phi_{k-1} r_{k-1} + V_k,$$

where

$$V_k = (v_0', v_1', \ldots, v_k')',$$

and Φ_{k-1} is an $s(k+1) \times (nk)$ matrix of the form

$$\Phi_{k-1} = \begin{pmatrix} C_0 & 0 \\ C_1 L_0 & 0 \\ \vdots & \vdots \\ C_{k-1} L_{k-2} & 0 \\ \multicolumn{2}{c}{C_k L_{k-1}} \end{pmatrix}.$$

We can thus use Cor. E.3.5, the equations above, and the data of the problem to compute

$$\hat{r}_{k-1}(Z_k) \quad \text{and} \quad E\{(r_{k-1} - \hat{r}_{k-1}(Z_k))(r_{k-1} - \hat{r}_{k-1}(Z_k))'\}.$$

Let us denote the linear least-squares estimates of x_{k+1} and x_k given Z_k by $\hat{x}_{k+1|k}$ and $\hat{x}_{k|k}$, respectively. We can now obtain $\hat{x}_{k|k} = \hat{x}_k(Z_k)$ and the corresponding error covariance matrix by using Cor. E.3.3, that is,

$$\hat{x}_{k|k} = L_{k-1}\hat{r}_{k-1}(Z_k),$$

$$E\{(x_k - \hat{x}_{k|k})(x_k - \hat{x}_{k|k})'\}$$
$$= L_{k-1} E\{(r_{k-1} - \hat{r}_{k-1}(Z_k))(r_{k-1} - \hat{r}_{k-1}(Z_k))'\} L_{k-1}'.$$

These equations may in turn be used to yield $\hat{x}_{k+1|k}$ and the corresponding error covariance again via Cor. E.3.3.

The Kalman Filtering Algorithm

The preceding method for obtaining the least-squares estimate of x_k is cumbersome when the number of measurements is large. Fortunately, the sequential structure of the problem can be exploited and the computations can be organized conveniently, as first proposed by Kalman [Kal60]. The main attractive feature of the Kalman filtering algorithm is that the estimate $\hat{x}_{k+1|k}$ can be obtained by means of a simple equation that involves the previous estimate $\hat{x}_{k|k-1}$ and the new measurement z_k but *does not involve any of the past measurements* $z_0, z_1, \ldots, z_{k-1}$.

Suppose that we have computed the estimate $\hat{x}_{k|k-1}$ together with the covariance matrix

$$\Sigma_{k|k-1} = E\big\{(x_k - \hat{x}_{k|k-1})(x_k - \hat{x}_{k|k-1})'\big\}. \tag{E.34}$$

At time k we receive the additional measurement

$$z_k = C_k x_k + v_k.$$

We may use now Cor. E.3.7 to compute the linear least-squares estimate of x_k given $Z_{k-1} = (z_0', z_1', \ldots, z_{k-1}')'$ *and* z_k. This estimate is denoted by $\hat{x}_{k|k}$ and, by Cor. E.3.7, it is given by

$$\hat{x}_{k|k} = \hat{x}_{k|k-1} + \hat{x}_k\big(z_k - \hat{z}_k(Z_{k-1})\big) - E\{x_k\}, \tag{E.35}$$

where $\hat{z}_k(Z_{k-1})$ denotes the linear least-squares estimate of z_k given Z_{k-1} and $\hat{x}_k\big(z_k - \hat{z}_k(Z_{k-1})\big)$ denotes the linear least-squares estimate of x_k given $\big(z_k - \hat{z}_k(Z_{k-1})\big)$.

We now calculate the term $\hat{x}_k\big(z_k - \hat{z}_k(Z_{k-1})\big)$ in Eq. (E.35). We have by Eqs. (E.31), (E.32), and Cor. E.3.3,

$$\hat{z}_k(Z_{k-1}) = C_k \hat{x}_{k|k-1}. \tag{E.36}$$

Also we use Cor. E.3.3 to obtain

$$E\Big\{\big(z_k - \hat{z}_k(Z_{k-1})\big)\big(z_k - \hat{z}_k(Z_{k-1})\big)'\Big\} = C_k \Sigma_{k|k-1} C_k' + N_k, \tag{E.37}$$

$$\begin{aligned} &E\Big\{x_k\big(z_k - \hat{z}_k(Z_{k-1})\big)'\Big\} \\ &\qquad = E\Big\{x_k\big(C_k(x_k - \hat{x}_{k|k-1})\big)'\Big\} + E\{x_k v_k'\} \\ &\qquad = E\big\{(x_k - \hat{x}_{k|k-1})(x_k - \hat{x}_{k|k-1})'\big\}C_k' + E\big\{\hat{x}_{k|k-1}(x_k - \hat{x}_{k|k-1})'\big\}C_k'. \end{aligned}$$

The last term in the right-hand side above is zero by Cor. E.3.2, so by using Eq. (E.34) we have

$$E\Big\{x_k\big(z_k - \hat{z}_k(Z_{k-1})\big)'\Big\} = \Sigma_{k|k-1} C_k'. \tag{E.38}$$

Using Eqs. (E.36)-(E.38) in Prop. E.3, we obtain

$$\hat{x}_k\big(z_k - \hat{z}_k(Z_{k-1})\big) = E\{x_k\} + \Sigma_{k|k-1}C_k'(C_k\Sigma_{k|k-1}C_k' + N_k)^{-1}(z_k - C_k\hat{x}_{k|k-1}),$$

and Eq. (E.35) is written as

$$\hat{x}_{k|k} = \hat{x}_{k|k-1} + \Sigma_{k|k-1}C_k'(C_k\Sigma_{k|k-1}C_k' + N_k)^{-1}(z_k - C_k\hat{x}_{k|k-1}). \tag{E.39}$$

By using Cor. E.3.3 we also have

$$\hat{x}_{k+1|k} = A_k\hat{x}_{k|k}. \tag{E.40}$$

Concerning the covariance matrix $\Sigma_{k+1|k}$, we have from Eqs. (E.30), (E.32), (E.33), and Cor. E.3.3,

$$\Sigma_{k+1|k} = A_k\Sigma_{k|k}A_k' + M_k, \tag{E.41}$$

where

$$\Sigma_{k|k} = E\big\{(x_k - \hat{x}_{k|k})(x_k - \hat{x}_{k|k})'\big\}.$$

The error covariance matrix $\Sigma_{k|k}$ may be computed via Cor. E.3.7 similar to $\hat{x}_{k|k}$ [cf. Eq. (E.35)]. Thus, we have from Eqs. (E.29), (E.37), (E.38)

$$\Sigma_{k|k} = \Sigma_{k|k-1} - \Sigma_{k|k-1}C_k'(C_k\Sigma_{k|k-1}C_k' + N_k)^{-1}C_k\Sigma_{k|k-1}. \tag{E.42}$$

Equations (E.39)-(E.42) with the initial conditions [cf. Eq. (E.33)]

$$\hat{x}_{0|-1} = E\{x_0\}, \qquad \Sigma_{0|-1} = S, \tag{E.43}$$

constitute the *Kalman filtering algorithm*. This algorithm recursively generates the linear least-squares estimates $\hat{x}_{k+1|k}$ or $\hat{x}_{k|k}$ together with the associated error covariance matrices $\Sigma_{k+1|k}$ or $\Sigma_{k|k}$. In particular, given $\Sigma_{k|k-1}$ and $\hat{x}_{k|k-1}$, Eqs. (E.39) and (E.42) yield $\Sigma_{k|k}$ and $\hat{x}_{k|k}$, and then Eqs. (E.41) and (E.40) yield $\Sigma_{k+1|k}$ and $\hat{x}_{k+1|k}$.

An alternative expression for Eq. (E.39) is

$$\hat{x}_{k|k} = A_{k-1}\hat{x}_{k-1|k-1} + \Sigma_{k|k} C_k N_k^{-1}(z_k - C_k A_{k-1}\hat{x}_{k-1|k-1}), \quad \text{(E.44)}$$

which can be obtained from Eqs. (E.39) and (E.40) by using the following equality

$$\Sigma_{k|k} C_k' N_k^{-1} = \Sigma_{k|k-1} C_k' (C_k \Sigma_{k|k-1} C_k' + N_k)^{-1}. \quad \text{(E.45)}$$

This equality may be verified by using Eq. (E.42) to write

$$\begin{aligned}\Sigma_{k|k} C_k' N_k^{-1} &= \big(\Sigma_{k|k-1} - \Sigma_{k|k-1} C_k' (C_k \Sigma_{k|k-1} C_k' + N_k)^{-1} C_k \Sigma_{k|k-1}\big) C_k' N_k^{-1} \\ &= \Sigma_{k|k-1} C_k' \big(N_k^{-1} - (C_k \Sigma_{k|k-1} C_k' + N_k)^{-1} C_k \Sigma_{k|k-1} C_k' N_k^{-1}\big),\end{aligned}$$

and then use in the above formula the following calculation

$$\begin{aligned}N_k^{-1} &= (C_k \Sigma_{k|k-1} C_k' + N_k)^{-1} (C_k \Sigma_{k|k-1} C_k' + N_k) N_k^{-1} \\ &= (C_k \Sigma_{k|k-1} C_k' + N_k)^{-1} (C_k \Sigma_{k|k-1} C_k' N_k^{-1} + I).\end{aligned}$$

When the system equation contains a control vector u_k,

$$x_{k+1} = A_k x_k + B_k u_k + w_k, \qquad k = 0, 1, \ldots, N-1,$$

it is straightforward to show that Eq. (E.44) takes the form

$$\begin{aligned}\hat{x}_{k|k} = {} & A_{k-1}\hat{x}_{k-1|k-1} + B_{k-1}u_{k-1} \\ & + \Sigma_{k|k} C_k' N_k^{-1}(z_k - C_k A_{k-1}\hat{x}_{k-1|k-1} - C_k B_{k-1} u_{k-1}),\end{aligned} \quad \text{(E.46)}$$

where $\hat{x}_{k|k}$ is the linear least-squares estimate of x_k given $z_0, z_1, \ldots, z_k$ and $u_0, u_1, \ldots, u_{k-1}$. The equations (E.41)-(E.43) that generate $\Sigma_{k|k}$ remain unchanged.

Steady-State Kalman Filtering Algorithm

Finally we note that Eqs. (E.41) and (E.42) yield

$$\Sigma_{k+1|k} = A_k \big(\Sigma_{k|k-1} - \Sigma_{k|k-1} C_k' (C_k \Sigma_{k|k-1} C_k' + N_k)^{-1} C_k \Sigma_{k|k-1}\big) A_k' + M_k, \quad \text{(E.47)}$$

with the initial condition $\Sigma_{0|-1} = S$. This equation is a matrix Riccati equation of the type considered in Section 4.1. Thus when A_k, C_k, N_k, and M_k are constant matrices,

$$A_k = A, \quad C_k = C, \quad N_k = N, \quad M_k = M, \qquad k = 0, 1, \ldots, N-1,$$

we have by invoking the proposition proved there, that $\Sigma_{k+1|k}$ tends to a positive definite symmetric matrix Σ that solves the algebraic Riccati equation

$$\Sigma = A\big(\Sigma - \Sigma C'(C\Sigma C' + N)^{-1}C\Sigma\big)A' + M,$$

assuming observability of the pair (A, C) and controllability of the pair (A, D), where $M = DD'$. Under the same conditions, we have $\Sigma_{k|k} \to \overline{\Sigma}$, where from Eq. (E.42),

$$\overline{\Sigma} = \Sigma - \Sigma C'(C\Sigma C' + N)^{-1}C\Sigma.$$

We may then write the Kalman filter recursion [cf. Eq. (E.44)] in the asymptotic form

$$\hat{x}_{k|k} = A\hat{x}_{k-1|k-1} + \overline{\Sigma}C'N^{-1}(z_k - CA\hat{x}_{k-1|k-1}). \tag{E.48}$$

This estimator is simple and convenient for implementation.

E.4 STABILITY ASPECTS

Let us consider now the stability properties of the steady-state form of the Kalman filter. From Eqs. (E.39) and (E.40), we have

$$\hat{x}_{k+1|k} = A\hat{x}_{k|k-1} + A\Sigma C'(C\Sigma C' + N)^{-1}(z_k - C\hat{x}_{k|k-1}). \tag{E.49}$$

Let e_k denote the "one-step prediction" error

$$e_k = x_k - \hat{x}_{k|k-1}.$$

By using Eq. (E.49), the system equation

$$x_{k+1} = Ax_k + w_k,$$

and the measurement equation

$$z_k = Cx_k + v_k,$$

we obtain

$$e_{k+1} = \big(A - A\Sigma C'(C\Sigma C' + N)^{-1}C\big)e_k + w_k - A\Sigma C'(C\Sigma C' + N)^{-1}v_k. \tag{E.50}$$

From the practical point of view it is important that the error equation (E.50) represents a stable system, that is, the matrix

$$A - A\Sigma C'(C\Sigma C' + N)^{-1}C \tag{E.51}$$

has eigenvalues strictly within the unit circle. This, however, follows by Prop. 4.1 of Section 4.1 under the observability and controllability assumptions given earlier, since Σ is the unique positive semidefinite symmetric solution of the algebraic Riccati equation

$$\Sigma = A\big(\Sigma - \Sigma C'(C\Sigma C' + N)^{-1}C\Sigma\big)A' + M.$$

Actually this proposition yields that the transpose of the matrix (E.51) has eigenvalues strictly within the unit circle, but this is sufficient for our purposes since the eigenvalues of a matrix are the same as those of its transpose.

Let us consider also the stability properties of the equation governing the estimation error

$$\tilde{e}_k = x_k - \hat{x}_{k|k}.$$

We have by a straightforward calculation

$$\tilde{e}_k = \big(I - \Sigma C'(C\Sigma C' + N)^{-1}C\big)e_k - \Sigma C'(C\Sigma C' + N)^{-1}v_k. \tag{E.52}$$

By multiplying both sides of Eq. (E.50) by $I - \Sigma C'(C\Sigma C' + N)^{-1}C$ and by using Eq. (E.52), we obtain

$$\begin{aligned} &\tilde{e}_{k+1} + \Sigma C'(C\Sigma C' + N)^{-1}v_{k+1} \\ &\quad = \big(A - \Sigma C'(C\Sigma C' + N)^{-1}CA\big)\big(\tilde{e}_k + \Sigma C'(C\Sigma C' + N)^{-1}v_k\big) \\ &\qquad + \big(I - \Sigma C'(C\Sigma C' + N)^{-1}C\big)\big(w_k - A\Sigma C'(C\Sigma C' + N)^{-1}v_k\big), \end{aligned}$$

or equivalently

$$\begin{aligned} \tilde{e}_{k+1} = {} & \big(A - \Sigma C'(C\Sigma C' + N)^{-1}CA\big)\tilde{e}_k \\ & + \big(I - \Sigma C'(C\Sigma C' + N)^{-1}C\big)w_k - \Sigma C'(C\Sigma C' + N)^{-1}v_{k+1}. \end{aligned} \tag{E.53}$$

Since the matrix (E.51) has eigenvalues strictly within the unit circle, the sequence $\{e_k\}$ generated by Eq. (E.50) tends to zero whenever the vectors w_k and v_k are identically zero for all k. Hence, by Eq. (E.52), the same is true for the sequence $\{\tilde{e}_k\}$. It follows from Eq. (E.53) that the matrix

$$A - \Sigma C'(C\Sigma C' + N)^{-1}CA \tag{E.54}$$

has eigenvalues strictly within the unit circle, and the estimation error sequence $\{\tilde{e}_k\}$ is generated by a stable system.

Let us finally consider the stability properties of the $2n$-dimensional system of equations with state vector $(x_k', \hat{x}_k')$:

$$x_{k+1} = Ax_k + BL\hat{x}_k, \tag{E.55}$$

$$\hat{x}_{k+1} = \overline{\Sigma}C'N^{-1}CAx_k + (A + BL - \overline{\Sigma}C'N^{-1}CA)\hat{x}_k. \tag{E.56}$$

This is the steady-state, asymptotically optimal closed-loop system that was encountered at the end of Section 5.2.

We assume that the appropriate observability and controllability assumptions stated there are in effect. By using the equation

$$\overline{\Sigma}C'N^{-1} = \Sigma C'(C\Sigma C' + N)^{-1},$$

shown earlier, we obtain from Eqs. (E.55) and (E.56) that

$$(x_{k+1} - \hat{x}_{k+1}) = \big(A - \Sigma C'(C\Sigma C' + N)^{-1}CA\big)(x_k - \hat{x}_k).$$

Since we have proved that the matrix (E.54) has eigenvalues strictly within the unit circle, it follows that

$$\lim_{k\to\infty} (x_{k+1} - \hat{x}_{k+1}) = 0, \tag{E.57}$$

for arbitrary initial states x_0 and $\hat{x}_0$. From Eq. (E.55) we obtain

$$x_{k+1} = (A + BL)x_k + BL(\hat{x}_k - x_k). \tag{E.58}$$

Since in accordance with the theory of Section 4.1 the matrix $(A + BL)$ has eigenvalues strictly within the unit circle, it follows from Eqs. (E.57) and (E.58) that we have

$$\lim_{k\to\infty} x_k = 0 \tag{E.59}$$

and hence from Eq. (E.57),

$$\lim_{k\to\infty} \hat{x}_k = 0. \tag{E.60}$$

Since the equations above hold for arbitrary initial states x_0 and $\hat{x}_0$ it follows that the system defined by Eqs. (E.55) and (E.56) is stable.

E.5 GAUSS-MARKOV ESTIMATORS

Suppose that we want to estimate a vector $x \in \Re^n$ given a measurement vector $z \in \Re^m$ that is related to x by

$$z = Cx + v, \tag{E.61}$$

where C is a given $m \times n$ matrix with rank m, and v is a random measurement error vector. Let us assume that v is uncorrelated with x, and has a known mean and a positive definite covariance matrix

$$E\{v\} = \overline{v}, \qquad E\{(v - \overline{v})(v - \overline{v})'\} = \Sigma_{vv}. \tag{E.62}$$

If the a priori probability distribution of x is known, we can obtain a linear least-squares estimate of x given z by using the theory of Section E.2 (cf. Cor. E.3.5). In many cases, however, the probability distribution of x is unknown. In such cases we can use the Gauss-Markov estimator, which is optimal within the class of linear estimators that satisfy certain restrictions, as described below.

Let us consider an estimator of the form

$$\hat{x}(z) = \hat{A}(z - \overline{v}),$$

where $\hat{A}$ minimizes

$$f(A) = \mathop{E}_{x,z}\{\|x - A(z - \overline{v})\|^2\} \tag{E.63}$$

over all $n \times m$ matrices A. Since x and v are uncorrelated, we have using Eqs. (E.61)-(E.63)

$$\begin{aligned} f(A) &= \mathop{E}_{x,v}\{\|x - ACx - A(v - \overline{v})\|^2\} \\ &= \mathop{E}_{x}\{\|(I - AC)x\|^2\} + \mathop{E}_{v}\{\|A(v - \overline{v})\|^2\}, \end{aligned}$$

where I is the $n \times n$ identity matrix. Since $f(A)$ depends on the unknown statistics of x, we see that the optimal matrix $\hat{A}$ also depends on these statistics. We can circumvent this difficulty by requiring that

$$AC = I.$$

Then our problem becomes

$$\begin{aligned} &\text{minimize} \quad \mathop{E}_{v}\{\|A(v - \overline{v})\|^2\} \\ &\text{subject to} \quad AC = I. \end{aligned} \tag{E.64}$$

Note that the requirement $AC = I$ is not only convenient analytically, but also makes sense conceptually. In particular, it is equivalent to requiring that the estimator $x(z) = A(z - \overline{v})$ be *unbiased* in the sense that

$$E\{x(z)\} = E\{x\} = \overline{x}, \qquad \text{for all } \overline{x} \in \Re^n.$$

This can be seen by writing

$$E\{x(z)\} = E\{A(Cx + v - \overline{v})\} = ACE\{x\} = AC\overline{x} = \overline{x}.$$

To derive the optimal solution $\hat{A}$ of problem (E.64), let a_i' denote the ith row of A. We have

$$\begin{aligned}
\|A(v-\overline{v})\|^2 &= (v-\overline{v})' \begin{pmatrix} a_1 & \cdots & a_n \end{pmatrix} \begin{pmatrix} a_1' \\ \vdots \\ a_n' \end{pmatrix} (v-\overline{v}) \\
&= \sum_{i=1}^n (v-\overline{v})' a_i a_i' (v-\overline{v}) \\
&= \sum_{i=1}^n a_i' (v-\overline{v})(v-\overline{v})' a_i.
\end{aligned}$$

Hence, the minimization problem (E.64) can also be written as

$$\begin{aligned}
&\text{minimize} \quad \sum_{i=1}^n a_i' \Sigma_{vv} a_i \\
&\text{subject to} \quad C' a_i = e_i, \qquad i = 1, \ldots, n,
\end{aligned}$$

where e_i is the ith column of the identity matrix. The minimization can be carried out separately for each i, yielding

$$\hat{x}_i = \Sigma_{vv}^{-1} C (C' \Sigma_{vv} C)^{-1} e_i, \qquad i = 1, \ldots, n,$$

and finally

$$\hat{A} = (C' \Sigma_{vv}^{-1} C)^{-1} C' \Sigma_{vv}^{-1}.$$

Thus, the Gauss-Markov estimator is given by

$$\hat{x}(z) = (C' \Sigma_{vv}^{-1} C)^{-1} C' \Sigma_{vv}^{-1} (z - \overline{v}). \tag{E.65}$$

Let us also calculate the corresponding error covariance matrix. We have

$$\begin{aligned}
E\big\{\big(x - \hat{x}(z)\big)\big(x - \hat{x}(z)\big)'\big\} &= E\big\{\big(x - \hat{A}(z-\overline{v})\big)\big(x - \hat{A}(z-\overline{v})\big)'\big\} \\
&= E\big\{\hat{A}(v-\overline{v})(v-\overline{v})'\hat{A}'\big\} \\
&= \hat{A} \Sigma_{vv} \hat{A}' \\
&= (C' \Sigma_{vv}^{-1} C)^{-1} C' \Sigma_{vv}^{-1} \Sigma_{vv} \Sigma_{vv}^{-1} C (C' \Sigma_{vv}^{-1} C)^{-1},
\end{aligned}$$

and finally

$$E\big\{\big(x - \hat{x}(z)\big)\big(x - \hat{x}(z)\big)'\big\} = (C' \Sigma_{vv}^{-1} C)^{-1}. \tag{E.66}$$

Finally, let us compare the Gauss-Markov estimator with the linear least-squares estimator of Cor. E.3.5. Assuming that Σ_{xx} is invertible, a straightforward calculation shows that the latter estimator can be written as

$$\hat{x}(z) = \overline{x} + \big(\Sigma_{xx}^{-1} + C' \Sigma_{vv}^{-1} C\big)^{-1} C' \Sigma_{vv}^{-1} (z - C\overline{x} - \overline{v}). \tag{E.67}$$

By comparing Eqs. (E.65) and (E.67), we see that the Gauss-Markov estimator is obtained from the linear least-squares estimator by setting $\overline{x} = 0$ and $\Sigma_{xx}^{-1} = 0$, i.e., a zero mean and infinite covariance for the unknown random variable x. Thus, the Gauss-Markov estimator may be viewed as a limiting form of the linear least-squares estimator. The error covariance matrix (E.66) of the Gauss-Markov estimator is similarly related with the error covariance matrix of the linear least-squares estimator.

E.6 DETERMINISTIC LEAST-SQUARES ESTIMATION

Suppose again that we want to estimate a vector $x \in \Re^n$ given a measurement vector $z \in \Re^m$ that is related to x by

$$z = Cx + v,$$

where C is a known $m \times n$ matrix of rank m. However, we know nothing about the probability distribution of x and v, and thus we can't use a statistically-based estimator. Then it is reasonable to select as our estimate the vector $\hat{x}$ that minimizes

$$f(x) = \|z - Cx\|^2,$$

that is, the estimate that fits best the data in a least-squares sense. We denote this estimate by $\hat{x}(z)$.

By setting to zero the gradient of f at $\hat{x}(z)$, we obtain

$$\nabla f|_{\hat{x}(z)} = 2C'\big(C\hat{x}(z) - z\big) = 0,$$

from which

$$\hat{x}(z) = (C'C)^{-1}C'z. \tag{E.68}$$

An interesting observation is that the estimate (E.68) is the same as the Gauss-Markov estimate given by Eq. (E.65), provided the measurement error has zero mean and covariance matrix equal to the identity, i.e., $\overline{v} = 0$, $\Sigma_{vv} = I$. In fact, if instead of $\|z - Cx\|^2$, we minimize

$$(z - \overline{v} - Cx)'\Sigma_{vv}^{-1}(z - \overline{v} - Cx),$$

then the deterministic least-squares estimate obtained is identical to the Gauss-Markov estimate. If instead of $\|z - Cx\|^2$ we minimize

$$(x - \overline{x})'\Sigma_{xx}^{-1}(z - \overline{x}) + (z - \overline{v} - Cx)'\Sigma_{vv}^{-1}(z - \overline{v} - Cx),$$

then the estimate obtained is identical to the linear least-squares estimate given by Eq. (E.67). Thus, we arrive at the interesting conclusion that the estimators obtained earlier on the basis of a stochastic optimization framework can also be obtained by minimization of a deterministic measure of fitness of estimated parameters to the data at hand.

APPENDIX F: Modeling of Stochastic Linear Systems

In this appendix we show how controlled linear time-invariant systems with stochastic inputs can be represented by the ARMAX model used in Section 5.3.

F.1 LINEAR SYSTEMS WITH STOCHASTIC INPUTS

Consider a linear system with output $\{y_k\}$, control input $\{u_k\}$, and an additional zero-mean random input $\{w_k\}$. We assume that $\{w_k\}$ is a stationary (up to second order) stochastic process. That is, $\{w_k\}$ is a sequence of random variables satisfying, for all $i, k = 0, \pm 1, \pm 2, \ldots,$

$$E\{w_k\} = 0, \qquad E\{w_0 w_i\} = E\{w_k w_{k+i}\} < \infty.$$

(All references to stationary processes in this section are meant in the limited sense just described.) By linearity, y_k is the sum of one sequence $\{y_k^1\}$ due to the presence of $\{u_k\}$ and another sequence $\{y_k^2\}$ due to the presence of $\{w_k\}$:

$$y_k = y_k^1 + y_k^2. \tag{F.1}$$

We assume that y_k^1 and y_k^2 are generated by some filters $B_1(s)/A_1(s)$ and $B_2(s)/A_2(s)$, respectively:

$$A_1(s)y_k^1 = B_1(s)u_k, \tag{F.2a}$$

$$A_2(s)y_k^2 = B_2(s)w_k. \tag{F.2b}$$

Operating on Eqs. (F.2a) and (F.2b) with $A_2(s)$ and $A_1(s)$, respectively, adding, and using Eq. (F.1), we obtain

$$\overline{A}(s)y_k = \overline{B}(s)u_k + v_k, \tag{F.3}$$

where $\overline{A}(s) = A_1(s)A_2(s)$, $\overline{B}(s) = A_2(s)B_1(s)$, and $\{v_k\}$, given by

$$v_k = A_1(s)B_2(s)w_k, \tag{F.4}$$

is a zero-mean, generally correlated, stationary stochastic process.

We are interested in the case where u_k is a control input applied after y_k has occurred and has been observed, so that in Eq. (F.2a) we have $B_1(0) = 0$. Then, we may assume that the polynomials $\overline{A}(s)$ and $\overline{B}(s)$ have the form

$$\overline{A}(s) = 1 + \overline{a}_1 s + \cdots + \overline{a}_{m_0} s^{m_0}, \qquad \overline{B}(s) = \overline{b}_1 s + \cdots + \overline{b}_{m_0} s^{m_0}$$

for some scalars $\overline{a}_i$ and $\overline{b}_i$, and some positive integer m_0.

To summarize, we have constructed a model of the form

$$\overline{A}(s)y_k = \overline{B}(s)u_k + v_k,$$

where $\overline{A}(s)$ and $\overline{B}(s)$ are polynomials of the preceding form and $\{v_k\}$ is some zero-mean, correlated, stationary stochastic process. We now need to model further the sequence $\{v_k\}$.

F.2 PROCESSES WITH RATIONAL SPECTRUM

Given a zero-mean, stationary scalar process $\{v_k\}$, denote by $V(k)$ the autocorrelation function

$$V(k) = E\{v_i v_{i+k}\}, \qquad k = 0, \pm 1, \pm 2, \ldots$$

We say that $\{v_k\}$ has *rational spectrum* if the transform of $\{V(k)\}$ defined by

$$S_v(\lambda) = \sum_{k=-\infty}^{\infty} V(k)e^{-jk\lambda}$$

exists for $\lambda \in [-\pi, \pi]$ and can be expressed as

$$S_v(\lambda) = \sigma^2 \frac{|C(e^{i\lambda})|^2}{|D(e^{j\lambda})|^2}, \qquad \lambda \in [-\pi, \pi], \tag{F.5}$$

where σ is a scalar, $C(z)$ and $D(z)$ are some polynomials with real coefficients

$$C(z) = 1 + c_1 z + \cdots + c_m z^m, \tag{F.6a}$$

$$D(z) = 1 + d_1 z + \cdots + d_m z^m, \tag{F.6b}$$

and $D(z)$ has no roots on the unit circle $\{z \mid |z| = 1\}$.

The following facts are of interest:

(a) If $\{v_k\}$ is an uncorrelated process with $V(0) = \sigma^2$, $V(k) = 0$ for $k \neq 0$, then

$$S_v(\lambda) = \sigma^2, \qquad \lambda \in [-\pi, \pi],$$

and clearly $\{v_k\}$ has rational spectrum.

(b) If $\{v_k\}$ has rational spectrum S_v given by Eq. (F.5), then S_v can be written as

$$S_v(\lambda) = \tilde{\sigma}^2 \frac{|\tilde{C}(e^{j\lambda})|^2}{|\tilde{D}(e^{j\lambda})|^2}, \qquad \lambda \in [-\pi, \pi],$$

where $\tilde{\sigma}$ is a scalar and $\tilde{C}(z)$, $\tilde{D}(z)$ are unique real polynomials of the form

$$\tilde{C}(z) = 1 + \tilde{c}_1 z + \cdots + \tilde{c}_m z^m,$$

$$\tilde{D}(z) = 1 + \tilde{d}_1 z + \cdots + \tilde{d}_m z^m,$$

such that:

(1) $\tilde{C}(z)$ has all its roots outside or on the unit circle, and if $C(z)$ has no roots on the unit circle, then the same is true for $\tilde{C}(z)$.

(2) $\tilde{D}(z)$ has all roots strictly outside the unit circle.

These facts are seen by noting that if $\rho \neq 0$ is a root of $D(z)$, then $|D(e^{j\lambda})|^2 = D(e^{j\lambda})D(e^{-j\lambda})$ contains a factor

$$(1 - \rho^{-1}e^{j\lambda})(1 - \rho^{-1}e^{-j\lambda}) = \rho^{-2}(\rho - e^{j\lambda})(\rho - e^{-j\lambda}).$$

A little reflection shows that the roots of $\tilde{D}(z)$ should be ρ or ρ^{-1} depending on whether ρ is outside or inside the unit circle. Similarly, the roots of $\tilde{C}(z)$ are obtained from the roots of $C(z)$. Thus the polynomials $\tilde{C}(z)$ and $\tilde{D}(z)$ as well as $\tilde{\sigma}^2$ can be uniquely determined. We may thus assume without loss of generality that $C(z)$ and $D(z)$ in Eq. (F.5) have no roots inside the unit circle.

There is a fundamental result here that relates to the realization of processes with rational spectrum. The proof is hard; see for example, [AsG75, pp. 75-76].

Proposition F.1: If $\{v_k\}$ is a zero-mean, stationary stochastic process with rational spectrum

$$S_v(\lambda) = \sigma^2 \frac{|C(e^{j\lambda})|^2}{|D(e^{j\lambda})|^2}, \qquad \lambda \in [-\pi, \pi],$$

where the polynomials $C(s)$ and $D(s)$ are given by

$$C(s) = 1 + c_1 s + \cdots + c_m s^m, \qquad D(s) = 1 + d_1 s + \cdots + d_m s^m,$$

and are assumed (without loss of generality) to have no roots inside the unit circle, then there exists a zero-mean, uncorrelated stationary process $\{\epsilon_k\}$ with $E\{\epsilon_k^2\} = \sigma^2$ such that for all k

$$v_k + d_1 v_{k-1} + \cdots + d_m v_{k-m} = \epsilon_k + c_1 \epsilon_{k-1} + \cdots + c_m \epsilon_{k-m}.$$

F.3 THE ARMAX MODEL

Let us now return to the problem of representation of a linear system with stochastic inputs. We had arrived at the model $\overline{A}(s)y_k = \overline{B}(s)u_k + v_k$. If the zero-mean stationary process $\{v_k\}$ has rational spectrum, the preceding analysis and proposition show that there exists a zero-mean, uncorrelated stationary process $\{\epsilon_k\}$ satisfying

$$D(s)v_k = C(s)\epsilon_k, \tag{F.7}$$

where $C(s)$ and $D(s)$ are polynomials, and $C(s)$ has no roots inside the unit circle. Operating on both sides of the equation $\overline{A}(s)y_k = \overline{B}(s)u_k + v_k$ with $D(s)$ and using the relation $D(s)v_k = C(s)\epsilon_k$, we obtain

$$A(s)y_k = B(s)u_k + C(s)\epsilon_k, \tag{F.8}$$

where $A(s) = D(s)\overline{A}(s)$ and $B(s) = D(s)\overline{B}(s)$. Since $\overline{A}(0) = 1$, $\overline{B}(0) = 0$, we can write Eq. (F.8) as

$$y_k + \sum_{i=1}^{m} a_i y_{k-i} = \sum_{i=1}^{m} b_i u_{k-i} + \epsilon_k + \sum_{i=1}^{m} c_i \epsilon_{k-i},$$

for some integer m and scalars a_i, b_i, c_i, $i = 1, \ldots, m$. This is the ARMAX model that we have used in Section 5.3.

References

[ABC65] Atkinson, R. C., Bower, G. H., and Crothers, E. J., 1965. An Introduction to Mathematical Learning Theory, Wiley, N. Y.

[ABF93] Arapostathis, A., Borkar, V., Fernandez-Gaucherand, E., Ghosh, M., and Marcus, S., 1993. "Discrete-Time Controlled Markov Processes with Average Cost Criterion: A Survey," SIAM J. on Control and Optimization, Vol. 31, pp. 282-344.

[ABG49] Arrow, K. J., Blackwell, D., and Girshick, M. A., 1949. "Bayes and Minimax Solutions of Sequential Design Problems," Econometrica, Vol. 17, pp. 213-244.

[AGK77] Athans, M., Ku, R., and Gershwin, S. B., 1977. "The Uncertainty Threshold Principle," IEEE Trans. on Automatic Control, Vol. AC-22, pp. 491-495.

[AHM51] Arrow, K. J., Harris, T., and Marschack, J., 1951. "Optimal Inventory Policy," Econometrica, Vol. 19, pp. 250-272.

[AKS58] Arrow, K. J., Karlin, S., and Scarf, H., 1958. Studies in the Mathematical Theory of Inventory and Production, Stanford Univ. Press, Stanford, CA.

[AdG86] Adams, M., and Guillemin, V., 1986. Measure Theory and Probability, Wadsworth and Brooks, Monterey, CA.

[AsG75] Ash, R. B., and Gardner, M. F., 1975. Topics in Stochastic Processes, Academic Press, N. Y.

[AnM79] Anderson, B. D. O., and Moore, J. B., 1979. Optimal Filtering, Prentice-Hall, Englewood Cliffs, N. J.

[AoL69] Aoki, M., and Li, M. T., 1969. "Optimal Discrete-Time Control Systems with Cost for Observation," IEEE Trans. Automatic Control, Vol. AC-14, pp. 165-175.

[AsW73] Aström, K. J., and Wittenmark, B., 1973. "On Self-Tuning Regulators," Automatica, Vol. 9, pp. 185-199.

[AsW84] Aström, K. J., and Wittenmark, B., 1984. Computer Controlled Systems, Prentice-Hall, Englewood Cliffs, N. J.

[AsW90] Aström, K. J., and Wittenmark, B., 1990. Adaptive Control, Addison-Wesley, Reading, MA.

[Ash70] Ash, R. B., 1970. Basic Probability Theory, Wiley, N. Y.

[Ash72] Ash, R. B., 1972. Real Analysis and Probability, Academic Press, N. Y.

[Ast83] Aström, K. J., 1983. "Theory and Applications of Adaptive Control – A Survey," Automatica, Vol. 19, pp. 471-486.

[AtF66] Athans, M., and Falb, P., 1966. Optimal Control, McGraw-Hill, N. Y.

[BGM94] Bertsekas, D. P., Guerriero, F., and Musmanno, R., 1994. "Parallel Label Corrrecting Methods for Shortest Paths," Lab. for Info. and Decision Systems Report LIDS-P-2250, Massachusetts Institute of Technology, J.O.T.A., to appear.

[BPS92] Bertsekas, D. P., Pallottino, S., and Scutella', M. G., 1992. "Polynomial Auction Algorithms for Shortest Paths," Lab. for Info. and Decision Systems Report LIDS Report P-2107, Massachusetts Institute of Technology; also Computational Optimization and Applications, Vol. 4, 1995, pp. 99-125.

[Bar81] Bar-Shalom, Y., 1981. "Stochastic Dynamic Programming: Caution and Probing," IEEE Trans. on Automatic Control, Vol. AC-26, pp. 1184-1195.

[BeD62] Bellman, R., and Dreyfus, S., 1962. Applied Dynamic Programming, Princeton Univ. Press, Princeton, N. J.

[BeG92] Bertsekas, D. P., and Gallager, R. G., 1992. Data Networks (2nd Edition), Prentice-Hall, Englewood Cliffs, N. J.

[BeR73] Bertsekas, D. P., and Rhodes, I. B., 1973. "Sufficiently Informative Functions and the Minimax Feedback Control of Uncertain Dynamic Systems," IEEE Trans. Automatic Control, Vol. AC-18, pp. 117-124.

[BeS78] Bertsekas, D. P., and Shreve, S. E., 1978. Stochastic Optimal Control: The Discrete Time Case, Academic Press, N. Y.

[BeT89] Bertsekas, D. P., and Tsitsiklis, J. N., 1989. Parallel and Distributed Computation: Numerical Methods, Prentice-Hall, Englewood Cliffs, N. J.

[BeT91] Bertsekas, D. P., and Tsitsiklis, J. N., 1991. "An Analysis of Stochastic Shortest Path Problems," Math. Operations Res., Vol. 16, pp. 580-595.

[Bel57] Bellman, R., 1957. Applied Dynamic Programming, Princeton University Press, Princeton, N. J.

[Ber70] Bertsekas, D. P., 1970. "On the Separation Theorem for Linear Systems, Quadratic Criteria, and Correlated Noise," Unpublished Report, Electronic Systems Lab., Massachusetts Institute of Technology.

[Ber72] Bertsekas, D. P., 1972. "On the Solution of Some Minimax Control Problems," Proc. 1972 IEEE Decision and Control Conf., New Orleans, LA.

[Ber75] Bertsekas, D. P., 1975. "Convergence of Discretization Procedures in Dynamic Programming," IEEE Trans. Automatic Control, Vol. AC-20, pp. 415-419.

[Ber76] Bertsekas, D. P., 1976. Dynamic Programming and Stochastic Control, Academic Press, N. Y.

[Ber82a] Bertsekas, D. P., 1982. "Distributed Dynamic Programming," IEEE Trans. Automatic Control, Vol. AC-27, pp. 610-616.

[Ber82b] Bertsekas, D. P., 1982. Constrained Optimization and Lagrange Multiplier Methods, Academic Press, N. Y.

[Ber91a] Bertsekas, D. P., 1991. Linear Network Optimization: Algorithms and Codes, MIT Press, Cambridge, MA.

[Ber91b] Bertsekas, D. P., 1991. "The Auction Algorithm for Shortest Paths," SIAM J. on Optimization, Vol. 1, pp. 425-447.

[Ber93] Bertsekas, D. P., 1993. "A Simple and Fast Label Correcting Algorithm for Shortest Paths," Networks, Vol. 23, pp. 703-709.

[Ber95] Bertsekas, D. P., 1995. Nonlinear Programming, Athena Scientific, Belmont, MA, to appear.

[BoV79] Borkar, V., and Varaiya, P. P., 1979. "Adaptive Control of Markov Chains, I: Finite Parameter Set," IEEE Trans. Automatic Control, Vol. AC-24, pp. 953-958.

[CCP70] Cannon, M. D., Cullum, C. D., and Polak, E., 1970. Theory of Optimal Control and Mathematical Programming, McGraw-Hill, N. Y.

[CDP92] Cerulli, R., De Leone, R., and Piacenti, G., 1992. "A Modified Auction Algorithm for the Shortest Path Problem," University of Salerno Report, J. of Optimization and Software, to appear.

[ChT89] Chow, C.-S., and Tsitsiklis, J. N., 1989. "The Complexity of Dynamic Programming," Journal of Complexity, Vol. 5, pp. 466-488.

[ChT91] Chow, C.-S., and Tsitsiklis, J. N., 1991. "An Optimal One–Way Multigrid Algorithm for Discrete–Time Stochastic Control," IEEE Trans. on Automatic Control, Vol. AC-36, 1991, pp. 898-914.

[Che72] Chernoff, H., 1972. "Sequential Analysis and Optimal Design," Regional Conference Series in Applied Mathematics, SIAM, Philadelphia, PA.

[Chu60] Chung, K. L., 1960. Markov Chains with Stationary Transition Probabilities, Springer-Verlag, Berlin and N. Y.

[CoL55] Coddington, E. A., and Levinson, N., 1955. Theory of Ordinary Differential Equations, McGraw-Hill, N. Y.

[DeG70] DeGroot, M. H., 1970. Optimal Statistical Decisions, McGraw-Hill, N. Y.

[DeP84] Deo, N., and Pang, C., 1984. "Shortest Path Problems: Taxonomy and Annotation," Networks, Vol. 14, pp. 275-323.

[DoS80] Doshi, B., and Shreve, S., 1980. "Strong Consistency of a Modified Maximum Likelihood Estimator for Controlled Markov Chains," J. of Applied Probability, Vol. 17, pp. 726-734.

[Dre65] Dreyfus, S. D., 1965. Dynamic Programming and the Calculus of Variations, Academic Press, N. Y.

[Dre69] Dreyfus, S. D., 1969. "An Appraisal of Some Shortest-Path Algorithms," Operations Research, Vol. 17, pp. 395-412.

[Eck68] Eckles, J. E., 1968. "Optimum Maintenance with Incomplete Information," Operations Res., Vol. 16, pp. 1058-1067.

[Elm78] Elmaghraby, S. E., 1978. Activity Networks: Project Planning and Control by Network Models, Wiley - Interscience, N. Y.

[Fal87] Falcone, M., 1987. "A Numerical Approach to the Infinite Horizon Problem of Deterministic Control Theory," Appl. Math. Opt., Vol. 15, pp. 1-13.

[FeM94] Fernandez-Gaucherand, E., and Markus, S. I., 1994. "Risk Sensitive Optimal Control of Hidden Markov Models," Proc. 33rd IEEE Conf. Dec. Control, Lake Buena Vista, Fla.

[Fel68] Feller, W., 1968. An Introduction to Probability Theory and Its Applications, Wiley, N. Y.

[For56] Ford, L. R., Jr., 1956. "Network Flow Theory," Report P-923, The Rand Corporation, Santa Monica, CA.

[For73] Forney, G. D., 1973. "The Viterbi Algorithm," Proc. IEEE, Vol. 61, pp. 268-278.

[Fox71] Fox, B. L., 1971. "Finite State Approximations to Denumerable State Dynamic Progams," J. Math. Anal. Appl., Vol. 34, pp. 665-670.

[GaP88] Gallo, G., and Pallottino, S., 1988. "Shortest Path Algorithms," Annals of Operations Research, Vol. 7, pp. 3-79.

[GoR85] Gonzalez, R., and Rofman, E., 1985. "On Deterministic Control Problems: An Approximation Procedure for the Optimal Cost, Parts I, II," SIAM J. Control Optimization, Vol. 23, pp. 242-285.

[GoS84] Goodwin, G. C., and Sin, K. S. S., 1984. Adaptive Filtering, Prediction, and Control, Prentice-Hall, Englewood Cliffs, N. J.

[GrA66] Groen, G. J., and Atkinson, R. C., 1966. "Models for Optimizing the Learning Process," Psychol. Bull., Vol. 66, pp. 309-320.

[GuF63] Gunckel, T. L., and Franklin, G. R., 1963. "A General Solution for Linear Sampled-Data Control," Trans. ASME Ser. D. J. Basic Engrg., Vol. 85, pp. 197-201.

[HMS55] Holt, C. C., Modigliani, F., and Simon, H. A., 1955. "A Linear Decision Rule for Production and Employment Scheduling," Management Sci., Vol. 2, pp. 1-30.

[HaL82] Hajek, B., and van Loon, T., 1982. "Decentralized Dynamic Control of a Multiaccess Broadcast Channel," IEEE Trans. Automatic Control, Vol. AC-27, pp. 559-569.

[Hes66] Hestenes, M. R., 1966. Calculus of Variations and Optimal Control Theory, Wiley, N. Y.

[Her89] Hernandez-Lerma, O., 1989. Adaptive Markov Control Processes, Springer-Verlag, N. Y.

[HoK61] Hoffman, K., and Kunze, R., 1961. Linear Algebra, Prentice-Hall, Englewood Cliffs, N. J.

[IEE71] IEEE Trans. Automatic Control, 1971. Special Issue on Linear-Quadratic Gaussian Problem, Vol. AC-16.

[JBE94] James, M. R., Baras, J. S., and Elliott, R. J., 1994. "Risk-Sensitive Control and Dynamic Games for Partially Observed Discrete-Time Nonlinear Systems," IEEE Trans. on Automatic Control, Vol. AC-39, pp. 780-792.

[Jac73] Jacobson, D. H., 1973. "Optimal Stochastic Linear Systems With Exponential Performance Criteria and their Relation to Deterministic Differential Games," IEEE Trans. Automatic Control, Vol. AC-18, pp. 124-131.

[Jaz70] Jazwinski, A. H., 1970. Stochastic Processes and Filtering Theory, Academic Press, N. Y.

[JoT61] Joseph, P. D., and Tou, J. T., 1961. "On Linear Control Theory," AIEE Trans., Vol. 80 (II), pp. 193-196.

[KGB82] Kimemia, J., Gershwin, S. B., and Bertsekas, D. P., 1982. "Computation of Production Control Policies by a Dynamic Programming Technique," in Analysis and Optimization of Systems, A. Bensoussan and J. L. Lions (eds.), Springer-Verlag, N. Y., pp. 243-269.

[KaD66] Karush, W., and Dear, E. E., 1966. "Optimal Stimulus Presentation Strategy for a Stimulus Sampling Model of Learning," J. Math. Psychology, Vol. 3, pp. 15-47.

[KaK58] Kalman, R. E., and Koepcke, R. W., 1958. "Optimal Synthesis of Linear Sampling Control Systems Using Generalized Performance Indexes," Trans. ASME, Vol. 80, pp. 1820-1826.

[Kal60] Kalman, R. E., 1960. "A New Approach to Linear Filtering and Prediction Problems," Trans. ASME Ser. D. J. Basic Engrg., Vol. 82, pp. 35-45.

[KeS60] Kemeny, J. G., and Snell, J. L., 1960. Finite Markov Chains, Van Nostrand-Reinhold, N. Y.

[Kim82] Kimemia, J., 1982. "Hierarchical Control of Production in Flexible Manufacturing Systems," Ph.D. Thesis, Dep. of Electrical Engineering and Computer Science, Massachusetts Institute of Technology.

[KuA77] Ku, R., and Athans, M., 1977. "Further Results on the Uncertainty Threshold Principle," IEEE Trans. on Automatic Control, Vol. AC-22, pp. 866-868.

[KuD92] Kushner, H. J., and Dupuis, P. G., 1992. Numerical Methods for Stochastic Control Problems in Continuous Time, Springer-Verlag, N. Y.

[KuL82] Kumar, P. R., and Lin, W., 1982. "Optimal Adaptive Controllers for Unknown Markov Chains," IEEE Trans. Automatic Control, Vol. AC-27, pp. 765-774.

[KuV86] Kumar, P. R., and Varaiya, P. P., 1986. Stochastic Systems: Estimation, Identification, and Adaptive Control, Prentice-Hall, Englewood Cliffs, N. J.

[Kum83] Kumar, P. R., 1983. "Optimal Adaptive Control of Linear-Quadratic-Gaussian Systems," SIAM J. on Control and Optimization, Vol. 21, pp. 163-178.

[Kum85] Kumar, P. R., 1985. "A Survey of Some Results in Stochastic Adaptive Control," SIAM J. on Control and Optimization, Vol. 23, pp. 329-380.

[Kus90] Kushner, H. J., 1990. "Numerical Methods for Continuous Control Problems in Continuous Time," SIAM J. on Control and Optimization, Vol. 28, pp. 999-1048.

[Las85] Lasserre, J. B., 1985. "A Mixed Forward-Backward Dynamic Programming Method Using Parallel Computation," J. Optimization Theory Appl., Vol. 45, pp. 165-168.

[Lev84] Levy, D., 1984. The Chess Computer Handbook, B. T. Batsford Ltd., London.

[LjS83] Ljung, L., and Soderstrom, T., 1983. Theory and Practice of Recursive Identification, MIT Press, Cambridge, MA.

[Lju86] Ljung, L., 1986. System Identification: Theory for the User, Prentice-Hall, Englewood Cliffs, N. J.

[Lov91] Lovejoy, W. S., 1991. "A Survey of Algorithmic Methods for Partially Observed Markov Decision Processes," Annals of Operations Research, Vol. 18, pp. 47-66.

[Lue69] Luenberger, D. G., 1969. Optimization by Vector Space Methods, Wiley, N. Y.

[Lue84] Luenberger, D. G., 1984. Linear and Nonlinear Programming, Addison-Wesley, Reading, MA.

[Med69] Meditch, J. S., 1969. Stochastic Optimal Linear Estimation and Control, McGraw-Hill, N. Y.

[Men73] Mendel, J. M., 1973. Discrete Techniques of Parameter Estimation – The Equation Error Formulation, Marcel Dekker, N. Y.

[Mik79] Mikhailov, V. A., 1979. Methods of Random Multiple Access, Candidate Engineering Thesis, Moscow Institute of Physics and Technology, Moscow.

[Mos68] Mossin, J., 1968. "Optimal Multi-Period Portfolio Policies," J. Business, Vol. 41, pp. 215-229.

[Nah69] Nahi, N., 1969. Estimation Theory and Applications, Wiley, N. Y.

[New75] Newborn, M., 1975. Computer Chess, Academic Press, N. Y.

[PBT95] Polymenakos, L. C., Bertsekas, D. P., and Tsitsiklis, J. N., 1995. "Efficient Algorithms for Continuous-Space Shortest Path Problems," Lab. for Info. and Decision Systems Report LIDS-P-2292, Massachusetts Institute of Technology.

[PaS82] Papadimitriou, C. H., and Steiglitz, K., 1982. Combinatorial Optimization: Algorithms and Complexity, Prentice-Hall, Englewood Cliffs, N. J.

[PaT87] Papadimitriou, C. H., and Tsitsiklis, J. N., 1987. "The Complexity of Markov Decision Processes," Math. Operations Res., Vol. 12, pp. 441-450.

[Pap74] Pape, V., 1974. "Implementation and Efficiency of Moore Algorithms for the Shortest Path Problem," Math. Progr., Vol. 7, pp. 212-222.

[Pap65] Papoulis, A., 1965. Probability, Random Variables and Stochastic Processes, McGraw-Hill, N. Y.

[Pea84] Pearl, J., 1984. Heuristics, Addison-Wesley, Reading, MA.

[Pic90] Picone, J., 1990. "Continuous Speech Recognition Using Hidden Markov Models," IEEE ASSP Magazine, July Issue, pp. 26-41.

[Pin95] Pinedo, M., 1995. Scheduling: Theory, Algorithms, and Systems, Prentice-Hall, Englewood Cliffs, N. J.

[PoB94] Polymenakos, L. C., and Bertsekas, D. P., 1994. "Parallel Shortest Path Auction Algorithms," Parallel Computing, Vol. 20, pp. 1221-1247.

[Pol71] Polak, E., 1971. Computational Methods in Optimization: A Unified Approach, Academic Press, N. Y.

[PrS94] Proakis, J. G., and Salehi, M., 1994. Communication Systems Engineering, Prentice-Hall, Englewood Cliffs, N. Y.

[Rab89] Rabiner, L. R., 1989. "A Tutorial on Hidden Markov Models and Selected Applications in Speech Recognition," Proc. of the IEEE, Vol. 77, pp. 257-286.

[Riv85] Rivest, R. L., 1985. "Network Control by Bayesian Broadcast," Lab. for Computer Science Report LCS/TM-285, Massachusetts Institute of Technology.

[Roc70] Rockafellar, R. T., 1970. Convex Analysis, Princeton University Press, Princeton, N. J.

[Ros83] Ross, S. M., 1983. Introduction to Stochastic Dynamic Programming, Academic Press, N. Y.

[Ros85] Ross, S. M., 1985. Probability Models, Academic Press, Orlando, Fla.

[Roy68] Royden, H. L., 1968. Principles of Mathematical Analysis, McGraw-Hill, N. Y.

[Rud64] Rudin, W., 1964. Real Analysis, McGraw-Hill, N. Y.

[Sar87] Sargent, T. J., 1987. Dynamic Macroeconomic Theory, Harvard Univ. Press, Cambridge, MA.

[Sca60] Scarf, H., 1960. "The Optimality of (s, S) Policies for the Dynamic Inventory Problem," Proceedings of the 1st Stanford Symposium on Mathematical Methods in the Social Sciences, Stanford University Press, Stanford, CA.

[Sha50] Shannon, C., 1950. "Programming a Digital Computer for Playing Chess," Phil. Mag., Vol. 41, pp. 356-375.

[Shi64] Shiryaev, A. N., 1964. "On Markov Sufficient Statistics in Non-Additive Bayes Problems of Sequential Analysis," Theory of Probability and Applications, Vol. 9, pp. 604-618.

[Shi66] Shiryaev, A. N., 1966. "On the Theory of Decision Functions and Control by an Observation Process with Incomplete Data," Selected Translations in Math. Statistics and Probability, Vol. 6, pp. 162-188.

[Shr81] Shreve, S. E., 1981. "A Note on Optimal Switching Between Two Activities," Naval Research Logistics Quarterly, Vol. 28, 185-190.

[Sim56] Simon, H. A., 1956. "Dynamic Programming Under Uncertainty with a Quadratic Criterion Function," Econometrica, Vol. 24, pp. 74-81.

[Skl88] Sklar, B., 1988. Digital Communications: Fundamentals and Applications, Prentice-Hall, Englewood Cliffs, N. Y.

[SmS73] Smallwood, R. D., and Sondik, E. J., 1973. "The Optimal Control of Partially Observable Markov Processes Over a Finite Horizon," Operations Res., Vol. 11, pp. 1071-1088.

[Sma71] Smallwood, R. D., 1971. "The Analysis of Economic Teaching Strategies for a Simple Learning Model," J. Math. Psychology, Vol. 8, pp. 285-301.

[Son71] Sondik, E. J., 1971. "The Optimal Control of Partially Observable Markov Processes," Ph.D. Dissertation, Department of Engineering-Economic Systems, Stanford University, Stanford, CA.

[StL89] Stokey, N. L., and Lucas, R. E., 1989. Recursive Methods in Economic Dynamics, Harvard University Press, Cambridge, MA.

[Str65] Striebel, C. T., 1965. "Sufficient Statistics in the Optimal Control of Stochastic Systems," J. Math. Anal. Appl., Vol. 12, pp. 576-592.

[Str76] Strang, G., 1976. Linear Algebra and its Applications, Academic Press, N. Y.

[ThW66] Thau, F. E., and Witsenhausen, H. S., 1966. "A Comparison of Closed-Loop and Open-Loop Optimum Systems," IEEE Trans. Automatic Control, Vol. AC-11, pp. 619-621.

[The54] Theil, H., 1954. "Econometric Models and Welfare Maximization," Weltwirtsch. Arch., Vol. 72, pp. 60-83.

[Tsi84a] Tsitsiklis, J. N., 1984. "Convexity and Characterization of Optimal Policies in a Dynamic Routing Problem," J. Optimization Theory Appl., Vol. 44, pp. 105-136.

[Tsi84b] Tsitsiklis, J. N., 1984. "Periodic Review Inventory Systems with Continuous Demand and Discrete Order Sizes," Management Sci., Vol. 30, pp. 1250-1254.

[Tsi87] Tsitsiklis, J. N., 1987. "Analysis of a Multiaccess Control Scheme," IEEE Trans. Automatic Control, Vol. AC-32, pp. 1017-1020.

[Tsi93] Tsitsiklis, J. N., 1993. "Efficient Algorithms for Globally Optimal Trajectories," Lab. for Info. and Decision Systems Report LIDS-P-2210, Massachusetts Institute of Technology, IEEE Trans. on Automatic Control, to appear.

[Vei65] Veinott, A. F., Jr., 1965. "The Optimal Inventory Policy for Batch Ordering," Operations Res., Vol. 13, pp. 424-432.

[Vei66] Veinott, A. F., Jr., 1966. "The Status of Mathematical Inventory Theory," Management Sci., Vol. 12, pp. 745-777.

[Vit67] Viterbi, A. J., 1967. "Error Bounds for Convolutional Codes and an Asymptotically Optimum Decoding Algorithm," IEEE Trans. on Info. Theory, Vol. IT-13, pp. 260-269.

[Wal47] Wald, A., 1947. Sequential Analysis, Wiley, N. Y.

[WeP80] Weiss, G., and Pinedo, M., 1980. "Scheduling Tasks with Exponential Service Times on Nonidentical Processors to Minimize Various Cost Functions," J. Appl. Prob., Vol. 17, pp. 187-202.

[WhH80] White, C. C., and Harrington, D. P., 1980. "Application of Jensen's Inequality to Adaptive Suboptimal Design," J. Optimization Theory Appl., Vol. 32, pp. 89-99.

[WhS89] White, C. C., and Scherer, W. T., 1989. "Solution Procedures for Partially Observed Markov Decision Processes," Operations Res., Vol. 30, pp. 791-797.

[Whi63] Whittle, P., 1963. Prediction and Regulation by Linear Least-Square Methods, English Universities Press, London.

[Whi69] White, D. J., 1969. Dynamic Programming, Holden-Day, San Francisco, CA.

[Whi78] Whitt, W., 1978. "Approximations of Dynamic Programs I," Math. Operations Res., Vol. 3, pp. 231-243.

[Whi79] Whitt, W., 1979. "Approximations of Dynamic Programs II," Math. Operations Res., Vol. 4, pp. 179-185.

[Whi82] Whittle, P., 1982. Optimization Over Time, Wiley, N. Y., Vol. 1, 1982, Vol. 2, 1983.

[Wit69] Witsenhausen, H. S., 1969. "Inequalities for the Performance of Suboptimal Uncertain Systems," Automatica, Vol. 5, pp. 507-512.

[Wit70] Witsenhausen, H. S., 1970. "On Performance Bounds for Uncertain Systems," SIAM J. on Control, Vol. 8, pp. 55-89.

[Wit71] Witsenhausen, H. S., 1971. "Separation of Estimation and Control for Discrete-Time Systems," Proc. IEEE, Vol. 59, pp. 1557-1566.

INDEX